Soundscape in Urban Forests

Soundscape in Urban Forests

Editors

Xin-Chen Hong
Jiang Liu
Guangyu Wang

MDPI • Basel • Beijing • Wuhan • Barcelona • Belgrade • Manchester • Tokyo • Cluj • Tianjin

Editors

Xin-Chen Hong
School of Architecture and
Urban-Rural Planning
Fuzhou University
Fuzhou, and
School of Architecture
Southeast University
Nanjing
China

Jiang Liu
School of Architecture and
Urban-Rural Planning
Fuzhou University
Fuzhou
China

Guangyu Wang
Faculty of Forestry
University of British
Columbia
Vancouver
Canada

Editorial Office
MDPI
St. Alban-Anlage 66
4052 Basel, Switzerland

This is a reprint of articles from the Special Issue published online in the open access journal *Forests* (ISSN 1999-4907) (available at: www.mdpi.com/journal/forests/special_issues/Soundscape_Urban).

For citation purposes, cite each article independently as indicated on the article page online and as indicated below:

LastName, A.A.; LastName, B.B.; LastName, C.C. Article Title. *Journal Name* **Year**, *Volume Number*, Page Range.

ISBN 978-3-0365-6274-2 (Hbk)
ISBN 978-3-0365-6273-5 (PDF)

Contents

About the Editors

Xin-Chen Hong

Dr. Xin-Chen Hong received his Ph.D. in Landscape Architecture from Fujian Agriculture and Forestry University (including a visiting process at University of British Columbia). He was a postdoctoral fellow at Southeast University. He is now a Fujian Higher-Level Talent and associate professor in the School of Architecture and Urban-Rural Planning at Fuzhou University, China. Dr. Hong's research interests are in the area of social sensing and computational social science. His recent projects include the National Natural Science Foundation of China, and the Program of Humanities and Social Science Research Program of Ministry of Education of China. Dr. Hong has published more than 60 articles in peer-reviewed scientific journals, such as *Sustainable Cities and Society, Urban Forestry & Urban Greening, Journal of Cleaner Production*, which included two ESI High Cited Papers. He serves as reviewers of *Nature Communications, Ecological Indicators, Urban Ecosystems, Frontiers in Public Health*, and many others.

Jiang Liu

Dr. Jiang Liu received his Ph.D. in Landscape Planning and Design at the University of Rostock in Germany in 2014. He is now a full Professor in the Department of Landscape Architecture in the School of Architecture and Urban–Rural Planning at Fuzhou University, China. With over 10 years' research experience in the soundscape field, his research focuses on the relationship between soundscape and landscape in different context, to reveal the fundamental mechanism of interactions between people and the physical as well as perceptual forms of soundscape and landscape. He has published over 40 journal publications in high-quality peer-reviewed journals, such as *Landscape and Urban Planning, Building and Environment, Urban Forestry and Urban Greening*, and *Science of the Total Environment*. He is also the director of Environmental Psychology and Health Innovation Research Team, and vice head and co-founder of the "Key Lab of Digital Technology in Simulation and Prediction of Territorial Space" at Fuzhou University.

Guangyu Wang

Dr. Guangyu Wang is an Associate Dean and Professor of the Faculty of Forestry at the University of British Columbia, and Director of the UBC's Multidisciplinary Institute of Natural Therapy (MINT), and Director National Park Research Centre. He has been working intensively in national park, recreation management and ecotourism development in the USA, Canada, and China. He brings extensive experience in protected areas research and project leadership while successfully managing USD 4 million in research projects over the last 10 years. He has completed several research projects on forest therapy and its contribution to both social and environmental wellness. Currently, his MINT team is mainly focused on the effectiveness of forest soundscape, lightscape, and viewscape on human health and wellbeing. He works very closely with Forest Therapy Hub, ANFT, Korea Forest Welfare Institute, China Forest Therapy Committee, BC Cancer Research Center, and the UBC School of Medicine.

He is the author of numerous books, chapters, technical reports, and scientific papers, including around 100 peer-reviewed international journals. He serves as an Associate Editor of *Journals of Forestry Research, Frontier of Ecology and Evolution, Forests*, and many others, and Director of Asia Pacific Forestry Education Coordination Mechanism.

Preface to "Soundscape in Urban Forests"

In recent years, we have witnessed an increased interest towards soundscape in urban forests. Soundscapes are one of the most important components of the landscape in urban green spaces, natural spaces and cultural landscapes, and soundscapes play an important role in the perception of residents, especially in quiet areas. Fortunately, urban forests are suitable habitats for many plants and animals, which play a key role in producing a soundscape. Therefore, there is increasing research attention focused on soundscapes as people become more concerned about health, urbanization, globalization etc.

We are honored to guest edit this Special Issue and thank all authors and reviewers for their excellent work and unyielding support. Meanwhile, our appreciation goes to the *Forests* editorial team for their enthusiastic dedication and professional editing. Contributions from our colleagues and scholars in different institutions are also sincerely appreciated. Meanwhile, many researches were funded by the National Natural Science Foundation of China (No. 52208052), and the Humanities and Social Science Research Program of Ministry of Education of China (No. 21YJCZH038).

Soundscapes have opened up whole new dimensions in forestry, urban planning and policy, but wide application of them still requires much effort to be made. By searching for multi-disciplinary techniques and strengthening cooperation between academics and industrialists across the world, we will continue to exploit the full potential of soundscape in order to better optimize the urban environment.

Xin-Chen Hong, Jiang Liu, and Guangyu Wang
Editors

Editorial

Soundscape in Urban Forests

Xin-Chen Hong [1,2], Jiang Liu [1,*] and Guang-Yu Wang [3,*]

1   School of Architecture and Urban-Rural Planning, Fuzhou University, Fuzhou 350108, China
2   School of Architecture, Southeast University, Nanjing 210096, China
3   Faculty of Forestry, University of British Columbia, Vancouver, BC V6T 1Z4, Canada
*   Correspondence: jiang.liu@fzu.edu.cn (J.L.); guangyu.wang@ubc.ca (G.-Y.W.)

1. Introduction

The World Health Organization (WHO) has made considerable efforts to reduce citizens' exposure to community noise in urban and suburban areas. Urban forests contribute to a healthy environment for the public in high-density cities, and the soundscapes in these areas are important health-related resources in urban areas, especially natural soundscapes [1]. The various soundscape standards proposed by the International Organization for Standardization (ISO) [2–4] represent a milestone in the recognition of soundscapes as a legitimate approach to manage and design urban sound environments, especially biophilic outdoor environments [5]. The effects of soundscapes in forested areas are not limited to psychological and physical rehabilitation, contributing to the visiting experience in urban forests together with other perceptions [6], but also have ecological significance, such as protecting the acoustic environment for animal communication [7]. These facts demonstrate that soundscapes in urban forests have profound connotations worth further exploration.

This Special Issue in *Forests* explored the role of soundscapes in urban forested areas. It is comprised of eleven papers involving soundscape studies conducted in urban forests from Asia and Africa. This collection contains six research fields: (1) the ecological patterns and processes of forest soundscapes; (2) the boundary effects and perceptual topology; (3) natural soundscapes and human health; (4) the experience of multi-sensory interactions; (5) environmental behavior and cognitive disposition; (6) soundscape resource management in forests.

2. Summary of Articles Included in the Special Issue

Soundscapes influencing individuals include physical stages, physiological stages, and psychological behavior stages. Our research collections focus on the experience of multi-sensory interactions in urban forests. Xu et al. [8] analyzed the audio-visual preference characteristics of exercisers in a park, and found that the forest and its landscape and birdsong avenues are most preferred by exercisers, though the audio-visual preferences of people with different exercise forms differ. Then, they analyzed children's multi-sensory landscape preferences in a park and explored the influence of multi-sensory experiences on children's behavioral experience [9]. They found that visual, auditory, tactile, and olfactory sensations were significantly correlated with children's behavioral experiences. Furthermore, to look into visitors' multi-sensory interactive experiences in an urban forest park [10], these contributors showed that urban forest park visitors' sensory experiences are dominated by visual and olfactory perceptions, followed by audio-visual and visual–tactile interactions. Gan et al. [11] tried to quantify and compare the contribution of audiovisual perception to visitors' overall park landscape preference. They found that visitors' visual landscape preference was generally higher than soundscape preference, while the influence of auditory preference on overall landscape preference was found to be greater than that of visual preference. Shao et al. [12] explored aural–visual interaction attributes that may

Citation: Hong, X.-C.; Liu, J.; Wang, G.-Y. Soundscape in Urban Forests. *Forests* **2022**, *13*, 2056. https://doi.org/10.3390/f13122056

Received: 28 November 2022
Accepted: 30 November 2022
Published: 3 December 2022

Publisher's Note: MDPI stays neutral with regard to jurisdictional claims in published maps and institutional affiliations.

influence people's perceived overall soundscape comfort in urban green spaces (UGSs). They suggested that a low level of environmental sound does not correspond to higher ratings on the overall soundscape comfort.

The natural properties of forests contribute to the occurrences of natural soundscapes which are beneficial to physiological stages. Our research collections focus on various scenes and contexts. Hong et al. [5] focused on the different forest structures, and explored the relationships between perceived soundscape and acoustical parameters, and observe physiological indicators. They suggested that L_{Aeq} and L_{10} are important factors that influence questionnaire responses, and electromyogram (EMG) signals were the most obvious and sensitive in physiological parameters. Wang et al. [13] explored the effects of soundscapes on human physiology and psychology from the perspective of auditory senses in national parks. The authors found that the sound of water had the most significant effects on the heart rate and respiratory rate of the subjects. Yu et al. [14] investigated the psycho-physiological effects of traffic sounds in urban green spaces. They found that traffic sounds had significant detrimental psychological and physiological effects. In terms of psychological responses, the peak sound level outperformed the equivalent sound level in determining the psychological impact of traffic sounds.

Furthermore, urban forested areas represent an important part of urban ecosystems. Our research collections focused on the ecological patterns and processes of forest soundscapes as well. Zhao et al. [15] explored the spatial and temporal characteristics of the urban forest area soundscape. They found that the power spectral density (PSD) and the soundscape diversity index (SDI) of the park were greatly affected by public recreation behaviors. Nwankwo et al. [16] explored how the soundscape of the natural environment could affect humans with respect to the different densities of vegetation, as well as the frequency of singing events and the sound pressure levels of common birds that generate natural sounds in a commonly visited urban park. Chen et al. [17] explored the spatial dependency of the soundscapes in Kulangsu, based on the spatiotemporal dynamics of soundscape and landscape perceptions. The authors found that the perceptions of soundscape and landscape were associated with significant spatiotemporal dynamics, and revealed the dominance of biological sounds in all sampling periods and human sounds in the evening.

3. The Researchers' Perspectives on Soundscapes in Urban Forests

To gather the perspectives of contributors for our Special Issue after completing their researches, we invited one contributor from each paper to answer the following open-ended question: "What are the current priorities and challenges for soundscape research in urban forested areas?". We required the contributors to respond with short comments, typically 100 to 150 words. We report below the full report of the contributors which sorted by title and family name initials.

3.1. Interviewee

Soundscape includes both the sound of natural environment and the played music, especially the natural sound which always attracts a tremendous audience. In the shaping of soundscapes, the sound characteristics of the natural environments are often skillfully used to craft a beautiful and diverse sense environment for human beings. It has been a long time since landscape architects have innovated natural sound into the traditional Chinese garden environment as a special landscape element, which have been abundantly documented in the history, such as rain drizzling down on the plantain leaves with a staggering sound, the sound of a reverberating breeze blowing across the lotus pond, and the whistling of the wind in the pines. In these time, research of soundscapes focuses more on high-quality built environments and the natural environments, which aim to satisfy people's auditory and perceptional demands, generally creating a relaxing, comfortable, and pleasant environmental atmosphere through landscape design.

(Prof. Dr. Yuning Cheng—Southeast University)

3.2. Interviewee

As a new perspective of experience and cognition, certain theoretical and practical achievements have been developed in the field of soundscape and multi-sensory research. However, the best way of maximizing the function of urban forest parks to serve children and improve their multi-sensory experience remains an open question. Compared with adults, children are natural sensorial explorers and are more sensitive to external sensory stimuli. Their sensory experience in urban forests is unique and energetic, which deeply influences the healthy growth and development of children. Therefore, it is vital to pay attention to specific groups of people in the study of urban forest soundscapes and analyze their experiences and preferences. Children's multi-sensory perceptions in forest parks are worth exploring, and more in-depth additional research is needed through increased research time, more case sites, and greater sample sizes.

(Prof. Dr. Muchun Li—South China University of Technology)

3.3. Interviewee

Natural sound, biological sound, and synthetic sound make up the majority of the soundscape, which is a crucial component of the composition of an environment. The urban forest soundscape contains a variety of biological sounds, including birds' and insects' sound. This type of soundscape has characteristics which vary over time and space. The activity of different biological communities, plant community structures, and other factors influence the soundscape diversity and characteristics of urban forest biological sound. The urban forest's acoustic environment can be improved by creating a habitat space to increase the richness of the soundscape. The relationship between biological activities, the diversity of soundscapes, and the vertical structure characteristics of urban forest plants can also be explored in future research.

(Prof. Dr. Qi Meng—Harbin Institute of Technology)

3.4. Interviewee

The current priorities of soundscape research in urban forested areas are as follows. Firstly, combing soundscape research with new tourism formats and tourist' requires in-depth experience. Future research should focus on diverse perspectives, such as "Soundscape + Culture" research, because different places have different soundscape characteristics and change processes, behind which is the expression of local cultural differences. Secondly, previous researches have tended to focus on a single perspective, such as vision. Focusing on a single research perspective is insufficient in the context of a normalized epidemic. More attention should be paid to multi-sensory dimensions such as "Soundscape + Olfactory".

The challenges of soundscape research is the quantitative analysis of the soundscape. As the concept of soundscape becomes more widespread, there is a need to experiment more with experimental methodologies of measuring the healing effects of soundscape to collect objective data, but also to focus on subjective factors such as changes in the psychological attributes of the study participants.

(Prof. Dr. Zhicai Wu—South China University of Technology)

3.5. Interviewee

Research on the interaction between auditory and non-auditory perceptions, including audiovisual interaction, and the sensory contribution mechanism of the urban forest landscape evaluation should be strengthened, and through this, soundscape perception can be expanded into a multi-sensory landscape experience. Moreover, the physiological and psychological healing functions and health promotion effects of urban forest multi-sensory landscapes on visitors should be further studied.

(Assoc. Prof. Dr. Yonghong Gan—Minnan Normal University)

3.6. Interviewee

The importance of multi-sensory perception in constructing human landscape experiences has been increasingly emphasized in contemporary urban life. The major priorities and challenges for soundscape research in urban forested areas is to broaden the understandings of environmental attributes with aural–visual interactive effects, and their impacts on human perceptions. These investigations should also consider whether their attributes and influences vary with sound-level ranges. Another important aspect that should be emphasized is the functional values of urban forested areas. Though numerous efforts have been made, the relations constructed between human perception and environmental aural–visual characteristics are only used as auxiliary tools in landscape practices. Future studies should further explore the effective ways of translating people's aural–visual landscape experience into design languages, which can further contribute to enriching human spatial experiences from the perspective of multi-sensory design interventions.

(Assoc. Prof. Dr. Yuhan Shao—Tongji University)

3.7. Interviewee

Combining the current major publication trends and research, I believe that the current priorities of urban forest soundscape research comprise mainly two aspects. On the one hand, the study not only focuses on human perception and soundscape experiences, but also includes psychological effects, such as recovery effects and stress reduction effects. On the other hand, the study of soundscape and other spaces is gradually integrated. The current related research mainly involves cultural, landscape, virtual, and color spaces. This extends the research direction and field of soundscape and enhances the application value of soundscape research results.

Research on urban soundscapes is also currently facing some challenges. Some studies have shown that noise in cities reduces people's sensitivity to sound changes, which is a large bias for researchers using urban samples. Second, new research techniques, such as virtual reality, as a way to improve experimental accuracy, have been slowly applied in the field of soundscape.

(Assoc. Prof. Dr. Jian Xu—South China University of Technology)

3.8. Interviewee

Urban green spaces provide important spaces for relaxation, recreation, social interaction, and sports. Many studies have proved that the soundscape had significant effects on visiting experiences. Our study revealed that soundscapes also had psychophysiological effects on people in urban green spaces. However, most existing studies focused on the short-term effects of soundscapes on people. However, for policy making and design practice, the long-term effects, including the effect on behavior and further health effects, are more important, which remains in question. Therefore, in my opinion, the quantitative investigation of the relationship between soundscape characteristics and the health effects is one of the most challenging problems for soundscape research in urban forested areas.

(Assoc. Prof. Dr. Boya Yu—Beijing Jiaotong University)

3.9. Interviewee

The soundscape in urban forested areas is complex, and factors such as forest structure, topography, biodiversity, seasonal changes, and environmental context should be considered. At the same time, the monitoring of forest soundscape should also change its ideas from specific monitoring points to the monitoring of the whole spatial and temporal dimensions, and from emphasizing the study of individual specific indicators to exploring the correlation of each indicator (e.g., biological vocalization is influenced by various factors such as microclimate, season, vegetation, etc.), so that the study of forest soundscape can be more systematic and perfect in the future.

(Dr. Weicong Fu—Fujian Agriculture and Forestry University)

3.10. Interviewee

The first point is the scientific evaluation of soundscapes in urban forested areas. According to the existing research methods, both physical data and perceptual data are closely related to the way people perceive the acoustic environment, which indicates that the measurement method, sampling method, sample size, interview form, and other soundscape contents may affect the objective quantification of soundscapes. The second point is the relationship between soundscape perception and visual aesthetics. Although the academic circle has conducted some studies on soundscapes and their interaction with visual landscapes, most of them are still limited to the theoretical level, and the existing results cannot effectively guide the planning, design, and optimization management of soundscapes in urban forested areas. The third point is the impact of soundscape perception on public health. Some studies have shown that the natural environment has positive effects on physical and emotional health, but more systematic empirical research and mechanism analysis are still lacking in the field of soundscape research.

(Dr. Peng Wang—Chinese Academy of Forestry)

3.11. Interviewee

I believe the most challenging work for soundscape research in either urban forests or other types of landscape is the database that could reflect both the highly spatiotemporal dynamic characteristics of soundscapes themselves and human collective values (or intersubjective values). At present, the data and evaluation in soundscape studies are mainly based on the short-term human responses to the soundscapes that could partially reflect and indicate the local perspective. However, they are very time-consuming and user- or weather-dependent, which is not very appropriate to be directly applied in landscape planning and management. To overcome these deficiencies, more efforts should be devoted to long-term acoustic monitoring (e.g., passive acoustic monitoring), which could be facilitated by intelligent sensors that could automatically record soundscapes and other associated environmental data, as well as Big Data sources that could reflect human perception and opinions. More effective and user-independent soundscape indicators and proxies that combine public intersubjective values and the acoustic environment are still needed.

(PhD Candidate Zhu Chen—Leibniz University Hannover)

4. Concluding Remarks

As mentioned by many of the contributors of this collection, multi-sensory experience in urban forests or green spaces have attracted increasing attention in the landscape field. The fields of landscape planning, design, and management also highlight the role of multi-sensory perception led by traditional visual perceptions in landscape experience, with an aim of creating high-quality landscapes to meet the current needs. Theories and technical approaches are becoming increasingly multi-disciplinary in light of how landscape experiences can affect human mental and/or physical status and well-being. However, there is still a long way to go to reveal the mechanisms on how soundscapes are formed in different landscapes, and how internal and external factors could affect the perception process which is related to the quality of landscape experience. Crucial influential factors, especially those previously mentioned, should be manipulated in planning and design processes should be identified for practical purposes. It is also good to see that the data sources used for soundscape researches are becoming diverse. Except for data acquired from traditional public investigation, on-site monitoring methods, etc., Big Data acquired from the Internet and human reaction parameters acquired from physiological apparatus are expanding the range of multi-sources. This will definitely facilitate soundscape research progress, and will clarify multi-sensory interaction mechanisms and their associated effects. We also believe more studies should address the following topics in the future:

1. Indicators for multi-sensory characteristics of landscape experience;
2. Sensory experiences with cultural and regional significance;
3. Theoretical and technical approaches for sensoryscape creation;
4. The innovative application of Internet of Things and multi-source data;
5. Landscape planning, design, and management based on multi-sensory experiences.

Author Contributions: Conceptualization, X.-C.H., J.L. and G.-Y.W.; writing—original draft preparation, X.-C.H.; writing—review and editing, X.-C.H. and J.L.; supervision, J.L. and G.-Y.W. All authors have read and agreed to the published version of the manuscript.

Funding: This project was supported by the National Natural Science Foundation of China (52208052, 51838003), the National Key Research & Development Program of China (2022YFF130130303), and the Humanities and Social Science Research Program of Ministry of Education of China (Grant No. 21YJCZH038).

Conflicts of Interest: The authors declare no conflict of interest.

References

1. Aletta, F.; Oberman, T.; Mitchell, A.; Erfanian, M.; Lionello, M.; Kachlicka, M.; Kang, J. Associations between Soundscape Experience and Self-Reported Wellbeing in Open Public Urban Spaces: A Field Study. *Lancet* **2019**, *394*, S17. [CrossRef]
2. *ISO 12913-1:2014*; Acoustics—Soundscape—Part 1: Definition and Conceptual Framework. ISO: Geneva, Switzerland, 2014.
3. *ISO 12913-2:2017*; Acoustics—Soundscape—Part 2: Data Collection and Reporting Requirements. ISO: Geneva, Switzerland, 2017.
4. *ISO 12913-2:2017*; Acoustics—Soundscape—Part 3: Data analysis. ISO: Geneva, Switzerland, 2017.
5. Hong, X.-C.; Cheng, S.; Liu, J.; Dang, E.; Wang, J.-B.; Cheng, Y. The Physiological Restorative Role of Soundscape in Different Forest Structures. *Forests* **2022**, *13*, 1920. [CrossRef]
6. Hong, X.-C.; Wang, G.-Y.; Liu, J.; Dang, E. Perceived loudness sensitivity influenced by brightness in urban forests: A comparison when eyes were opened and closed. *Forests* **2013**, *11*, 1242. [CrossRef]
7. Hong, X.-C.; Wang, G.-Y.; Liu, J.; Song, L.; Wu, E.T. Modeling the impact of soundscape drivers on perceived birdsongs in urban forests. *J. Clean. Prod.* **2021**, *292*, 125315. [CrossRef]
8. Xu, J.; Li, M.; Gu, Z.; Xie, Y.; Jia, N. Audio-Visual Preferences for the Exercise-Oriented Population in Urban Forest Parks in China. *Forests* **2022**, *13*, 948. [CrossRef]
9. Xu, J.; Chen, L.; Liu, T.; Wang, T.; Li, M.; Wu, Z. Multi-Sensory Experience and Preferences for Children in an Urban Forest Park: A Case Study of Maofeng Mountain Forest Park in Guangzhou, China. *Forests* **2022**, *13*, 1435. [CrossRef]
10. Xu, J.; Xu, J.; Gu, Z.; Chen, G.; Li, M.; Wu, Z. Network Text Analysis of Visitors' Perception of Multi-Sensory Interactive Experience in Urban Forest Parks in China. *Forests* **2022**, *13*, 1451. [CrossRef]
11. Gan, Y.; Zheng, Y.; Zhang, L. Audio-Visual Analysis of Visitors' Landscape Preference for City Parks: A Case Study from Zhangzhou, China. *Forests* **2022**, *13*, 1376. [CrossRef]
12. Shao, Y.; Hao, Y.; Yin, Y.; Meng, Y.; Xue, Z. Improving Soundscape Comfort in Urban Green Spaces Based on Aural-Visual Interaction Attributes of Landscape Experience. *Forests* **2022**, *13*, 1262. [CrossRef]
13. Wang, P.; He, Y.; Yang, W.; Li, N.; Chen, J. Effects of Soundscapes on Human Physiology and Psychology in Qianjiangyuan National Park System Pilot Area in China. *Forests* **2022**, *13*, 1461. [CrossRef]
14. Yu, B.; Bai, J.; Wen, L.; Chai, Y. Psychophysiological Impacts of Traffic Sounds in Urban Green Spaces. *Forests* **2022**, *13*, 960. [CrossRef]

15. Zhao, Y.; Xu, S.; Huang, Z.; Fang, W.; Huang, S.; Huang, P.; Zheng, D.; Dong, J.; Chen, Z.; Yan, C.; et al. Temporal and Spatial Characteristics of Soundscape Ecology in Urban Forest Areas and Its Landscape Spatial Influencing Factors. *Forests* **2022**, *13*, 1751. [CrossRef]
16. Nwankwo, M.; Meng, Q.; Yang, D.; Liu, F. Effects of Forest on Birdsong and Human Acoustic Perception in Urban Parks: A Case Study in Nigeria. *Forests* **2022**, *13*, 994. [CrossRef]
17. Chen, Z.; Zhu, T.-Y.; Liu, J.; Hong, X.-C. Before Becoming a World Heritage: Spatiotemporal Dynamics and Spatial Dependency of the Soundscapes in Kulangsu Scenic Area, China. *Forests* **2022**, *13*, 1526. [CrossRef]

forests

Article

The Physiological Restorative Role of Soundscape in Different Forest Structures

Xin-Chen Hong [1,2,†], Shi Cheng [1,†], Jiang Liu [2], Emily Dang [3], Jia-Bing Wang [2] and Yuning Cheng [1,*]

1   School of Architecture, Southeast University, Nanjing 210096, China
2   School of Architecture and Urban-Rural Planning, Fuzhou University, Fuzhou 350025, China
3   Department of Forest Resources Management, University of British Columbia, Vancouver, BC V6T 1Z4, Canada
*   Correspondence: cyn999@126.com
†   These authors contributed equally to this work.

Abstract: Natural soundscape is considered a dominant type of hearing in forested areas and contributes to health and recovery effects from exposure to the biophilic outdoor environment. This study focuses on the different forest structures, and aims to explore the relationship between perceived soundscape and acoustical parameters, observe physiological indicators, and model the physiological restorative role of soundscape. Questionnaires and measuring equipment were used to gather psychophysical and physiological information at 20 observation sites in urban forested areas. Back-propagation neural network techniques were conducted to determine the forecasting model from psychophysical to physiological parameters. Our results suggested that L_{Aeq} and L_{10} are important factors that influence questionnaire responses. Our findings also showed that electromyogram (EMG) signals were the most obvious and sensitive in physiological parameters. Additionally, we found that L_{10-90} played the most important role among all physical parameters in the physiological restorativeness soundscape model. This can facilitate the understanding of the physiological restorative role of soundscape in different forest structures when proposing suitable forest-based health care strategies.

Keywords: soundscape; restorativeness; acoustic parameters; physiological parameters; pleasantness

Citation: Hong, X.-C.; Cheng, S.; Liu, J.; Dang, E.; Wang, J.-B.; Cheng, Y. The Physiological Restorative Role of Soundscape in Different Forest Structures. *Forests* **2022**, *13*, 1920. https://doi.org/10.3390/f13111920

Academic Editor: Chi Yung Jim

Received: 28 September 2022
Accepted: 14 November 2022
Published: 15 November 2022

Publisher's Note: MDPI stays neutral with regard to jurisdictional claims in published maps and institutional affiliations.

1. Introduction

In 2020, the public found themselves in an unprecedented situation. With COVID-19 spreading, lockdowns were enforced, and outdoor activities subsequently declined. This led to a rise in both psychological stress and mortality by suicide [1]. There has been a growing demand by the public, especially in high-density cities, for relaxation and entertainment. This public demand contributes to an important effect for public health and work efficiency, and thus the public pays increasing attention to this issue [2,3]. Fortunately, urban forested areas, which are an important part of the urban green infrastructure, are natural places that provide relaxation, entertainment and perceived restoration to the public [4–6]. Furthermore, forest landscapes contribute ecosystem services to the public [7].

Urban forested areas potentially play a key role in the construction of healthy cities [8]. The World Health Organization (WHO) announced goals for healthy cities, which aim to continuously improve the health and quality of life of city dwellers [9]. Various studies have explored physiological and psychological relationships, including taste [10], touch [11], smell [12], vision [13], hearing [14] and other senses [15,16]. In general, vision is considered the most important driver for sensory and cognition effects in environmental exposure. However, the second important driver, hearing, also plays a key role in cognition and behavior. This includes tracking functions, such as spatial cognition without visuals [17]; positioning and connecting functions, such as judging sound sources and audiovisual relationships [18]; and focusing and memory functions, such as understanding of the environment combining visual information [19]. Thus, hearing potentially occupies an important position in the perception of urban forested areas.

Soundscape is described in the ISO as 'acoustic environment as perceived or experienced and/or understood by a person or people, in context' [20]. In urban forested areas, natural soundscape is considered a dominant type as it performs in natural sound occurrences [21], perceived geophony and biophony [22], and birdsong identification [23]. Exposure to the biophilic outdoor environment contributes to health and recovery, and is applied especially in forest-based health care [24]. Previous studies used questionnaire responses to explore the perception of forest soundscape in national parks [25], urban parks [18], and forest parks [26]. The Perceived Restorativeness Soundscape Scale (PRSS) was developed and tested to assess a soundscape's potential to provide psychological restoration [27]. The PRSS focuses on the dimensions of psychological restoration, such as curiosity, interest, concentration, and demand for soundscape in context. However, the dimensions of physiological restoration for forest soundscapes are lacking, especially in different forest structures. Another potential research gap is how to simulate advanced mental processes that elicit physiological responses and contribute to the modelling of psychophysical parameters to physiological parameters in forested areas.

This study was conducted to fill these gaps and aims to: (1) explore the relationship between perceived soundscape and acoustical parameters in different forest structures; (2) observe physiological indicators in different forest structures; and (3) model the physiological restorative role of soundscapes in forested areas.

2. Methodology

2.1. Study Area

Our study was conducted in the arboretum (8,469,323 m^2) of Fuzhou National Forest Park (57,439,074 m^2) in Fuzhou, Fujian, China. Fuzhou National Forest Park is located to the north of downtown Fuzhou. It has a subtropical oceanic monsoon climate with an average annual rainfall of about 1438.5 mm. The average wind speed is 1.8 m/s, relative humidity is 75%, and average annual sunshine is 1848 h. The arboretum consists of well-maintained paths, various tree species, and high forest coverage (65.54%). We found that the arboretum was a suitable site for soundscape research as it contains potential sources of both natural sounds and man-made sounds.

Based on different forest structures and previous research [28,29], 20 observation sites were chosen in the arboretum (see Figure 1), comprising five in bamboo forests, five in broad-leaved forests, five in coniferous forests, and five in coniferous/broad-leaved mixed forests. The acoustic environmental conditions at each site were measured for 5 min, and included L_{Aeq}, L_{10}, L_{90} and L_{10}-L_{90}. The measured L_{Aeq} ranged from 43.9 dBA to 76.8 dBA, L_{10} from 47.7 dBA to 78.6 dBA, L_{90} from 41.8 dBA to 63.1 dBA, and L_{10}-L_{90} from 1.9 dBA to 22.1 dBA.

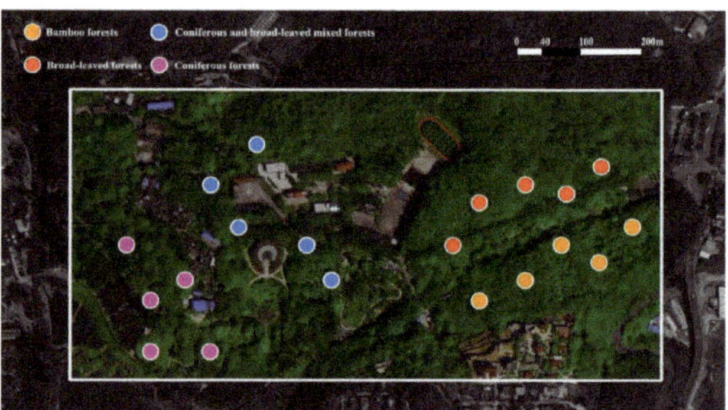

Figure 1. Observation sites in study area.

2.2. Physiological and Soundscape Information

2.2.1. Physiological Parameters

To observe participant responses to stress inducement, we examined the variations in the physiological parameters between baseline value (BL) with eye mask and earmuffs, and pre-test value (Pr) after stress inducement. Pr-BL represented the variation in physiological parameters contributed by stress inducement.

Previous studies have shown that four physiological parameters potentially reflect the role of physiological restoration [30–32]. The first parameter is electromyogram (EMG) and is influenced by frontal muscle activity. Frontal muscle activity decrease when participants are exposed to positive influences from environmental scenes, and increase when exposed to negative influences [33]. This influences EMG levels. The second parameter is electrodermal activity (EDA), a representative measure for mood changes in biometrics research [34]. EDA is affected by exocrine sweat gland activity based on the sympathetic nervous system increasing secretion from sweat glands [35]. The third parameter is photoplethysmography (PPG), which uses low-intensity infrared (IR) light to detect blood volume changes in the microvascular bed of tissue [36]. The fourth parameter is respiration (RESP), the movement of respiratory gases (such as oxygen and carbon dioxide) into and out of the lungs [37].

Physiological parameters were measured using ErgoLAB [38,39], a wearable polygraph with 2048 Hz sampling rate, 16-bit resolution and a wireless communication frequency of 2.4 GHz. We found that each of the physiological parameters had different signal accuracies: 0.183 μV for EMG with 16-bit resolution, 0.01 μs for EDA, 1% for PPG, and 1 rpm for RESP.

To observe the role of physiological restoration in different forest structures, we gathered the pre-test value (Pr) for physiological parameters in bamboo forests, broad-leaved forests, coniferous forests, and coniferous/broad-leaved mixed forests. We then gathered the post-test value (Po) in the same forest structures. The absolute value of Pr-Po reflected how relaxed the participants were and the degree of physiological experienced restoration. Pr-Po was represented by ΔEMG, ΔEDA, ΔPPG, and ΔRESP, respectively.

Five scales were selected to represent the degree to which soundscapes affect physiological restorative role (PRR) [40]. These included 'extremely restorative', 'very restorative', 'moderately restorative', 'slightly restorative', and 'not restorative at all'. To match the five scales, the intervals of ΔEMG, ΔEDA, ΔPPG, and ΔRESP were derived. In green space, EMG, EDA and RESP values are the same order of magnitude, with maximum around or less than 10 [41]. PPG value is another order of magnitude with maximum around or less than 40. Then, the maximum values were split into twenty parts to observe the variation of parameters. The interval length of ΔEMG, ΔEDA, and ΔRESP was 0.5, and that of ΔPPG was 2. The value of EMG, EDA and RESP dropped gradually without stress inducement in general [42]. For PPG, due to be affected by factors other than stress inducement, we took a symmetric interval distribution. Thus, the scales corresponding to the intervals of ΔEMG, ΔEDA, and ΔRESP were $[-\infty, -1.5)$, $[-1.5, -1.0)$, $[-1.0, -0.5)$, $[-0.5, 0.0)$, and $[0.0, +\infty]$. ΔPPG was $[-\infty, -3.0)$, $[-3.0, -1.0)$, $[-1.0, 1.0)$, $[1.0, 3.0)$, and $[3.0, +\infty]$.

2.2.2. Soundscape Parameters

In this study, questionnaires and measuring equipment were used to gather soundscape parameters in different forest structures [43]. Questionnaires were conducted to inquire about the pleasantness of perceived soundscape (PL): not pleasant at all (+1), slightly pleasant (+2), moderately pleasant (+3), very pleasant (+4), and extremely pleasant (+5).

Soundscape parameters were collected via measurements from Type-1 sound level meters (AWA 6228+) at 1.5 m height. This included measuring L_{Aeq}, L_{10}, L_{90}, and L_{10}-L_{90}. L_{Aeq} was the A-weighted equivalent sound pressure level. L_{10} and L_{90} were statistical levels that represented the levels that exceeded 10% and 90%, respectively. L_{10}-L_{90} measured temporal variability and represented the difference between L_{10} and L_{90}.

2.2.3. Stress Inducement

Stress inducement came mainly through mathematical calculations for participants and consisted of two parts. The first part involved asking participants to add two three-digit random numbers. The results were a four-digit number, such as '571 + 815 = 1386'. The second part involved asking them to multiply a two-digit and a one-digit random number. The results were a three-digit number, such as '89 × 5 = 445'. There were ten sets in total, with five sets in each part.

2.3. Physiological Restorativeness Soundscape Modeling

To simulate psychophysical processes that elicited physiological responses, a back-propagation neural network was created to determine the forecasting model from psychophysical parameters to physiological parameters [44]. For the back-propagation neural network, L_{Aeq}, L_{10} L_{90}, L_{10}-L_{90}, and PL were selected as input variables, while ΔEMG, ΔEDA, ΔPPG, ΔRESP, and PRR were selected as output variables.

There were two hidden layers in the physiological restorativeness soundscape model (PRS model), which included 5 neurons and 4 neurons in the first and second hidden layers, respectively. Hyperbolic tangent functions were used for all neurons in each hidden layer.

2.4. Procedure

2.4.1. Participants and Equipment Measuring

Physiological and soundscape information were gathered on weekdays with sunny weather between 9:00 and 17:00 in the months of February and March 2020. Young adults make up the majority of urban forest visitors [45]. Thus, we randomly recruited staff and graduate students from local universities in Fuzhou. A total of 48 participants (male = 25, female = 23, average 29.5 ± 5.1) with normal hearing abilities were recruited to respond to questionnaires and gather physiological information in a sitting position. Before the test began, all participants were required to sign a consent form outlining the details of the study, including content, purpose and methodology. Furthermore, participants could quit the study at any point if they felt uncomfortable during the process.

In this study, the measuring process included five steps with a total duration of 15 min (See Figure 2). Due to the limited number of ErgoLAB devices (24 sets), all participants were divided into two groups and required to complete the measuring process separately by single group. In the group, half of the participants were tested at the same time.

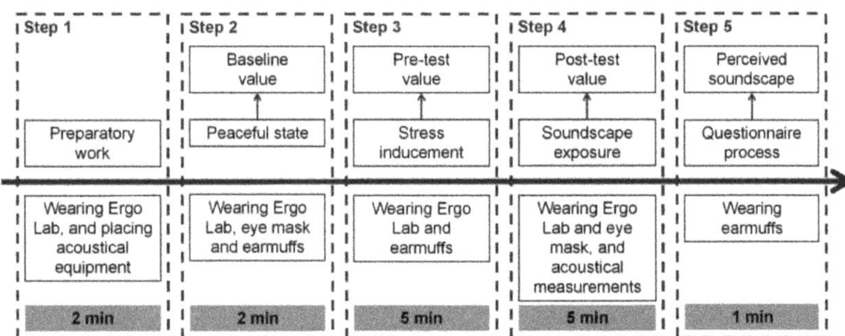

Figure 2. Measurements time stamp for gathering parameters.

Step 1: Preparatory work. We spent 2 min on participants putting on ErgoLAB and placing acoustical equipment at an observation site.

Step 2: Peaceful statement. Participants spent 2 min on maintaining a peaceful state, wearing eye masks and earmuffs. Meanwhile, we used ErgoLAB to gather their BL values in this step.

Step 3: Stress inducement. We spent 5 min on inducing stress in participants. Meanwhile, we used ErgoLAB to gather their Pr values in this step. Participants were required to wear earmuffs before soundscape exposure (Step 4). This contributed to a reduction in auditory short-term memory effects resulting from non-experimental procedures.

Step 4: Soundscape exposure. Participants were exposed to the soundscape at an observation site for 5 min. Meanwhile, we used ErgoLAB to gather their Po values, and used sound-level meters to gather acoustical information. To focus on the physiological restorative role of soundscape, participants were required to wear eye masks during soundscape exposure. This contributed to a reduction in memory attenuation caused by visual distraction, and an increase in the level of their auditory attention [46].

Step 5: Questionnaire process. Participants wearing earmuffs spent 1 min on filling out questionnaires. We gathered the values of perceived soundscape in this step.

After this, we conducted tests to analyze reliability and validity for physiological and psychological parameters. Our results suggested that Cronbach's alpha of physiological and psychological parameters was 0.87, and Cronbach's alpha of each parameter ranged from 0.71 to 0.93. Then, we found that KMO of physiological and psychological parameters was more than 0.75, suggesting an acceptable reliability and validity [28].

2.4.2. Statistical Analyses

To explore the physiological restorative role of soundscape in different forest structures, various statistical analyses were used. Pearson's correlation was conducted to analyze the relationship between acoustic parameters and perceived soundscape. T-test was conducted to analyze: (1) EMG, EDA, PPG, and RESP at tranquility and stress-inducement state; and (2) EMG, EDA, PPG, and RESP during pre-test and post-test in different forest structures. Principal components analysis (PCA) was used to determine the different contents of PRR in psychophysical and physiological parameters. The statistical analysis was carried out in SPSS 26.0. To forecast psychophysical parameters to physiological parameters, a back-propagation neural network was conducted using Matlab R2021a.

3. Results

3.1. Relationship between Perceived Soundscape and Acoustical Parameters

Figure 3 shows the distribution of soundscape pleasantness degree (PL) and acoustical parameters, as well as L_{10}, L_{90}, L_{10}-L_{90} and L_{Aeq}, at the study sites. Figure 3a shows that L_{10} was distributed during interval [51.1, 62.2] dBA, and the distribution of PL was concentrated at response 4 ('very pleasant'). As L_{10} further increased, the distribution of PL fluctuated from response 2 to 4 ('slightly pleasant' to 'very pleasant'). Figure 3b shows that L_{90} was distributed during interval [43.0, 54.8] dBA, and the distribution of PL was during interval [3, 4]. As L_{90} further increased, the distribution of PL fluctuated from response 2 to 3 ('slightly pleasant' to 'moderately pleasant'). Figure 3c shows that L_{10}-L_{90} was distributed during interval [5.3, 14.7] dBA, and the distribution of PL was concentrated at response 4 ('very pleasant'). As L_{90} further increased, the distribution of PL fluctuated during interval [2, 3]. Furthermore, Figure 3d shows that L_{Aeq} was distributed during interval [44.7, 55.9] dBA, and the distribution of PL was concentrated at response 4 ('very pleasant'). The distribution of PL decreased from [3, 4] to [2, 3] as L_{Aeq} increased from [55.9, 62.9] dBA to more than 62.9 dBA.

In general, there was a negative tendency between PL and acoustical parameters. We conducted the Pearson correlation analysis to explore the different relationships between these parameters in different forest structures (See Table 1). Based on a total of 960 sets of data, our results showed that the value of perceived soundscape significantly correlated with all acoustical parameters in bamboo forests, and with L_{10}, L_{90} and L_{10}-L_{90} in other forest structures.

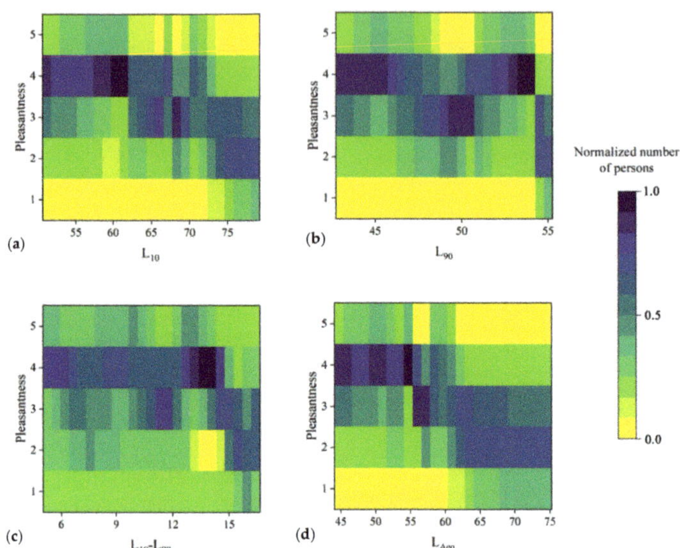

Figure 3. Distribution of soundscape pleasantness level and (**a**) L_{10}, (**b**)L_{90}, (**c**) L_{10}-L_{90} and (**d**) L_{Aeq}, in urban forests.

Table 1. Relationship between acoustic parameters and perceived pleasantness in different forest structures, where Pearson correlation coefficients are shown in each cell.

	Bamboo Forests	Broad-Leaved Forests	Coniferous Forests	Coniferous and Broad-Leaved Mixed Forests
L_{10}	−0.825 **	−0.666 *	−0.821 **	−0.689 *
L_{90}	−0.562 **	−0.441	−0.314	−0.279
L_{10}-L_{90}	−0.720 *	−0.595 *	−0.967 **	−0.793 **
L_{Aeq}	−0.847 **	−0.753 **	−0.676 *	−0.709 *

* $p < 0.05$, ** $p < 0.01$.

3.2. Physiological Indicators in Different Forest Structures

3.2.1. Effect of Stress Inducement for Physiological Indicators

Table 2 shows the variations of electromyography (EMG), electrodermal activity (EDA), photoplethysmography (PPG) and respiration (RESP) at tranquility and stress inducement. Our results showed that the values of EMG, EDA, PPG, and RESP rose at stress inducement based on the difference between Pr and BL. This suggested that the process of stress inducement increased physiological activity such as prefrontal muscle contraction, vigorous activity of exocrine sweat glands, accelerated pulse, and shortness of breath. Furthermore, results of paired-sample t-tests showed significant changes for all physiological indicators, which suggested that the process of stress inducement was effective for physiological indicators in this study.

Table 2. The *t*-test of EMG, EDA, PPG, and RESP at tranquility and stress inducement.

	EMG	EDA	PPG	RESP
Baseline value (BL)	2.847	0.925	39.104	10.052
Pre-test value (Pr)	4.735	1.607	39.890	10.536
Pr-BL	1.888	0.681	0.786	0.484
t value	9.874 **	7.105 **	2.482 **	2.561 **

** $p < 0.01$.

3.2.2. Variation Degree of Physiological Indicators

We conducted t-tests of EMG, EDA, PPG, and RESP in different forest structures based on stress inducement (see Table 3) during pre-test and post-test conditions. For EMG, EDA, and RESP, there were negative tendencies in different forest structures. These physiological indicators showed the most obvious drop in values while in bamboo forests. The decline of EMG and EDA were not obvious in broad-leaved forests. Furthermore, for the PPG of participants, results showed a negative tendency in bamboo forests.

Table 3. The *t*-test of EMG, EDA, PPG, and RESP at pre-test and post-test in different forest structures.

		Bamboo Forests	Broad-Leaved Forests	Coniferous Forests	Coniferous and Broad-Leaved Mixed Forests
EMG	Pr	4.477	4.572	4.979	4.912
	Po	3.800	4.335	4.413	4.436
	ΔEMG	−0.677	−0.237	−0.566	−0.476
	t value	−4.171 **	−2.058 *	−3.637 *	−2.403 *
EDA	Pr	1.562	1.638	1.616	1.611
	Po	1.017	1.404	1.227	1.277
	ΔEDA	−0.546	−0.234	−0.389	−0.334
	t value	−4.628 **	−1.561	3.343 **	−1.764
PPG	Pr	39.854	39.781	40.229	39.698
	Po	38.083	39.708	39.208	40.917
	ΔPPG	−1.770	−0.724	−0.020	1.219
	t value	−3.313 **	−0.125	−1.226	1.204
RESP	Pr	10.57	10.438	10.565	10.577
	Po	10.034	10.276	10.134	10.495
	ΔRESP	−0.536	−0.161	−0.430	−0.076
	t value	−3.062 **	−1.058	−2.832 **	−0.361

* $p < 0.05$, ** $p < 0.01$; Pre-test value (Pr), Post-test value (Po), Po-Pr value (ΔEMG, ΔEDA, ΔPPG, ΔRESP).

For ΔEMG and ΔEDA (See Figure 4), most participants recorded 'moderately restorative' in different forest structures. The total proportion of answers that included 'moderately restorative' and above was more than 65%. This suggested that all forest structures played a role in EMG and EDA for the participants. Furthermore, our results showed that 'slightly restorative' and 'not restorative at all' amounted to a small proportion of answers when in bamboo forests, compared to other forest structures. Few participants answered 'extremely restorative' in broad-leaved forests, which was consistent with the above results for the decline of EMG.

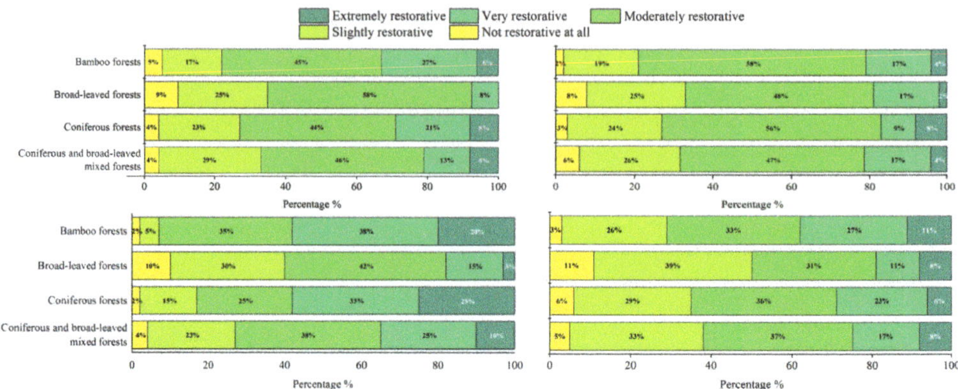

Figure 4. Proportion of ΔEMG (**top left**), ΔEDA (**top right**), ΔPPG (**bottom left**), and ΔRESP (**bottom right**) in different forests Scheme 4. Our results showed that 'slightly restorative' and 'not restorative at all' accounted for more than 40% of responses in broad-leaved forests. In bamboo forests, we found that only 7% of participants answered 'slightly restorative' and 'not restorative at all' for ΔPPG, which suggested a consistency of above results for the decline of physiological indicators. Furthermore, our findings showed the restorative effect of RESP was limited in urban forests, with an average that was more than 38%.

3.3. Modelling the Physiological Restorative Role of Soundscape

3.3.1. Relationship between Psychophysical and Physiological Parameters

Psychophysical and physiological datasets were combined to create a PRR model. The model could be applied to different forest structures to explore the relationship between psychophysical and physiological parameters.

Principal components analysis (PCA) was conducted to reduce the dimensionality of psychophysical and physiological parameters, and to combine the original variables into potential restorative factors [47]. Table 4 shows the PCA results of the psychophysical and physiological dataset. Two components obtained by PCA showcased the differences between psychophysical and physiological parameters: component 1 showed that 66.22% of the variance in functional parameter was due to its large capacity for loading most of the psychophysical and physiological parameters; component 2 showed that 21.68% of the variance in background sound was due to a high factor of loading L_{Aeq} and L_{90}. As restorative factors for the public, these components affected human perception and response to the soundscape in different forest structures. Thus, we suggested a potential interaction between psychophysical parameters and physiological parameters.

Table 4. Summary of principal component analysis (PCA) on physiological restorative parameters.

	Varimax-Rotated Component (Explained Variance, %)	
	Functional Parameter (66.22)	**Background Sound (21.68)**
L_{Aeq}	0.645	0.666
L_{10}	0.704	0.555
L_{90}	0.146	0.961
L_{10}-L_{90}	0.912	0.225
PL	−0.842	−0.290
PRR	−0.880	−0.192
ΔEMG	0.880	0.314
ΔEDA	0.899	0.296
ΔPPG	0.943	0.247
ΔRESP	0.935	0.320

3.3.2. Back-Propagation Neural Network for PRS Model

In our physiological restorativeness soundscape model (PRS model), 960 samples were used and divided into three randomly chosen sets: training set (672 samples, 70.0%), test set (144 samples, 15.0%) and validation set (144 samples, 15.0%). Three-fold cross validation was also conducted.

Table 5 shows the accuracy of the PRS model based on the validation set. The accuracy percentage of both the training and testing sets was more than 90%. After training the PRS model, classified soundscape data results indicated that the accuracy of ΔEMG, ΔEDA, ΔPPG, ΔRESP, and PR were 81.1%, 87.2%, 95.6%, 92.9% and 86.2%, respectively. As shown in Figure 5, ΔEMG, ΔEDA, and ΔRESP maintained stable accuracy during interval $[-2.5, 2.5]$, while ΔPPG and PRR maintained stable accuracy during interval $[-5.0, 10.0]$ and $[-2.5, 3.5]$, respectively. Accuracy decreased when values were outside these intervals.

Table 5. Accuracy of parameters in the PRR model based on the validation set.

	Parameters	Percent Correct (%)
Training	ΔEMG	84.1
	ΔEDA	80.9
	ΔPPG	95.7
	ΔRESP	94.8
	PRR	88.3
	Overall percent	90.8
Testing	ΔEMG	81.1
	ΔEDA	87.2
	ΔPPG	95.6
	ΔRESP	92.9
	PRR	86.2
	Overall percent	90.1

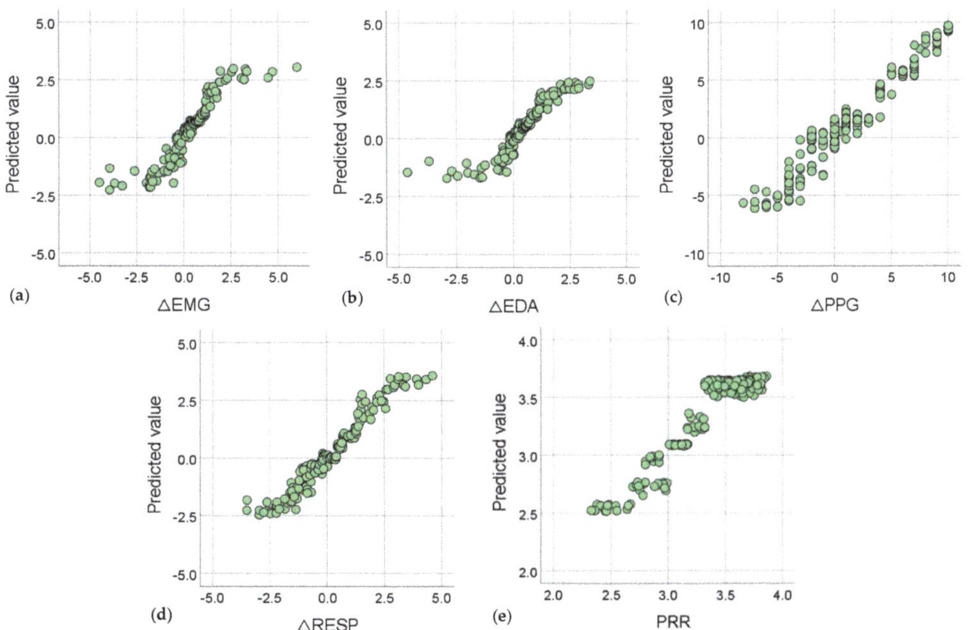

Figure 5. Prediction error for (**a**) ΔEMG, (**b**) ΔEDA, (**c**) ΔPPG, (**d**) ΔRESP and (**e**) PRR in ANNs.

Figure 6 showcases the importance of input variables for determining outputs. Results showed that PL impacted accuracy the most and accounted for more than 35% of the independent variable importance. This suggested that perception was a main driver for physiological parameters.

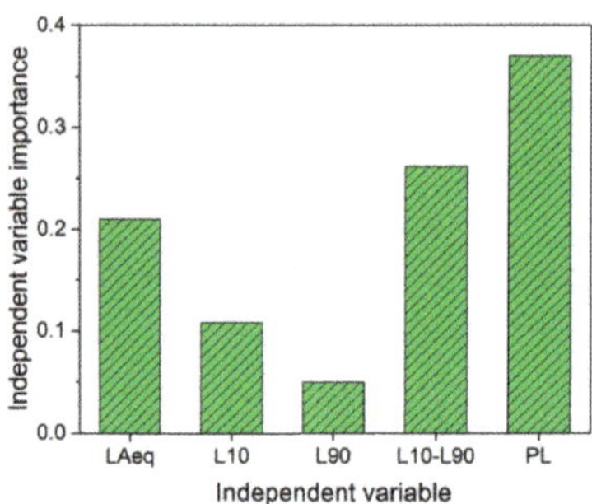

Figure 6. Importance of input indicators in determining the output value of soundscape.

4. Discussion

When Figure 3a,d were combined, our findings showed that L_{Aeq} and L_{10} were important drivers that influenced questionnaire responses in urban forests. This was similar to previous studies conducted in urban green areas [48,49]. We also found that L_{Aeq} and L_{10} displayed overlapping intervals, since there were fewer sources of mechanical noise in urban forests and animals did not need to raise their volume to communicate [50]. As our observation sites were in areas with some tourist activity, the maximum L_{Aeq} was higher when compared to previous studies [51], which suggests more various soundscape exposure contributing to more significant PRR in our research. Furthermore, our results showed a similarity between the distribution of L_{90} and L_{10}-L_{90} (see Figure 3b,c). Most of these values fluctuated and dropped after values reached the threshold, which suggested that the threshold may conduct enough physiological and psychological phenomena on an individual to change them from a steady state to an excited state. Table 4 showed that L_{90} and L_{10}-L_{90} influenced the data for the parameters of physiological restoration from different components. Therefore, there should be more attention and control over L_{90} and L_{10}-L_{90}, especially in forest-based health care [52]. In practice, L_{90} that is lower than 54.8 dBA is beneficial in the creation of quiet areas in urban areas, while L_{10}-L_{90} lower than 14.7 dBA is effective in weakening the negative effects of eventual and unexpected sound events [53,54].

Our findings showed that EMG was the most sensitive physiological parameter in our data set (see Table 3 and Figure 4). This suggested an optimal effect of physiological restorativeness because of the dual sensory channels of input and output in the muscular system, contributing to the cognition and response of participants in forest soundscapes [55,56]. However, the accuracy of EMG was lower than the other parameters in the artificial neural network (ANN). This suggested that EMG was influenced by other environmental drivers and the individual's senses. We also found that the coefficients ranking of parameters in the principal components analysis (PCA) and the accuracy ranking of parameters in the ANN testing group were potentially consistent (see Tables 4 and 5): ΔEMG > ΔEDA > ΔRESP > ΔPPG. As shown in Figure 5, our ANN results suggested that L_{10-90} played the most important role of all physical parameters in determining the output value of soundscape. L_{10-90} was also consistent with the highest coefficient in the functional parameter. These results suggested that PCA can be used as a pre-experiment method for the creation of a model for the physiological restorative role of soundscape (PRR model). Figure 6 showed that soundscape questionnaire responses were the most important for input indicators and suggested that physiological responses were based on the cognitive basis of soundscape [57]. Furthermore, soundscape pleasantness contributed to the enhancement of attention to sound sources [58], which promoted a physiological response. Thus, we found that questionnaires were one of the most important methods for gathering physiological information. Furthermore, previous research suggests different absorption and radiation of the leaves and woods among different forest structures [55,59]. This helps us to understand the physiological restorative role of soundscapes in different forest structures strategies when proposing suitable forest-based health care.

In our study, some limitations may have been presented. Although we tried to minimize the effect of vision on physiological responses by using eye masks, participants were potentially affected by somatosensory effects including the variation of temperature and humidity. Additionally, audio-visual interaction was expected in urban forests, but we did not consider audio-visual drivers in this study.

5. Conclusions

Urban forested areas contribute to favorable exposure to the biophilic outdoor environment, which is beneficial to public health and recovery. This study revealed that psychophysical parameters jointly function in the physiological restorative role of soundscape in urban forested areas. Our findings showed that: (1) L_{Aeq} and L_{10} were important drivers that influence questionnaire responses; (2) EMG was the most sensitive physiological parameter; and (3) L_{10}-L_{90} played the most important role of all physical parameters in the PRS model.

We suggest that the biophilic outdoor environment may offer physiological restorative potential for therapy after COVID-19. Furthermore, other potential drivers such as audio-visual interaction in forested areas may be considered in future studies to further explore physiological restorative patterns in different forest structures.

Author Contributions: Conceptualization, X.-C.H. and S.C.; methodology, X.-C.H. and J.L.; software, E.D.; formal analysis, X.-C.H. and J.-B.W.; investigation, X.-C.H. and S.C.; data curation, X.-C.H.; writing—original draft preparation, X.-C.H., S.C. and J.L.; writing—review and editing, E.D. and J.-B.W.; supervision, Y.C. All authors have read and agreed to the published version of the manuscript.

Funding: This project was supported by the National Natural Science Foundation of China (52208052, 51838003), the National Key Research & Development Program of China (2019YFD1100405), and the Program of Humanities and Social Science Research Program of Ministry of Education of China (Grant No. 21YJCZH038).

Informed Consent Statement: Informed consent was obtained from all subjects involved in the study.

Conflicts of Interest: The authors declare no conflict of interest.

References

1. Leske, S.; Klves, K.; Crompton, D.; Arensman, E.; Leo, D.D. Real-time suicide mortality data from police reports in queensland, australia, during the COVID-19 pandemic: An interrupted time-series analysis. *Lancet Psychiatry* **2021**, *8*, 58–63. [CrossRef]
2. Stec, K. Yoga and relaxation for promoting public health: A review of the practice and supportive research. *Biomed. Hum. Kinet.* **2020**, *12*, 133–140. [CrossRef]
3. Lu, M.; Hu, S.; Mao, Z.; Liang, P.; Xin, S.; Guan, H. Research on work efficiency and light comfort based on EEG evaluation method—Sciencedirect. *Build. Environ.* **2020**, *183*, 107122. [CrossRef]
4. Escobedo, F.; Giannico, V.; Jim, C.; Sanesi, G.; Lafortezza, R. Urban forests, ecosystem services, green infrastructure and nature-based solutions: Nexus or evolving metaphors? *Urban For. Urban Green.* **2019**, *37*, 3–12. [CrossRef]
5. Qiu, M.; Sha, J.; Utomo, S. Listening to Forests: Comparing the Perceived Restorative Characteristics of Natural Soundscapes before and after the COVID-19 Pandemic. *Sustainability* **2021**, *13*, 293. [CrossRef]
6. Wang, P.; He, Y.; Yang, W.; Li, N.; Chen, J. Effects of Soundscapes on Human Physiology and Psychology in Qianjiangyuan National Park System Pilot Area in China. *Forests* **2022**, *13*, 1461. [CrossRef]
7. Shengyan, D.; Fude, S.; Lexiang, Q.; Xinxiang, C.; Shuang, L.; Haomin, L. Forest landscape patterns dynamics of Yihe-Luohe river basin. *J. Geogr. Sci.* **2003**, *13*, 153–162. [CrossRef]
8. Han, L.; Shi, L.; Yang, F.; Xiang, X.Q.; Gao, L. Method for the evaluation of residents' perceptions of their community based on landsenses ecology. *J. Clean. Prod.* **2020**, *281*, 124048.

9. Boulos, M.K.; Al-Shorbaji, N.M. On the internet of things, smart cities and the who healthy cities. *Int. J. Health Geogr.* **2014**, *13*, 10. [CrossRef]

10. Ding, R.; Logemann, J.A.; Larson, C.R.; Rademaker, A.W. The effects of taste and consistency on swallow physiology in younger and older healthy individualsa surface electromyographic study. *J. Speech Lang. Hear. Res.* **2014**, *46*, 977–989. [CrossRef]

11. Yackinous, C.; Guinard, J.X. Relation between prop taster status and fat perception, touch, and olfaction. *Physiol. Behav.* **2001**, *72*, 427–437. [CrossRef]

12. Xiao, J.; Tait, M.; Kang, J. A perceptual model of smellscape pleasantness. *Cities* **2018**, *76*, 105–115. [CrossRef]

13. Hunt, D.F.; Hunt, H.; Park, J.H. Bioenergetic costs and state influence distance perception. *Physiol. Behav.* **2017**, *180*, 103–106. [CrossRef] [PubMed]

14. Skoe, E.; Krizman, J.; Spitzer, E.; Kraus, N. The auditory brainstem is a barometer of rapid auditory learning. *Neuroence* **2013**, *243*, 104–114. [CrossRef]

15. Bingham, G.P.; Winona, S.C.; Zhu, Q. Information about relative phase in bimanual coordination is modality specific (not amodal), but kinesthesis and vision can teach one another. *Hum. Mov. Sci.* **2018**, *60*, 98–106. [CrossRef]

16. Liu, J. Towards a unified model of human information behavior: An equilibrium perspective. *J. Doc.* **2017**, *73*, 666–688. [CrossRef]

17. Viaud-Delmon, I.; Warusfel, O. From ear to body: The auditory-motor loop in spatial cognition. *Front. Neurosci.* **2014**, *9*, 283. [CrossRef]

18. Guo, X.; Liu, J.; Albert, C.; Hong, X.-C. Audio-visual interaction and visitor characteristics affect perceived soundscape restorativeness: Case study in five parks in China. *Urban For. Urban Green.* **2022**, *77*, 127738. [CrossRef]

19. Steffens, J. When do we judge sounds? relevant everyday situations for the estimation of ecological validity of indoor soundscape experiments. *J. Acoust. Soc. Am.* **2013**, *133*, 3371. [CrossRef]

20. ISO/TS 12913-1:2014. Acoustics–Soundscape–Part 1: Definition and Conceptual Framework; International Organization for Standardization ISO: Geneva, Switzerland, 2014.

21. Hong, X.-C.; Wang, G.-Y.; Liu, J.; Dang, E. Perceived loudness sensitivity influenced by brightness in urban forests: A comparison when eyes were opened and closed. *Forests* **2013**, *11*, 1242. [CrossRef]

22. Miller, Z.D.; Hallo, J.C.; Sharp, J.L.; Powell, R.B.; Lanham, J.D. Birding by ear: A study of recreational specialization and soundscape preference. *Hum. Dimens. Wildl.* **2014**, *19*, 498–511. [CrossRef]

23. Hu, F.; Wang, Z.; Sheng, G.; Lia, X.; Chen, C.; Geng, D.; Hong, X.; Xu, N.; Zhu, Z.; Zhang, Z.; et al. Impacts of national park tourism sites: A perceptual analysis from residents of three spatial levels of local communities in Banff national park. *Environ. Dev. Sustain.* **2021**, *24*, 3126–3145. [CrossRef]

24. Simkin, J.; Ojala, A.; Tyrvinen, L. Restorative effects of mature and young commercial forests, pristine old-growth forest and urban recreation forest—A field experiment. *Urban For. Urban Green.* **2020**, *48*, 126567. [CrossRef]

25. Xu, X.; Wu, H. Audio-visual interactions enhance soundscape perception in China's national parks. *Urban For. Urban Green.* **2020**, *61*, 127090. [CrossRef]

26. Calleja, A.; Díaz-Balteiro, L.; Iglesias-Merchan, C.; Solio, M. Acoustic and economic valuation of soundscape: An application to the 'retiro' urban forest park. *Urban For. Urban Green.* **2017**, *27*, 272–278. [CrossRef]

27. Payne, S.R. The production of a perceived restorativeness soundscape scale. *Appl. Acoust.* **2013**, *74*, 255–263. [CrossRef]

28. Hong, X.; Wang, G.; Liu, J.; Lan, S. Cognitive persistence of soundscape in urban parks. *Sustain. Cities Soc.* **2019**, *51*, 17–26. [CrossRef]

29. Kim, Y.H.; Hwang, I.H.; Hong, J.Y.; Lee, S.C. Effects of vegetation on soundscape of an urban religious precinct: Case study of Myeong-dong cathedral in Seoul. *Build. Environ.* **2019**, *155*, 389–398. [CrossRef]

30. Chang, C.Y.; Hammitt, W.E.; Chen, P.K.; Machnik, L.; Su, W.C. Psychophysiological responses and restorative values of natural environments in taiwan. *Landsc. Urban Plan.* **2008**, *85*, 79–84. [CrossRef]

31. Gathright, J.; Yamada, Y.; Morita, M. Comparison of the physiological and psychological benefits of tree and tower climbing. *Urban For. Urban Green.* **2006**, *5*, 141–149. [CrossRef]

32. Shu, S.; Ma, H. Restorative effects of urban park soundscapes on children's psychophysiological stress. *Appl. Acoust.* **2020**, *164*, 107293. [CrossRef]

33. Korpela, K.M.; Klemettila, T.; Hietanen, J.K. Evidence for rapid affective evaluation of environmental scenes. *Environ. Behav.* **2002**, *34*, 634–650. [CrossRef]

34. Kim, J.; Cha, S.H.; Koo, C.; Tang, S.-K. The effects of indoor plants and artificial windows in an underground environment. *Build. Environ.* **2018**, *138*, 53–62. [CrossRef]

35. Martinsen, O.G.; Grimnes, S. *Bioimpedance and Bioelectricity Basics*; Academic Press: Pittsburgh, PA, USA, 2011.

36. Elsadek, M.; Sun, M.; Sugiyama, R.; Fujii, E. Cross-cultural comparison of physiological and psychological responses to different garden styles. *Urban For. Urban Green.* **2019**, *38*, 74–83. [CrossRef]

37. Calfee, C.S.; Delucchi, K.; Parsons, P.E.; Thompson, B.T.; Ware, L.B.; Matthay, M.A.; Nhlbi Ards Network. Subphenotypes in acute respiratory distress syndrome: Latent class analysis of data from two randomised controlled trials. *Lancet Respir. Med.* **2014**, *2*, 611–620. [CrossRef]

38. Liu, L.; Qu, H.; Ma, Y.; Wang, K.; Qu, H. Restorative benefits of urban green space: Physiological, psychological restoration and eye movement analysis. *J. Environ. Manag.* **2022**, *301*, 113930. [CrossRef]

39. Liu, Y.; Hu, M.; Zhao, B. Audio-visual interactive evaluation of the forest landscape based on eye-tracking experiments. *Urban For. Urban Green.* **2019**, *46*, 126476. [CrossRef]
40. Ye, H.; Zhu, X. Study on the Effect of Spatial Characteristics and Health Restoration of Urban Parks in Harbin City in Autumn—Taking Zhaolin Park as an Example. *J. Hum. Settl. West China* **2018**, *33*, 73–79.
41. Wang, Z.; Li, Y.; An, J.; Dong, W.; Li, H.; Ma, H.; Wang, J.; Wu, J.; Jiang, T.; Wang, G. Effects of Restorative Environment and Presence on Anxiety and Depression Based on Interactive Virtual Reality Scenarios. *Int. J. Environ. Res. Public Health* **2022**, *19*, 7878. [CrossRef]
42. Huang, S.; Qi, J.; Li, W.; Dong, J.; Bosch, C.K.V.D. The Contribution to Stress Recovery and Attention Restoration Potential of Exposure to Urban Green Spaces in Low-Density Residential Areas. *Int. J. Environ. Res. Public Health* **2021**, *18*, 8713. [CrossRef]
43. ISO/TS 12913-2:2018. Acoustics—Soundscape—Part 2: Data Collection and Reporting Requirements; International Organization for Standardization ISO: Geneva, Switzerland, 2018.
44. Yu, L.; Kang, J. Modeling subjective evaluation of soundscape quality in urban open spaces: An artificial neural network approach. *J. Acoust. Soc. Am.* **2009**, *126*, 1163–1174. [CrossRef] [PubMed]
45. Yang, J.; Yu, Y.; You, L. Segmentation by Visitor Motivation in Fuzhou National Forest Park: A Factor-Cluster Approach. *Scientia Silvae Sin.* **2015**, *51*, 106–116.
46. Perfect, T.J.; Andrade, J.; Eagan, I. Eye closure reduces the cross-modal memory impairment caused by auditory distraction. *J. Exp. Psychol. Learn. Mem. Cogn.* **2011**, *37*, 1008–1013. [CrossRef]
47. Nichols, S.M.; Bradley, D.L. Characterization of very low frequency ambient noise by principal component analysis. *J. Acoust. Soc. Am.* **2016**, *140*, 3352. [CrossRef]
48. Zhang, X.; Ba, M.; Kang, J.; Meng, Q. Effect of soundscape dimensions on acoustic comfort in urban open public spaces. *Appl. Acoust.* **2018**, *133*, 73–81. [CrossRef]
49. Ma, K.W.; Mak, C.M.; Hai, M.W. Effects of environmental sound quality on soundscape preference in a public urban space. *Appl. Acoust.* **2021**, *171*, 107570. [CrossRef]
50. Farina, A. *Soundscape Ecology: Principles, Patterns, Methods and Applications*; Springer: Berlin/Heidelberg, Germany, 2021.
51. Hong, X.; Liu, J.; Wang, G.; Jiang, Y.; Wu, S.; Lan, S. Factors influencing the harmonious degree of soundscapes in urban forests: A comparison of broad-leaved and coniferous forests. *Urban For. Urban Green.* **2019**, *39*, 18–25. [CrossRef]
52. Nilsson, K. Papers from sessions on forests, trees and human health and wellbeing. *Urban For. Urban Green.* **2006**, *5*, 109. [CrossRef]
53. Cerwén, G.; Mossberg, F. Implementation of Quiet Areas in Sweden. *Int. J. Environ. Res. Public Health* **2019**, *16*, 134. [CrossRef]
54. Hong, X.-C.; Wang, G.-Y.; Liu, J.; Song, L.; Wu, E.T. Modeling the impact of soundscape drivers on perceived birdsongs in urban forests. *J. Clean. Prod.* **2021**, *292*, 125315. [CrossRef]
55. Hong, X.-C.; Zhu, Z.-P.; Liu, J.; Geng, D.-H.; Wang, G.-Y.; Lan, S.-R. Perceived occurrences of soundscape influencing pleasantness in urban forests: A comparison of broad-leaved and coniferous forests. *Sustainability* **2019**, *11*, 4789. [CrossRef]
56. Payne, S.R.; Guastavino, C. Exploring the validity of the perceived restorativeness soundscape scale: A psycholinguistic approach. *Front. Psychol.* **2018**, *9*, 2224. [CrossRef] [PubMed]
57. Dijk, P.V.; Başkent, D.; Gaudrain, E.; Kleine, E.D.; Wagner, A.; Lanting, C. Physiology, psychoacoustics and cognition in normal and impaired hearing. In *Advances in Experimental Medicine Biology*; Springer International Publishing: Berlin/Heidelberg, Germany, 2017.
58. Erfanian, M.; Mitchell, A.; Aletta, F.; Kang, J. Psychological well-being and demographic factors can mediate soundscape pleasantness and eventfulness: A large sample study. *J. Environ. Psychol.* **2021**, *77*, 101660. [CrossRef]
59. Embleton, T.F.W. Sound propagation in homogeneous deciduous and evergreen woods. *J. Acoust. Soc. Am.* **1963**, *35*, 1119. [CrossRef]

Article

Audio-Visual Preferences for the Exercise-Oriented Population in Urban Forest Parks in China

Jian Xu [1,2,3,4], Muchun Li [1,*], Ziyang Gu [1], Yongle Xie [1] and Ningrui Jia [1]

[1] Department of Tourism Management, South China University of Technology, Guangzhou 510006, China; jianxu@scut.edu.cn (J.X.); 201930460128@mail.scut.edu.cn (Z.G.); 201930460463@mail.scut.edu.cn (Y.X.); 201930460203@mail.scut.edu.cn (N.J.)
[2] State Key Laboratory of Subtropical Building Science, Guangzhou 510641, China
[3] Key Laboratory of Digital Village and Sustainable Development of Culture and Tourism, Guangzhou 510641, China
[4] Guangdong Tourism Strategy and Policy Research Center, Guangzhou 510641, China
*   Correspondence: limch@scut.edu.cn; Tel.: +86-181-288-6885

Abstract: The purpose of this study is to explore the audio-visual preferences of exercisers in urban forest parks in China and to make practical suggestions for park landscape design. Taking Beigushan Forest Park in Lianyungang City, Jiangsu Province as a case, based on field research and questionnaire survey, this study analyzed the audio-visual preference characteristics of exercisers in the park, revealed the correlation between audio-visual preference and exercisers' behaviors and individual characteristics, and explored the influence of audio-visual preferences on exercise feelings by establishing a structural equation model. It was found that (1) the forest and its avenue landscape and birdsong are most preferred by exercisers; (2) the audio-visual preferences of people with different exercise forms differ, for example, people who slowly walk, run, and briskly walk have stronger preferences for natural soundscape and visual landscape, while people who use fitness equipment have stronger inclusiveness for human activity sound and prefer public facility-based landscapes. In addition, some individual characteristics such as exercise intensity and exercise frequency significantly affect exercisers' audio-visual preferences; (3) visual landscape preferences have a greater direct impact on exercise feelings, with natural waterscape having the greatest direct impact, but overall soundscape preferences do not have a high degree of direct impact on exercise feelings, with natural sound still having a strong positive impact. These findings provide a more quantitative basis for the landscape design of urban forest parks from the perspective of exercise behavior.

Keywords: urban forest park; exercise behaviors; audio-visual preferences; correlation analysis

Citation: Xu, J.; Li, M.; Gu, Z.; Xie, Y.; Jia, N. Audio-Visual Preferences for the Exercise-Oriented Population in Urban Forest Parks in China. *Forests* **2022**, *13*, 948. https://doi.org/10.3390/f13060948

Academic Editors: Xin-Chen Hong, Jiang Liu and Guang-Yu Wang

Received: 21 May 2022
Accepted: 14 June 2022
Published: 17 June 2022

Publisher's Note: MDPI stays neutral with regard to jurisdictional claims in published maps and institutional affiliations.

1. Introduction

As a new type of park arising from urbanization in the era of ecological civilization [1], urban forest parks have multiple functions such as recreation, recuperation, and avoiding the heat, which help improve air quality [2], reduce noise [3], and provide a pleasant environment for people to promote physical and mental health development [4]. Forest parks can meet the needs of urban residents for environment, leisure, and sports, thus attracting an increasing number of urban residents for physical exercise close to nature [5]. In China, where urbanization is accelerating, the relationship between the health level and quality of life of urban residents and urban forest parks has become increasingly important [6], especially in the context of COVID-19, where the demand for parks and outdoor green spaces has increased rather than decreased [7]. Studies have shown that physical exercise in green spaces can release stress and enhance the body's ability to fight infectious diseases [7–9]. Therefore, it can be predicted that in the future, the demand for forest parks and exercise activities will continue to grow and more urban residents can benefit from the services of forests and parks [10]. In this context, exercisers in urban forest parks and their audio-visual preferences become the focus of this study.

In the studies on exercise behavior in urban forest parks, many scholars have devoted their research to exploring the main factors that influence the attractiveness and satisfaction of exercise in parks. For example, Li et al. found that open activity space with waterscape, landscape sketch, can attract more people toward exercise activities [11]; McCormack et al. summarized qualitative studies about the relationship between park use and physical exercise and found that safety, aesthetics, park facilities, and landscape maintenance were important factors influencing satisfaction with park use [12]. Although some scholars have also emphasized the importance of individual perceptual factors in influencing park use satisfaction [13,14], the majority of scholars have focused on objective factors such as facilities and public services provided by parks, and have not paid enough attention to people's underlying psychological motivations and preferences. Therefore, an investigation of people's exercise satisfaction from the perspective of their visual and auditory preferences at the psychological level would provide a relevant complement to the current research on the factors influencing exercise satisfaction in parks.

People's perceptions of the landscape initially originate from human intuitive experiences, in which people rely on their eyes to obtain 87% of the information from the outside world, and 75–90% of human activities is visually induced [15]. Studies have shown that individual visual aesthetic preferences influence people's perceptions of ecological and aesthetic values, which further influence their behavioral choices. [16]. For example, Ma et al. explored the influence of the degree of visual landscape heterogeneity on landscape aesthetic quality and public visual perception effects [17]; Zhang et al. used eye-tracking to explore the visual preferences of different types of visitors to trail landscapes and revealed the reasons for the differences in visitors' landscape gaze time [18]. Visual factors also affect people's perception and evaluation of soundscapes, and the relationship between these two is the focus of this study. In this regard, many scholars have made significant research contributions. Cassina et al., proposed a linear model for predicting perceived tranquility in different environments based on visual and acoustic features [19]. Romero et al. found that visual factors such as ocean visibility can affect the perception of the soundscape quality in the areas with road traffic [20], and they also found there are color associations between people and different urban soundscapes [21]. Preis et al. found that the addition of visual information increases the noise annoyance assessment [22]. Moreover, numerous studies have also demonstrated that people's visual preferences affect the perception of landscape and environmental behavior [23,24].

Soundscape is the acoustic environment perceived by an individual, group, or community in a given scene [25] and is highly relevant to people's health [26]. With the increasing concern about health, urbanization, and globalization, more and more studies are focusing on soundscape. It has been shown that people's perceptions and preferences of soundscapes can play a key role in the construction of related landscapes in urban forest parks [27]. Currently, most of the soundscape studies focus on people's perception. Among them, scholars have found that people's perception of soundscape is related to the type of soundscape, people's personal preferences and sensitivities, and demographic indicators related to soundscape [28–32]. For example, Fang et al. found that five main dimensions of social, demographic, and behavioral attributes (age and familiarity of site, educational and economic condition, companion and type of recreational use, gender, and length of stay) were associated with people's soundscape perceptions and preferences [33]. In addition, scholars have expanded their research in related fields, such as Hong et al. who analyzed the relationship between each soundscape element that has an impact on forest park soundscapes and its physical stimulus amount and people's soundscape preferences [34]. Subdivided into the field of soundscape preference research, some scholars have found that natural sounds are more preferred by people [35]; some scholars have put people's soundscape preferences in the context of COVID-19 and found that individual characteristics such as age, occupation, education level, and life happiness are the main factors affecting soundscape preferences [36]; in addition, the frequency of visits to destinations also affects people's preferences for beautiful soundscapes [37]. Although the above studies have addressed different influencing factors of soundscape preference, few studies have focused on soundscape preference among a population with specific behaviors; therefore, the variability in soundscape preference cannot be explained in a more behavioral characteristic sense.

At present, many research results in the field of audio-visual perception are directed to the applied science fields such as medicine and engineering, but there are still few reports in the natural science fields such as landscape and ecology, as well as natural and social interdisciplinary subjects. Among the studies on landscape and ecological environment that focus on audio-visual perception, there are mostly studies on people's single-sensory perception and preference, but there is a lack of studies on multisensory preference and its interaction. Therefore, this study investigates the audio-visual preferences of the exercisers, which will be helpful to explore and improve the research on the audio-visual field of the exercisers in urban forest parks. In this study, 406 exercisers in Beigushan Forest Park, Lianyungang, Jiangsu Province, were surveyed according to the subjective evaluation method. The purpose of this study is to explore the audio-visual preferences of exercisers in China's urban forest parks, and to reveal the correlation between these preferences and exercisers' behaviors and individual characteristics, so as to put forward practical suggestions for park landscape design. Unlike most previous studies, this study is novel in that it focuses on audio-visual preferences among a specific behavioral population and reveals differences in audio-visual preferences from an environmental behavioral perspective. However, the restrictions on pedestrian flow in the park under the influence of the epidemic and the impact of some precautionary measures on people's landscape evaluation pose certain challenges to this study. Overall, the study helps to maximize the usefulness of natural resources and provide auxiliary visual and acoustic landscape design for urban forest park designers and planners, while providing better exercise experience for exercise groups and improving people's quality of life and happiness.

This paper is divided into four parts. The method part mainly introduces the study area, questionnaire design, and field research. In the result part, firstly, the reliability and validity test results of the collected questionnaires and the statistical results of the respondents' personal characteristics are analyzed, then the audio-visual preference characteristics of the exercisers are revealed, and the correlation between the audio-visual preferences and exercisers' behaviors as well as their personal characteristics is discussed through correlation analysis. Finally, the influence of audio-visual preferences on the exercise feeling is explored by establishing a structural equation model. On the basis of comparing and summarizing the similarities and differences between this study and the existing scholars' research, this discussion section explores the landscape design of urban forest parks from the perspective of exercise behavior. In the conclusion section, the full text and its important points are reviewed, while the limitations and future work of this study are also summarized.

This study explores the following issues to be addressed:

1. What are the audio-visual preferences of exercisers in urban forest parks?
2. What is the correlation between audio-visual preferences and exercise behavior choices and individual characteristics?
3. What is the effect of each audio-visual preference on people's exercise feelings?

2. Materials and Methods

2.1. Study Area

Beigushan Forest Park in Lianyungang City, Jiangsu Province, China, was selected as the case site for the study. The location map of Beigushan Forest Park is shown in Figure 1. The park has three major ecosystems: marine, forest, and wetland, with a forest coverage of 86.3%. Beigushan Forest Park is a mountainous forest park, which is the most common type of forest park in China. The average annual temperature is 14 °C, which is similar to China's average annual temperature of 13 °C. The park has good accessibility and experiences strong demand from residents; the trail around the mountain was officially opened to the public in March 2017 with new facilities. It is 5000 m in length, which circles around Beigu Mountain, and won the "2017 Jiangsu Most Beautiful Running Route" award. The exercisers account for 19% of visitors to this park, which is comparable to the percentage of Chinese nationals exercising in urban forest parks [38], and is highly representative as a case study for this study.

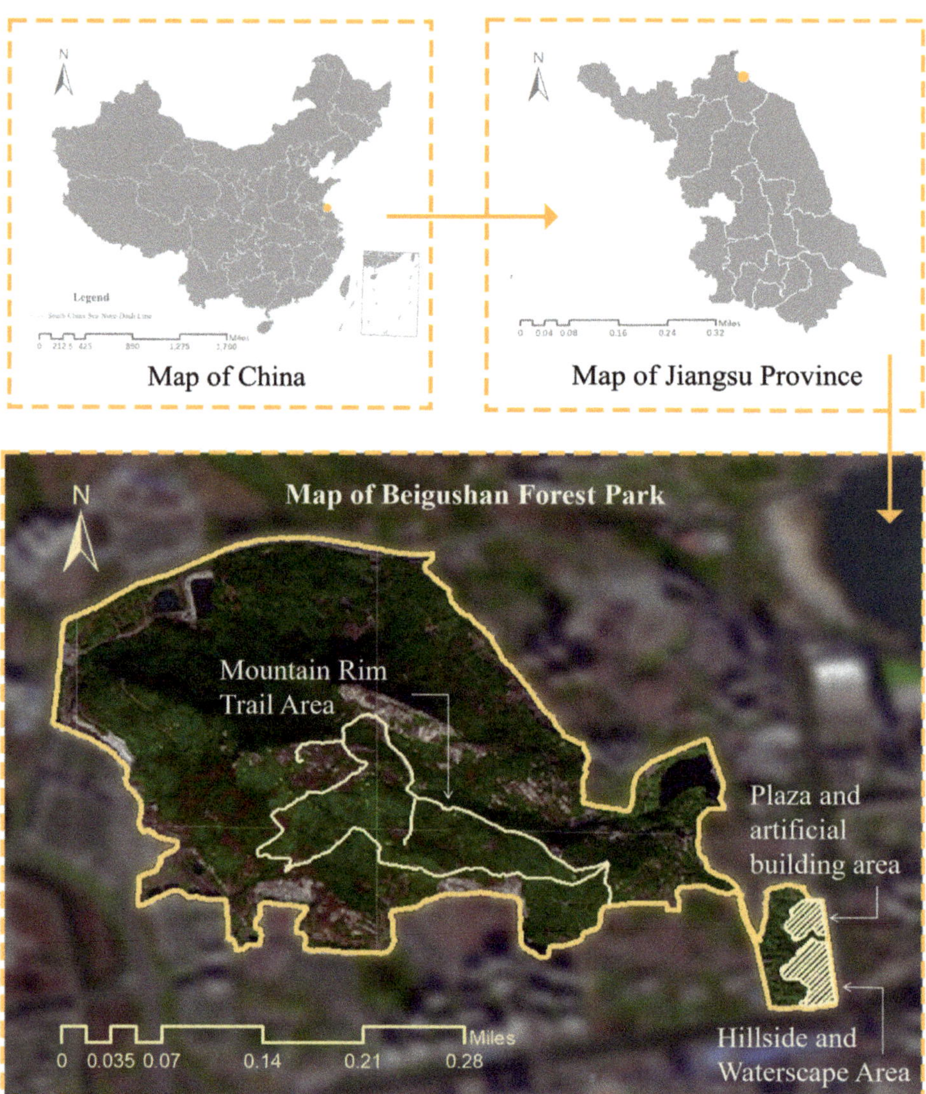

Figure 1. Geographical location of Beigushan Forest Park in Lianyungang.

Before conducting the questionnaire survey, through fieldwork, Beigushan Forest Park can be divided into a plaza and artificial building area, a hillside and waterscape area, and a mountain rim trail area. There are 17 main types of landscapes in Beigushan Forest Park, including natural waterscape (streams, ponds, and lakes), topographic landscape (lawns, avenues, hillsides, and lakesides), natural vegetation (shrubs, ornamental flowers, and forests), artificial landscape (rockery, parterres, fountains, sculptures, bridges, and pavilions), artificial facilities (fitness equipment, squares, and public buildings). We often hear 16 different kinds of sounds, including natural sounds (sound of wind, birdsong, cry of insects, rustle of leaves, and water flow sound), human activity sounds (conversational voice, sound of children playing, footstep, and exercise sound), and artificial sounds (traffic sound, entertainment sound, device music, construction noise, machine noise, and broadcast).

2.2. Questionnaire Design

Respondents were asked to fill out the questionnaires created by "Questionnaire Star" using the tablet PCs provided to them by the researchers ("Questionnaire Star" is a professional, unlimited free online questionnaire, assessment, voting platform, focusing on providing users with a powerful, user-friendly online questionnaire design; free to use the program, it provides powerful, fast, easy to use, and low-cost obvious advantages [39]. The questionnaire star program has released a total of 154 million questionnaires, which can fully meet the number of questionnaire research and question type setting requirements; https://www.wjx.cn/, accessed on 25 January 2022). In addition, the researchers also prepared a certain number of paper questionnaires for the elderly who cannot use electronic devices skillfully.

The questionnaire consisted of four parts, with 18 questions in total. The first part was designed to collect demographic information about the respondents, such as age, gender, number of participants in exercise activities, distance of residence from the target park, and activities performed in the park other than exercise; the second part focused on the exercise profile of the respondents, including exercise mode, exercise time, exercise duration, exercise frequency, driving factors for exercising, reasons for choosing the park as an exercise site, specific location of exercise, and exercise frequency. The second part focused on the respondents' exercise patterns, exercise time, exercise duration, exercise frequency, exercise site, exercise intensity, overall feelings of exercise, and willingness to exercise in parks in the future. In the third and fourth sections, the three types that constitute a soundscape as defined by Kraus [40] (abiotic natural sounds from the physical environment, nonhuman biological sounds emitted by all organisms in a given habitat, and anthropomorphic sounds emitted by stationary and moving man-made objects) were used as the basis for classifying the types of soundscapes in the questionnaire. On this basis, the scales used in the study of landscape perception by scholars such as Zheng Zhao [41] and the scales involved in the study of urban forest park soundscape by Wei Zhao [42] and Banu Chitra [43], respectively, were used, and the scales used in this study were appropriately adjusted and modified by combining the ISO [44] definition and classification of soundscape and the actual situation of Beigushan Forest Park in Lianyungang. The third part focused on the understanding of individual visual landscape preferences, mainly using a five-point Likert scale (strongly dislike (−2), dislike (−1), average (0), like (1), and like very much (2)) to illustrate their overall preferences for visual landscapes. In the fourth section, a selection of the frequency of occurrence of 16 soundscapes and a five-point rating of the soundscapes (from very dislike (−2) to very like (2)) were included to illustrate the overall preference of respondents for common soundscapes. After a pre-research test with 20 people, the average response time was 4 min and 39 s, all of whom had no objections to answer the questionnaire questions. The relevant contents of the questionnaire are shown in Table 1.

Table 1. Exerciser landscape and soundscape preference system.

Question	Sub-Factors	Type	Key Findings	Reference
Q5 Activities other than exercise	Family bonding type activities	Activity type	Exercisers may engage in activities in the park other than those related to exercise activities, and these activities may influence, to some extent, exercisers' preferences for landscape and soundscape.	[33]
	Social activities			
	Leisure activities			
	Quiet-type activities			
	Group activities			
	Just exercise			
Q6 Based on different types of exercise	Slow walking	Exercise form	The common types of exercise in the park are listed, and the correlation with landscape type and soundscape type is explored from the perspective of exercise type.	[45]
	Jogging			
	Brisk walking			
	Using fitness equipment			
	Square dance			
	Gymnastics			
	Chinese Martial arts			
	Ball games			
Q10 Different types of areas based on parks	Square open space	Exercise site	Focus on the main types of sites present in the park, and study the preference of the exercisers for landscape types and soundscape types through the analysis of site types.	[40]
	Lawn footpath			
	Forest footpath			
	Lakeside footpath			
	Fitness equipment venue			
Q14 Evaluation of the landscape	Natural waterscape	Landscape	A comprehensive overview of the existing landscape in the park from five aspects: natural waterscape, topographic landscape, natural vegetation, artificial landscape, and artificial facilities, based on which to study the preference of exercisers for a certain landscape or landscape type.	[40]
	Topographic landscape			
	Natural vegetation			
	Artificial landscape			
	Artificial facilities			
Q15 Evaluation of Soundscape	Natural sounds	Soundscape	The soundscape is divided into three categories: natural, human activity and artificial sound, and the preference of the exercise population in the park for a certain soundscape is explored.	[33]
	Human activity sounds			
	Artificial sounds			

2.3. Field Research

The sampling sites covered three subdivisions in the park, including three sampling sites in the mountain rim trail area, two sampling sites in the hillside and water area, and one sampling site in the plaza and artificial building area, for a total of six sampling sites, the specific locations of which are shown in Figure 2. The survey was conducted during the daytime in April 2021 under sunny weather, and each survey lasted for eight hours (from about 9:00 to 17:00); when the average temperature is about 17 °C, the climate is suitable, the vegetation is abundant, the residents are willing to travel more, and the number of exercise activities is higher. Before each part of research, attention was paid to temperature, relative humidity and wind speed, and similar weather was used for the research to avoid differences in audio-visual preferences of respondents due to climate effects. Additionally, to reduce any bias due to the selection of respondents at a specific time, each sampling site was surveyed twice on different days. The research was conducted using anonymous random interviews, where respondents were first explained the purpose and procedures of the survey, which did not mention positive or negative sounds, noise pollution, etc. They were then informed that their responses would be anonymous. To avoid distractions from other participants during their stay on the site, those who wished to participate in the survey were given a tablet containing the questionnaire and invited to fill it out individually. Respondents were invited to go to a secluded place near their location while the questionnaire was being filled out, and the ambient sound was tested using a decibel meter to ensure that there were no significant sound disturbances in the surroundings. Due to the short duration of the questionnaire and the fact that only a limited number of sounds may be present during a given time period, participants were asked to respond for the length of time chosen in the questionnaire, based on their long-term experience in the park. In addition to this, for the issue of hearing impairment, specific questioning was conducted prior to the study and observations were made to ensure that respondents did not have any significant hearing impairment.

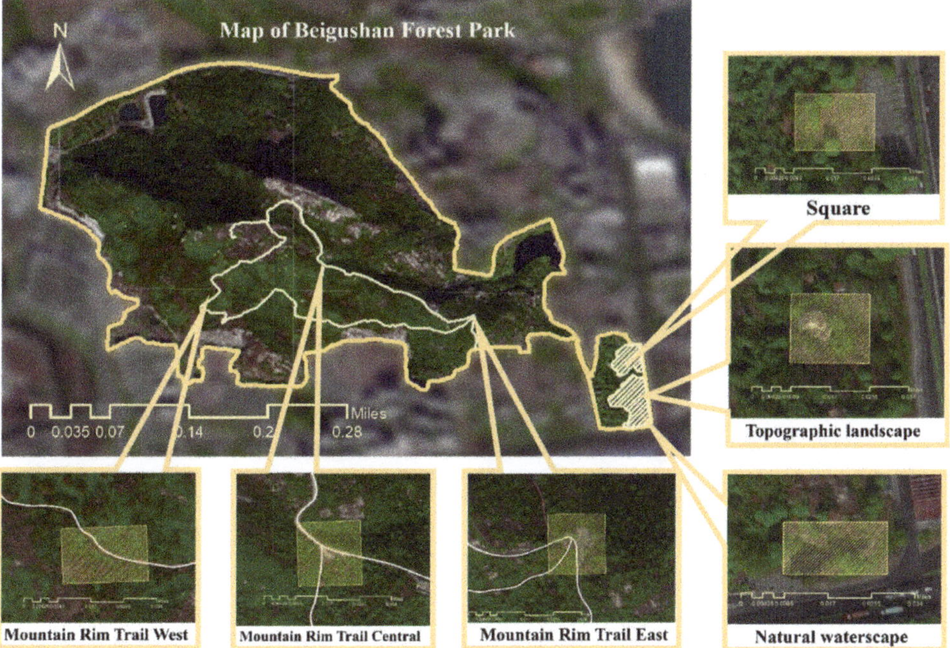

Figure 2. Sampling point distribution map.

In order to reduce the influence of the order effect on the accuracy of the questionnaire results due to the single form of questions, the researcher randomly switched the order of scoring questions to improve the accuracy of the questionnaire results before interviewing the respondents, and used the method of setting irrelevant interfering items to filter the questionnaire (eliminating the questionnaire with irrelevant interfering items), so as to guarantee the authenticity and credibility of the questionnaire results to the greatest extent. For the final collection of 406 questionnaires, the number of actual valid questionnaires was 344, and the questionnaire efficiency was 84.7%.

The research framework is shown in Figure 3.

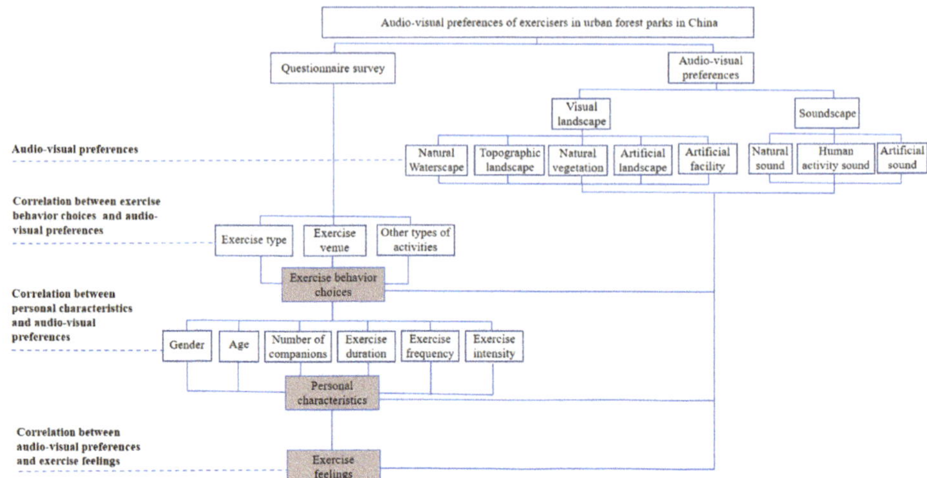

Figure 3. Research framework.

3. Results

3.1. Data Testing and Demographic Analysis

After the 344 valid questionnaires were collected and sorted, SPSS 26.0 software (IBM, Armonk, NY, USA) was used to test the reliability and validity of the questionnaire data. In this study, Cronbach's alpha was used to analyze the reliability of the questionnaire, and $\alpha \geq 0.7$ represents reliable results [46]. The results of the questionnaire were calculated to meet this reliability criterion: natural water features (0.8), topographic landscapes (0.862), natural vegetation (0.772), artificial landscapes (0.891), and public facilities (0.849); natural sounds (0.839), activity sounds (0.936), and artificial sounds (0.921). The reliability of the overall perception factor was 0.914. Thus, it can be seen that the reliability of the questionnaire meets the survey requirements. In this study, the validity of KMO was tested by factor analysis, and KMO = 0.92, which satisfied the condition of factor analysis (KMO $\geq$ 0.6), indicating that the validity of the questionnaire also met the requirements. Bartlett's ball test approximated a chi-square value of 9331.808, corresponding to a probability value of 0.000 ($p < 0.01$), indicating that the questionnaire measures significant correlation of the question items and that the data are valid.

The personal characteristics of the respondents are shown in Figure 4. The proportions of respondents were 48% and 52% for men and women, respectively, which were relatively equal (Figure 4a), but there were fewer respondents over the age of 60, and the respondents were mainly the young and middle-aged group (Figure 4b). In terms of travel mode (Figure 4c), respondents traveled in a variety of ways, and exercising with three to five friends was the composition of the largest number of exercisers. In terms of the distance of the respondents' addresses from the park (Figure 4d), 500–3000 m (36% for 500–1500 m and 28% for 1500–3000) accounted for the majority, and the majority of exercisers were living near the park (Figure 4e) About 58% of the respondents exercised for 1–2 h, and the overall frequency of exercise was low (Figure 4f), generally concentrated on once a month, 2–3 times a month and 1–2 times a week, and more respondents (39%) exercised during the time period of 18:00–21:00 (Figure 4g), with 61% of respondents only exercising lightly (Figure 4h), exercisers generally exercised less intensely.

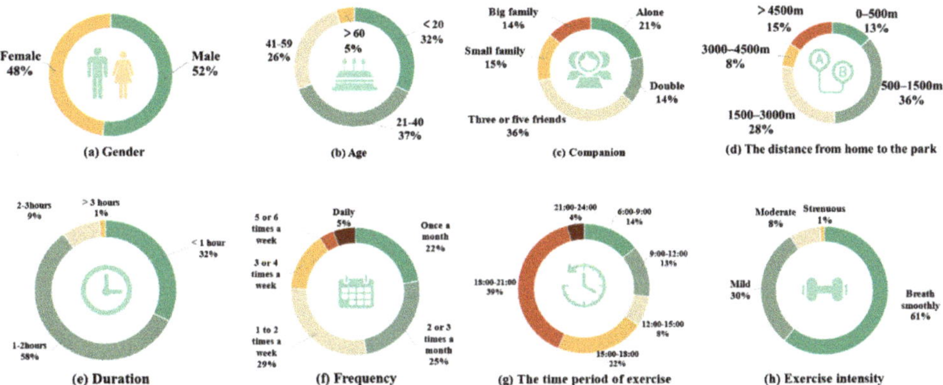

Figure 4. Statistical results of respondents' personal characteristics (**a–h**): gender, age, companion, the distance from home to the park, duration, frequency, and the time period of exercise and exercise intensity.

3.2. Audio-Visual Preference Characteristics of Exercisers in Urban Forest Park

As shown in Figure 5, the most preferred visual landscapes for people exercising in urban forest parks are, in order of preference: avenues, forests, streams, lakesides, bridges and pavilions, ornamental flowers, and lawns, and less preferred landscapes are rockery, public buildings, and sculptures. In terms of overall categories, people prefer topographic landscapes and natural landscapes, and have a lower preference for artificial landscapes.

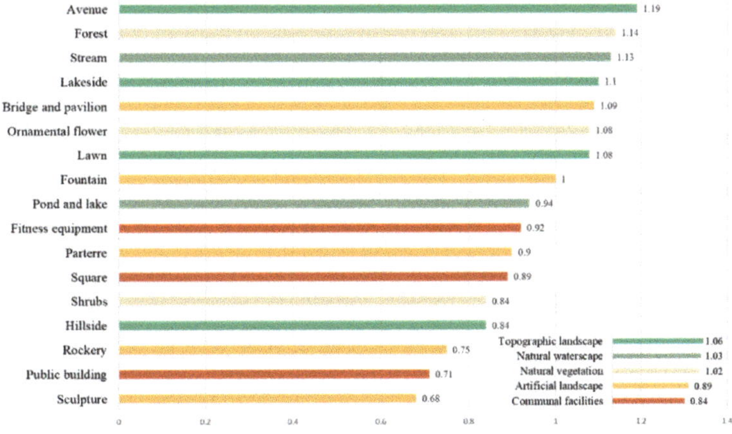

Figure 5. Average landscape preferences of exercisers in urban forest park.

As shown in Figure 6, the most preferred sounds for the exercisers were birdsong, water flow sound, and rustling of leaves, and the least preferred sounds for landscapes were construction noise, machine noise, traffic sounds, and broadcasts. The exercise population prefers nature-related landscapes and soundscapes more, and were less fond of sounds and landscapes generated or created by people.

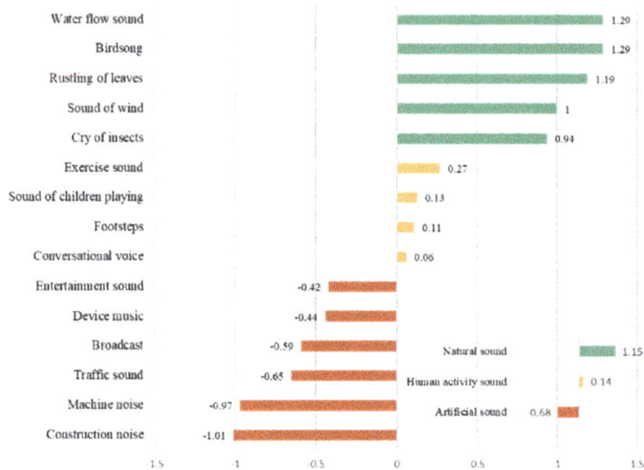

Figure 6. Average soundscape preferences of exercisers in urban forest park.

3.3. The Effect of Exercisers' Exercise Style and Venue Choice on Audio-Visual Preference

3.3.1. Effects of Exercise Modality and Exercise Site Selection on Soundscape Preference

As the results of the correlation analysis (Table 2) show, different exercise methods affect people's preference for soundscape. Those who chose jogging and brisk walking had a higher potential preference for natural sounds, while this group tended to choose exercise sites that were close to the natural landscape in the form of lawn, forest, and lakeside footpath.

Table 2. Correlation between different exercise forms and soundscape preferences.

Sound Classification		Exercise Options							
		Slow Walking	Jogging	Brisk Walking	Use Fitness Equipment	Square Dance	Gymnastics	Chinese Martial Arts	Ball Games
Natural sound	Sound of wind	0.058	0.122	−0.060	−0.027	0.067	−0.011	−0.072	−0.063
	Birdsong	0.158 **	0.102	0.101	0.043	0.095	0.013	0.063	−0.067
	Cry of insects	0.110 *	0.070	−0.031	0.043	0.082	0.017	−0.004	−0.067
	Rustle of leaves	0.117 *	0.141 **	0.130 *	0.013	0.079	−0.017	−0.032	−0.071
	Water flow sound	0.157 **	0.023	0.086	0.024	0.032	0.007	0.123 *	0.010
Human activity sound	Conversational voice	0.023	0.071	0.021	0.062	−0.034	−0.083	−0.049	−0.019
	Sound of children playing	0.024	−0.009	0.009	0.075	−0.016	−0.048	−0.025	0.020
	Footsteps	0.046	0.058	−0.004	0.013	−0.063	−0.141 **	−0.073	0.014
	Exercise sound	−0.042	0.101	−0.009	0.096	0.059	−0.029	0.096	0.036
Artificial sound	Traffic sound	0.028	−0.033	0.005	0.015	−0.066	−0.019	−0.039	0.017
	Entertainment sound	−0.017	−0.024	0.008	0.017	−0.031	0.059	−0.005	−0.024
	Device music	−0.069	−0.051	−0.016	0.005	0.019	0.033	−0.012	0.017
	Construction noise	0.043	−0.007	0.000	−0.010	−0.148 **	0.012	0.021	0.019
	Machine noise	0.020	0.006	−0.037	0.012	−0.095	−0.005	−0.001	0.056
	Broadcast	0.022	−0.088	−0.011	0.078	−0.084	0.012	0.013	0.054

* Correlation is significant at the 0.05 level. ** Correlation is significant at the 0.01 level.

Since the venues for dance and gymnastics are limited by the mountainous terrain and are far from the mountains and forests, those who choose these activities have less exposure to natural sounds [47]. There is a positive correlation between the preference for device music and the crowd of square dancers and gymnasts, but they have a stronger aversion to activity sound and artificial noise. People using fitness equipment showed a higher acceptance for activity sound on a potential level.

The data in Table 3 indicate that there is a positive correlation between the choice of footpath in forests and lawns and preference for natural sounds, especially sounds in the forest, and a negative correlation between preference for activity sounds and artificial noise.

There was a positive correlation between the choice of lakeside footpath and preference for water flow sounds. People exercising in square open spaces and fitness equipment venues had a higher tolerance for the sound of human activity and the sound of playing music. Almost all people in different exercise areas have different levels of aversion to traffic, construction, and machine sounds.

Table 3. Correlation between different exercise site selections and soundscape preferences.

Sound Classification		Field Options				
		Square Open Space	Lawn Footpath	Forest Footpath	Lakeside Footpath	Fitness Equipment Venue
Natural sound	Sound of wind	−0.083	0.021	0.146 **	0.047	0.082
	Birdsong	−0.097	0.095	0.168 **	0.032	0.077
	Cry of insects	−0.077	0.054	0.184 **	−0.011	0.059
	Rustle of leaves	0.060	0.089	0.195 **	0.050	0.147 **
	Water flow sound	−0.118 *	0.036	0.052	0.135 *	0.165 **
Human activity sound	Conversational voice	0.080	−0.052	−0.034	0.025	0.096
	Sound of children playing	0.012	0.018	−0.021	0.020	0.048
	Footsteps	0.051	−0.113 *	−0.075	0.002	0.025
	Exercise sound	0.074	−0.156 **	−0.089	−0.075	0.127 *
Artificial sound	Traffic sound	−0.024	−0.012	0.006	−0.032	−0.018
	Entertainment sound	−0.010	−0.029	0.060	−0.056	0.044
	Device music	0.024	−0.032	0.011	−0.099	0.036
	Construction noise	−0.113 *	−0.023	−0.026	−0.054	−0.079
	Machine noise	−0.070	−0.056	−0.052	−0.077	−0.053
	Broadcast	−0.003	0.013	−0.053	−0.006	0.035

* Correlation is significant at the 0.05 level. ** Correlation is significant at the 0.01 level.

3.3.2. Effects of Exercise Modality and Exercise Site Selection on Visual Landscape Preference

As shown in Table 4, in terms of visual landscape, there is a significant positive correlation between slow walking, jogging, and brisk walking crowds and preference for natural landscape, with the people who run showing a lower preference for artificial landscape compared to the other two categories. In contrast, the fitness crowd and the square dancing crowd have a higher preference for fitness equipment and squares. There is a positive correlation between the choice of carrying out gymnastics activities, Chinese martial arts, ball games, and the preference for artificial landscapes, but a lower degree of relationship with the preference for natural landscapes, in which those who perform ball games do not show a positive preference for natural landscapes.

The data in Table 5 illustrate that there is a strong positive correlation between the choice of fitness equipment site and the preference for natural landscape, artificial landscape, and communal facilities. Weak correlations exist between the choice of being on a lawn or forest trail and the preference for artificial landscapes, while positive correlations exist with natural landscapes. There is a strong positive correlation between the choice of lakeside footpath and preference for fountains, bridges and pavilions, and parterres.

3.3.3. Effects of Choice of Activity Type Other Than Exercise on Audio-Visual Preference

Considering that exercise is not the only purpose for which people visit urban forest, and that most exercisers engage in concurrent activities, such as family, social, and leisure activities, it is necessary to explore the visual landscape and soundscape preferences for these activities as well.

Table 4. Correlation between different exercise forms and visual landscape preferences.

Landscapes		Exercise Options							
		Slow Walking	Jogging	Brisk Walking	Use Fitness Equipment	Square Dance	Gymnastics	Chinese Martial Arts	Ball Games
Natural waterscape	Stream	0.054	0.034	0.111 *	0.018	0.087	0.011	0.073	0.023
	Pond and lake	0.113 *	0.084	0.091	0.028	0.106	0.015	−0.015	−0.020
Topographic landscape	Lawn	0.059	0.039	0.064	0.045	0.145 **	0.033	0.045	−0.056
	Avenue	0.075	0.057	0.125 *	0.094	0.142 *	0.019	0.011	−0.081
	Hillside	0.081	0.037	0.031	0.009	0.082	0.054	−0.003	−0.025
	Lakeside	0.085	0.005	0.059	0.008	0.109	0.009	0.062	0.008
Natural vegetation	Shrubs	−0.050	0.076	0.107 *	−0.096	0.034	0.028	−0.043	−0.034
	Ornamental flower	0.048	0.006	0.021	0.032	0.123 *	0.055	0.111 *	−0.005
	Forest	0.040	0.079	0.155 **	0.039	0.097	−0.069	0.080	0.033
Artificial landscape	Rockery	0.051	0.001	0.038	−0.022	0.099	0.066	0.111 *	−0.033
	Parterre	0.069	0.018	0.045	0.010	0.118 *	0.077	0.129 *	0.055
	Fountain	0.05	−0.029	0.012	0.013	0.080	0.090	0.064	0.071
	Sculpture	0.049	−0.030	0.005	0.003	0.056	0.079	0.081	0.075
	Bridge and pavilion	0.081	−0.037	0.005	0.067	0.140 **	0.087	0.046	0.060
Communal facilities	Fitness equipment	−0.047	0.075	0.061	0.139 **	0.180 **	0.062	0.121 *	0.007
	Square	0.004	0.066	−0.040	0.049	0.133 *	−0.004	0.063	0.074
	Public building	−0.018	0.054	−0.078	0.021	0.082	0.044	−0.016	0.049

* Correlation is significant at the 0.05 level. ** Correlation is significant at the 0.01 level.

Table 5. Correlation between different exercise site selections and visual landscape preferences.

Landscapes		Field Options				
		Square Open Space	Lawn Footpath	Forest Footpath	Lakeside Footpath	Fitness Equipment Venue
Natural waterscape	Stream	0.066	0.017	0.108 *	0.100	0.129 *
	Pond and lake	0.015	0.002	0.120 *	0.110	0.105
Topographic landscape	Lawn	0.008	0.183	0.033	0.026	0.136 *
	Avenue	0.005	0.062	0.118 *	−0.050	0.092
	Hillside	0.050	0.051	0.069	0.041	0.150 **
	Lakeside	0.056	−0.021	−0.009	0.115	0.092
Natural vegetation	Shrubs	−0.046	−0.018	0.110 *	−0.036	0.069
	Ornamental flower	0.016	0.009	0.031	−0.006	0.086
	Forest	−0.012	0.043	0.110 *	−0.029	0.077
Artificial landscape	Rockery	0.006	−0.010	0.009	0.013	0.117 *
	Parterre	0.084	0.002	0.003	0.039	0.129 *
	Fountain	−0.013	−0.022	−0.006	0.068	0.145 **
	Sculpture	0.058	0.003	−0.073	0.039	0.121
	Bridge and pavilion	−0.028	0.007	−0.057	0.096	0.130 *
Communal facilities	Fitness equipment	0.088	−0.048	−0.006	−0.036	0.238 **
	Square	0.114 *	−0.039	−0.038	−0.028	0.227 **
	Public Building	0.039	−0.063	−0.044	0.025	0.172 **

* Correlation is significant at the 0.05 level. ** Correlation is significant at the 0.01 level.

The data in Tables 6 and 7 illustrate that there is a positive correlation between those who perform family activities and all visual landscape preferences after exercise, but such activities show a weaker correlation with natural water features, hillside, and rockery preferences, while there is a positive correlation with natural sound preferences, which are more averse to noise. There is a positive correlation between the choice of social activities, leisure activities and group activities and preference for communal facilities, with a higher tolerance for activity sound. In contrast, there was a positive correlation between the choice of quiet–type activities and the preference for avenues, hillsides, and forests, and a negative correlation between the preference for artificial sound and activity sound.

Table 6. Correlation between the choices of other types of activities other than exercise and visual landscape preferences.

Landscapes		Other Activities besides Exercise					
		Family Activities	Social Activities	Leisure Activities	Quiet–Type Activities	Group Activities	Just Exercise
Natural waterscape	Stream	0.123 **	0.025	0.060	0.008	0.086	−0.046
	Pond and lake	0.107 *	0.013	0.067	−0.006	0.038	−0.020
Topographic landscape	Lawn	0.234 **	−0.010	−0.009	0.000	0.017	−0.035
	Avenue	0.176 **	−0.022	−0.034	0.058	0.021	−0.037
	Hillside	0.102	0.012	0.059	0.038	0.021	−0.007
	Lakeside	0.144 **	0.047	0.015	−0.016	0.025	−0.043
Natural vegetation	Shrubs	0.124	−0.009	0.038	−0.099	−0.036	−0.020
	Ornamental flower	0.196 **	0.043	0.014	−0.140 **	−0.008	−0.044
	Forest	0.122 *	−0.024	0.042	0.023	0.012	−0.087
Artificial landscape	Rockery	0.093	0.070	0.082	−0.110 *	0.040	0.002
	Parterre	0.194 **	0.026	0.069	−0.081	0.090	−0.001
	Fountain	0.180 **	0.054	0.088	−0.115 *	0.120 *	−0.006
	Sculpture	0.133 *	0.112 *	0.106 *	−0.062	0.082	−0.050
	Bridge and pavilion	0.170 **	0.072	0.078	−0.016	0.103	−0.001
Communal facilities	Fitness equipment	0.174 **	0.044	0.095	−0.042	0.155 **	0.001
	Square	0.166 **	0.017	0.120 *	−0.012	0.133 *	−0.020
	Public Building	0.137 *	0.086	0.090	−0.017	0.144 **	−0.070

* Correlation is significant at the 0.05 level. ** Correlation is significant at the 0.01 level.

Table 7. Correlation between the choices of other types of activities other than exercise and soundscape preferences.

Sound Classification		Other Activities besides Exercise					
		Family Activities	Social Activities	Leisure Activities	Quiet–Type Activities	Group Activities	Just Exercise
Natural sound	Sound of wind	0.040	0.064	0.068	0.031	0.081	0.023
	Birdsong	0.142 **	−0.012	0.009	0.023	0.035	0.000
	Cry of insects	0.160 **	0.019	0.018	0.060	0.045	−0.087
	Rustle of leaves	0.141 **	0.025	0.060	0.035	0.094	−0.024
	Water flow sound	0.171 **	0.002	0.072	0.050	0.085	−0.020
Human activity sound	Conversational voice	0.006	0.120 *	0.122 *	−0.056	0.041	−0.036
	Sound of children playing	0.086	0.091	0.104	−0.048	0.082	−0.054
	Footsteps	0.081	0.046	0.151 **	−0.031	0.026	−0.070
	Exercise sound	0.017	0.008	0.072	−0.031	0.015	0.037
Artificial sound	Traffic sound	−0.093	0.028	−0.022	−0.026	−0.004	−0.003
	Entertainment sound	−0.105	0.025	−0.070	−0.058	0.038	−0.004
	Device music	−0.069	0.071	−0.028	−0.047	0.029	0.004
	Construction noise	−0.020	0.022 *	0.023	−0.048	0.024	−0.105
	Machine noise	−0.051	−0.021 *	0.036	−0.045	0.009	−0.092
	Broadcast	0.012	0.068	0.003	−0.029	−0.006	−0.106 *

* Correlation is significant at the 0.05 level. ** Correlation is significant at the 0.01 level.

3.4. The Influence of Individual Characteristics of Urban Forest Park Exercisers on Audio-Visual Preferences

To explore the differences in soundscape and visual landscape preferences under other exercise-related factors, we conducted an ANOVA between soundscape and visual landscape preferences under different individual characteristic indicators and plotted radar plots. If significant differences were presented ($p < 0.05$ or $p < 0.01$), the specific differences were described by specifically comparing the mean size; if no significance was presented, it means that there were no significant differences in audio-visual preferences under different individual characteristics. The analysis revealed that distance from home to the park and time period did not have significant effects on audio-visual preferences ($p > 0.05$), so the effects of the major individual characteristic factors of gender, age, number of companions, frequency, and exercise intensity were mainly explored.

3.4.1. The Relationship between Gender and Audio-Visual Preference

As shown in Figure 7a, females generally preferred natural sounds more than males, while males had a higher acceptance for activity and artificial sounds. Males and females showed significant differences in their preferences for traffic sound ($p = 0.006$ **), construction noise ($p = 0.038$ *), and mechanical noise ($p = 0.007$ **), with females showing more significant aversions to these three types of sounds. Figure 7b shows that females preferred visual landscapes in the park more than males, significantly in terms of preference for ornamental flowers ($p = 0.049$ *), fountains ($p = 0.02$ *), and bridge and pavilion ($p = 0.039$ *).

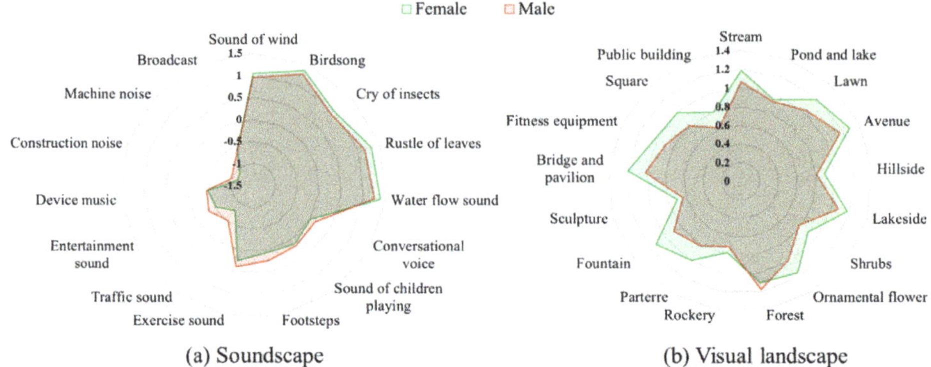

(a) Soundscape   (b) Visual landscape

Figure 7. Radar map of soundscape (**a**) and visual landscape (**b**) preferences distribution of exercisers at different genders.

3.4.2. The Relationship between Age and Audio-Visual Preference

ANOVA results show that people aged 40–59 years have a higher preference for natural sounds, particularly birdsong ($p = 0.009$ **), cry of insects ($p = 0.047$ *), and water flow sound ($p = 0.018$ *). Older people are more tolerant of activity sounds, significantly for the sound of children playing ($p = 0.02$ *) and entertainment sounds ($p = 0.032$ *). The visualized mean data are shown in Figure 8a.

According to the ANOVA results, people aged 40–59 years showed a more significant preference for streams ($p = 0.006$ **), ponds and lakes ($p = 0.001$ **), lawns ($p = 0.000$ **), avenues ($p = 0.000$ **), hillsides ($p = 0.002$ **), lakesides ($p = 0.000$ **), ornamental flowers ($p = 0.001$ **), forests ($p = 0.000$ **), parterres ($p = 0.001$ **), and fitness equipment ($p = 0.02$ *) compared to other age groups. This is also shown in Figure 8b, where middle-aged people have a higher preference for the overall park landscape compared to other age groups.

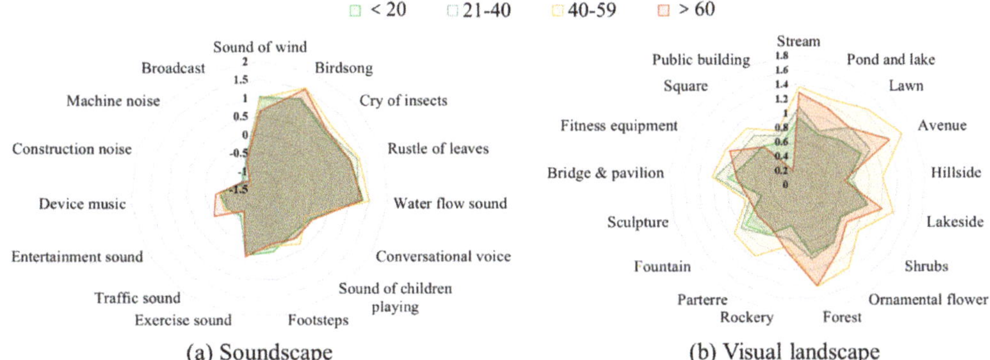

(a) Soundscape   (b) Visual landscape

Figure 8. Radar map of soundscape (**a**) and visual landscape (**b**) preferences distribution of exercisers at different ages.

3.4.3. The Relationship between Companion Number and Audio-Visual Preference

According to the results of the ANOVA, those who went with small families showed a significant preference for the four natural sounds of birdsong (p = 0.005 **), insects (p = 0.045 **), leaves (p = 0.046 *), water flow (p = 0.028 **) and the sound of children playing (p = 0.001 **) compared to the rest of the population. In addition, according to Figure 9a, people traveling with small families were more averse to artificial noise and entertainment equipment, but no significant differences were found in the ANOVA. Those who were accompanied by a companion showed a higher preference for soundscapes compared to those who were alone, while being more averse to noise.

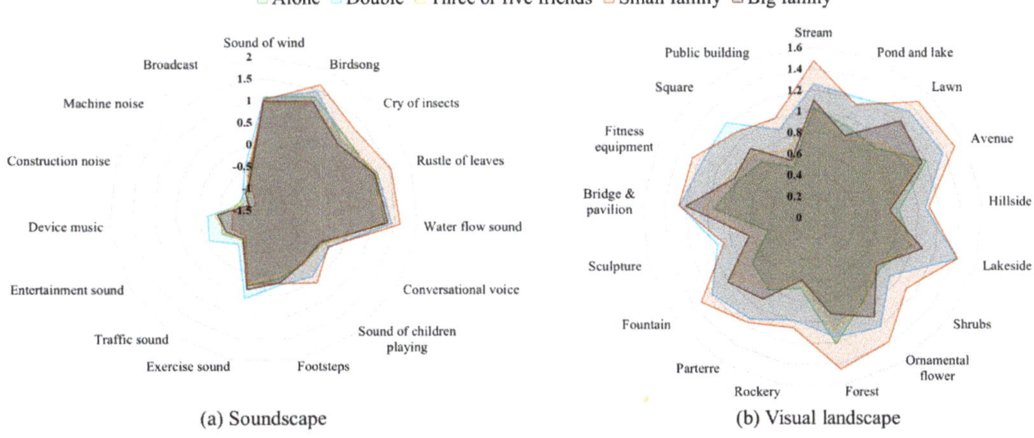

(a) Soundscape   (b) Visual landscape

Figure 9. Radar map of soundscape (**a**) and visual landscape (**b**) preferences distribution of exercisers at different companions.

According to the results of the ANOVA and in conjunction with Figure 9b, the visual landscape preferences of the exercisers showed significant differences for different numbers of companions, except for the bridges and pavilions (p = 0.132). Those who traveled in small families showed more significant preferences for streams (p = 0.003 **), ponds and lakes (p = 0.008 **), lawns (p = 0.000 **), avenues (p = 0.011 *), hillsides (p = 0.004 **), lakesides (p = 0.001 **), shrubs (p = 0.047 *), ornamental flowers (p = 0.006 **), forests (p = 0.007 **), rockeries (p = 0.01 *), parterres (p = 0.003 **), fountains (p = 0.001 **), fitness equipment

(p = 0.002 **), and public buildings (p = 0.02 *), while those who traveled in pairs showed more significant preferences for sculptures (p = 0.029 *) and squares (p = 0.003 **).

3.4.4. The Relationship between Exercise Frequency and Audio-Visual Preference

According to the variance results and in conjunction with the results shown in Figure 10a, people who exercise more frequently show a significant preference for natural sounds such as wind (p = 0.012 *), birdsong (p = 0.04 *), cry of insects (p = 0.014 *), rustle of leaves (p = 0.007 **), and water flow (p = 0.024 *), and a more significant tolerance for noise: device music (p = 0.024 *), entertainment sound (p = 0.022 *), and traffic sound (p = 0.045 *). People with lower activity frequencies show opposite trends in sound preference to those with higher activity frequencies.

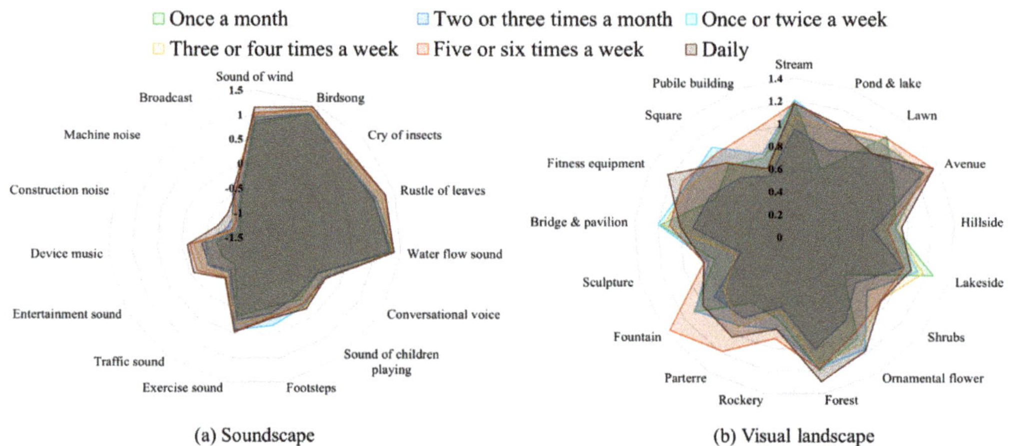

Figure 10. Radar map of soundscape (**a**) and visual landscape (**b**) preferences distribution of exercisers at different exercise frequencies.

The results of the ANOVA showed that those who exercised more frequently showed a significant preference for ponds and lakes (p = 0.015 *), avenues (p = 0.021 *), shrubs (p = 0.045 *), parterres (p = 0.015 *), and fitness equipment (p = 0.027 *), while those who exercised less frequently showed a lower preference for all these landscapes.

3.4.5. The Relationship between Exercise Intensity and Audio-Visual Preference

According to the ANOVA results, the light exercisers showed a significant preference for natural sounds such as wind (p = 0.047 *), birdsong (p = 0.022 *) and leaves (p = 0.000 **) compared to the strenuous exercisers, while the strenuous exercisers showed a more significant tolerance for noise such as traffic (p = 0.015 *), construction (p = 0.03 *), and machine noise (p = 0.009 **). These trends are also shown in Figure 11a.

Using ANOVA, we found that the light exercise group showed a significant preference for the categories of streams (p = 0.001 **), ponds and lakes (p = 0.015 *), avenues (p = 0.025 *), lakesides (p = 0.003 **), parterres (p = 0.05 *), and squares (p = 0.021 *). Figure 11b also reflects that the heavy exercisers showed a lower preference for these types of landscapes compared to the light exercisers.

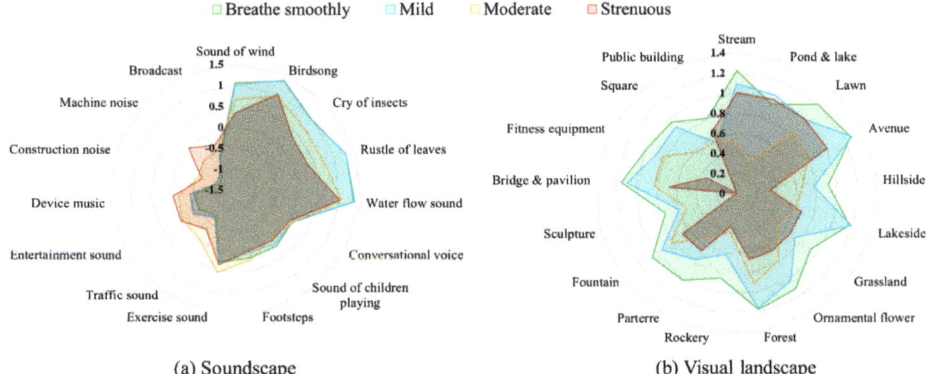

(a) Soundscape   (b) Visual landscape

Figure 11. Radar map of soundscape (**a**) and visual landscape (**b**) preferences distribution of exercisers at different exercise intensity.

3.5. The Effect of Audio-Visual Preference on Exercise Perception among Urban Forest Park Exercisers

Structural equation modeling was first developed by Swedish statisticians as a multivariate statistical analysis method for analyzing the complex structure of relationships between multi-indicator variables, SEM, which has similar aims to regression analysis but has two advantages over regression analysis [48]. Firstly, SEM is able to take into account the estimated residuals of the observed variables, which gives a more realistic picture of the sample information [49]. Secondly, SEM allows the reader to understand the relationship between variables in a more intuitive way by presenting the results in a simple graphical output [50].

In this paper, in order to explore the influence of soundscape and visual landscape preferences on the perceptions of exercising people in the park, methods that can reveal the relationships that exist between multiple variables need to be used. Combining the properties and advantages of SEM, we ultimately used SEM to explore the influence of relationships between the variables. The visual and acoustic landscapes were divided into previously classified categories and the mean scores were used to calculate the scores for each dimension. The mean score is the most commonly used dimensional induction treatment. For the accuracy of the model, a multicollinearity test was performed to ensure that there was no multicollinearity between the variables. If tolerance ≤ 0.1 or VIF ≥ 10, it exist multicollinearity. According to what is shown in Table 8, none of the observed variables in the model are subject to multicollinearity.

Table 8. Collinearity diagnostics.

	Collinearity Diagnostics	
	Tolerance	**VIF**
Waterscape	0.393	2.545
Topography	0.329	3.042
Natural vegetation	0.321	3.120
Manufactured Landscapes	0.330	3.030
Communal facilities	0.422	2.369
Natural sound	0.588	1.701
Human activity sound	0.984	1.016
Artificial sound	0.882	1.133

Model plotting was performed in SPSS AMOS software (IBM, Armonk, NY, USA), and relevant data were imported for computational analysis; the final model plot is shown in Figure 12. The oval in the figure represents the latent variable and the rectangle represents the observed variable, which is the measurement item in the questionnaire. Each measurement term must have a residual term, which is a circular term from e1–e9 in the figure.

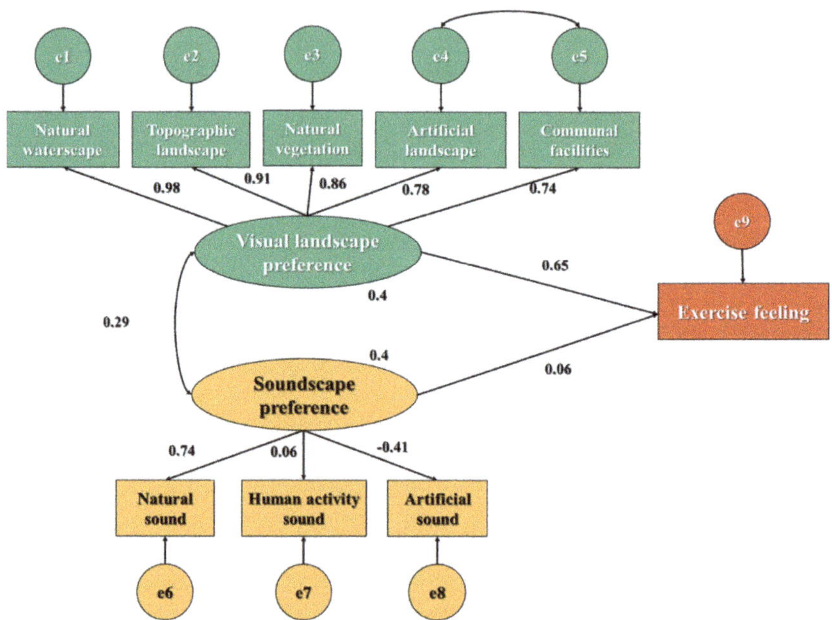

Figure 12. Structural equation model of audio-visual preferences and exercise feelings.

In the model, soundscape and visual landscape preferences are latent variables, and the dependent variable exercise perception is the observed variable. Natural sound, human activity sound, and artificial sound are used as observed variables for soundscape preference, and natural waterscape, topographic landscape, natural vegetation, artificial landscape, and communal facilities are used as observed variables for visual landscape preference. According to the principles of SEM, residual terms need to be added to the latent variables and double arrows added between the exogenous latent variables. The model fit indices were as follows, CMIN/DF = 2.595(<3), GFI = 0.958(>0.9), AGFI = 0.925(>0.9), RMR = 0.023(<0.05), CFI = 0.973(>0.9), reflecting the overall goodness of fit of the model. The validation level α = 0.05 and each estimated parameter is significant. The results of the optimal model path coefficient estimation are shown in Table 9.

Correlation analysis of each visual landscape preference and exercise perceptions found that they were all significantly correlated. Figure 12 of the structural equation model shows that the overall visual landscape preference (0.65) has a greater direct effect on the exercise population's perception in the park and has the highest direct effect status for natural waterscape (0.98), followed by topographic landscape (0.91), and less direct effect for artificial landscape (0.78) and communal facilities (0.74). Additionally, the linear regression of the single term shows that sculptures have a significant negative effect on exercise perception.

Table 9. Path analysis of modified structural equation model.

Path	Estimate	S.E.	C.R.	*p*
Waterscape ← Visual landscape preference	0.976	0.054	17.979	<0.001
Topography ← Visual landscape preference	0.913	0.046	19.991	<0.001
Natural vegetation ← Visual landscape preference	0.863	0.045	19.257	<0.001
Manufactured Landscapes ← Visual landscape preference	0.777	0.050	15.508	<0.001
Communal facilities ← Visual landscape preference	0.741	0.057	13.029	<0.001
Natural sound ← Soundscape preference	0.740	0.041	18.003	<0.001
Human activity sound ← Soundscape preference	0.060	0.031	1.938	0.047
Artificial sound ← Soundscape preference	−0.414	0.083	−4.971	<0.001
Exercise feeling ← Visual landscape preference	0.649	0.095	6.807	<0.001
Exercise feeling ← Soundscape preference	0.059	0.101	0.579	0.034

The correlation analysis between soundscape preference and exercise perception found that most of them had significant correlation with perception, except for conversational sounds, footsteps, music, and entertainment equipment sounds that belonged to personal activity sounds. The structural equation model showed that the overall soundscape preference (0.06) did not have a high degree of direct effect on perception, and the activity sound had a relatively minor effect, which was also consistent with the correlation analysis results. However, noise still plays a negative influence in it. There is still a strong positive influence of natural sound, and in the regressions of individual items, bird song and water flow sound are found to have a more significant influence, which has some connection with landscape preference.

4. Discussion

This discussion section explores the landscape design of urban forest parks from the perspective of exercise behavior on the basis of comparing and summarizing the similarities and differences between this study and existing scholars' research views.

4.1. Audio-Visual Preference Characteristics of Exercisers in Urban Forest Parks

This study found that the exercise population had a higher preference for natural soundscapes and visual landscapes and a lower preference for man-made landscapes and human-generated sounds, and developed an aversion to noise in particular, which is consistent with the findings of Jeon et al. [51–53]. Among them, forests and footpaths and bird songs in it were most preferred by the exercising population, followed by streams and the water flow sounds they produce. Significant aversion was shown toward construction noise, traffic sounds, and machine noise, confirming the findings of Fang [33]. Park designers can install trails within the natural landscape or in the surrounding areas to increase the frequency of exercisers' contact with the natural landscape and optimize people's exercise experience in forest parks. At the same time, noise needs to be controlled. Some studies have shown that the use of bird calls to mask noise may improve people's soundscape perception [29]. On the one hand, park designers can consider artificially setting up some bird nests or bird feeders in the forest to attract birds to nest, thereby increasing bird calls; on the other hand, sound insulation panels can be installed next to noise sources for noise reduction.

4.2. The Effect of Exercise Style and Individual Characteristics on Audio-Visual Preferences
4.2.1. Effects of Exercise Modality Choices on Audio-Visual Preferences

Among the different exercise activities, it is worth attention that the crowd of square dancers prefer squares and fitness equipment sites, probably because square dancing requires flat and open sites. They also prefer ornamental flowers and parterres compared to the crowd of other activities. The researchers' observation and questioning found that most of this group were middle-aged and older women who preferred flowers. There was a positive correlation between the square dancing and gymnastics crowd and the preference for music sound played by the equipment, but they had a strong aversion to activity sound and artificial noise. After questioning, it was found that they did not like the activity sound and artificial noise of people around them to interfere with the music sound they played. People using fitness equipment showed a higher acceptance for activity sound at the potential level, because there are often more people gathered at the fitness equipment in the park, and there are many children playing around.

4.2.2. Effects of Exercise Site Selections on Audio-Visual Preferences

Those who choose forest and lawn footpaths are more concerned about the purity of the surrounding natural landscape and do not want artificial landscapes in the natural landscape, while being very averse to fire announcements in the forest. Those who choose lakeside footpath are more interested in small bridges and pavilions, fountains, parterres, waterscapes, and water sounds. Compared with other groups, those who choose the square open space are more interested in artificial landscape, and those who choose the fitness equipment site are very receptive to the landscape in the park. Both of these two types of exercise groups have a higher tolerance for activity sound as well as music sound.

Chen et al. showed that when the surrounding conditions can satisfy park users to engage in active health behaviors, they will still engage in exercise behaviors even if the landscape preference is weak, while when the conditions are not satisfied, they will not engage in exercise behaviors despite the strong landscape preference [54], which partly explains the differences in landscape preferences among people with different exercise behaviors. Park designers can try to provide adequate exercise conditions by installing some fitness equipment and trails at the interface area between natural and man-made landscapes. When planning forest footpaths, the purity of the surrounding natural landscape can be ensured by minimizing the involvement of man-made landscape and sound elements along the trails.

4.2.3. Effects of Factors Other Than Exercise on Audio-Visual Preferences

Most of the population will also perform some other activities, and there is a relationship between these activities and audio-visual preferences. Those who engage in family activities prefer visual landscapes, but have a lower preference for natural waterscapes, hillsides, and rockeries, perhaps because parents are concerned that their children may be harmed in these areas. Park designers may consider putting signs and hints in prominent places to indicate potential hazards. Moreover, noise is more uncomfortable for them, probably because parents are more concerned about their children's feelings, so designers can consider creating special areas for children's activities away from noise.

The recreational crowd with public participation type of activities prefers to be near communal facilities and artificial landscapes, while having a higher tolerance for activity sound. Some studies have shown that people who come to play with children and gather with family and friends prefer to stay in areas that can accommodate group activities, such as lawns and recreation areas [55]. Designers can consider providing them with some resting benches or tables and chairs in areas close to human activity sites. Quiet-type activity groups prefer to be in less crowded avenues, hillsides, and forests, away from artificial landscapes and places with dense activity sounds, and designers can consider installing resting facilities in these areas for them to use.

4.2.4. Effects of Individual Characteristics on Exercisers' Audio-Visual Preferences

Among the different individual characteristics, the distance from home to the park and the time period of exercise had a low degree of influence on audio-visual preference, while gender, age, number of companions, exercise frequency, and intensity factors had an influence on audio-visual preference.

The present study found that gender has an effect on soundscape preference, which fits with Hedblom's findings [56]. Among them, female exercisers preferred natural sounds more than males, probably because females are more sensitive to some sounds and can easily perceive sounds that males ignore [57]. Moreover, similar to Gozalo's view in this study is that men have a higher acceptance for activity and artificial sounds, probably because it is usually women who take care of children and they are closer to artificial sound sources and thus need to endure the distress caused by this part of the sound [58]. In addition, females prefer visual landscapes in parks more than males, although males have a greater preference for forest landscapes.

The effect of age on audio-visual preferences also deserves our attention. The present study showed that as residents age, their evaluation of natural sounds increases and their tolerance for musical and activity sounds increases, and these results are generally consistent with Zhou's findings [59], which may result from the decline of human perception of high-frequency sounds with age [60]. Compared to middle-aged and young adults, older adults have a stronger perception of natural sounds. Some studies have shown that older adults are more likely to derive a sense of calm from natural sounds [56]. This study found that middle-aged exercisers had the highest preference for various park landscapes, and in line with Paneerchelvam et al.'s study, it was found that older adults enjoyed parks more than younger people, with a particular preference for natural landscapes [61]. Park designers may consider installing facilities within or around natural landscapes that are convenient for the elderly, such as handrails along the footpaths or adding anti-slip features to the trails.

This study found that the number of companions showed an effect on both auditory and visual landscape preferences. The audio-visual preference characteristics were similar for people on small family outings and those in the parent-child category. Accompanied exercisers are more likely to resent noise than those who are alone, possibly because the presence of noise tends to interfere with the activity they are carrying out or their interaction with their partner. People who traveled in pairs and small families had a stronger preference for the visual landscape in the park, and those who moved alone and in large groups appeared to be less concerned about the visual landscape.

The present study found a significant effect of exercise frequency on audio-visual preference, which was not mentioned in previous studies. People who exercised more frequently had a greater preference for the sounds of birds, insects, and water flow, and had a higher tolerance for noise, probably because this group of people was more accustomed to the surrounding soundscape environment. People with low exercise frequency are more sensitive to the noise, and are susceptible to the negative effects of noise. People with high exercise frequency also have more preference for visual landscape in the park, while people with low frequency do not perceive visual landscape significantly.

Exercise intensity also had a significant effect on exercisers' audio-visual preferences, and a correlation between greenway built environment and exercise intensity has been demonstrated by Dong et al. [62]. The present study found a higher degree of preference for natural sounds and natural landscapes among people with mild exercise intensity by further investigation. However, the phenomenon that the moderate and heavy exercise population found in this study had lower preference for each soundscape and visual landscape still needs further discussion. The author believes that this group is more concerned with the process of exercise itself to motivate themselves to achieve the desired exercise effect, so they may not pay attention to the surrounding sound and visual scenery.

4.3. Effect of Audio-Visual Preferences on Exercisers' Perception of Exercise

In this study, natural landscapes and soundscapes had the effect of enhancing people's exercise perceptions, which is consistent with Stigsdotter and Watts. Natural landscape spaces representing the tranquil type were rated as having the most restorative potential [63]. The inclusion of natural soundscapes in the environment helps to create a tranquil atmosphere and stimulate positive and pleasant emotions [64]. Interestingly, the overall soundscape preference did not have a high degree of direct effect on exercise perception, a finding that needs to be verified in further discussions. It is possible that the heavier breathing sounds produced during exercise or the preference of some exercisers to wear headphones to listen to music during exercise may have affected their ability to capture the surrounding sounds.

5. Conclusions

This study was based on a field questionnaire in Beigushan Urban Forest Park, Lianyungang City, Jiangsu Province, China. The survey examined the different preferences for sound and visual landscapes among people exercising in urban forest parks. In terms of overall audio-visual preferences, natural soundscapes and visual landscapes were generally preferred, but artificially generated sounds and constructed landscapes were preferred to a relatively lesser extent. In terms of specific factors, the study analyzed the relationship between exercise style, exercise venue, other activities, and audio-visual preferences, and showed that these factors differed in terms of people's audio-visual preferences. People's audio-visual preferences for their surroundings influence their choice of exercise venue. Other types of activities outside of exercise also have an effect on the audio-visual preference of exercisers, for example, the crowd of family activities are more concerned about the possible harm caused by rockeries, hillsides, waterscapes, and noise to children. The study also used ANOVA to investigate whether different individual characteristics had a significant effect on audio-visual preferences. The results show that audio-visual preferences vary by gender, age, number of companions, frequency, and intensity of exercise, which provides important information for park planners to consider when designing their parks. However, when studying the effect of age on audio-visual preferences, due to the fact that some elderly people and children are not proficient in the use of electronic devices and require substantial manpower to assist in the research, and that this group has a weaker perception of the landscape, a large amount of insensitive data were screened out, resulting in a small sample size for these two groups. The sample size of the elderly and children will be further increased in the future for more in-depth supplementary research.

By constructing a structural variance model between soundscape preference, visual landscape preference and exercise perception, it was found that visual landscape has a more direct impact on exercise perception than soundscape for exercisers, and that soundscape planning can be taken into account when planning parks while ensuring visual landscape planning.

Author Contributions: Conceptualization, J.X., M.L., Z.G. and Y.X.; data curation, J.X. and Z.G.; funding acquisition, J.X. and M.L.; methodology, J.X. and Z.G.; project administration, J.X. and M.L.; resources, J.X., Z.G., Y.X. and N.J.; supervision, M.L.; validation, J.X. and M.L.; visualization, Z.G; writing—original draft, J.X., Z.G., Y.X. and N.J.; writing—review and editing, J.X., Z.G. and Y.X. All authors have read and agreed to the published version of the manuscript.

Funding: Project funding was provided by "The Study on Rural Tourism Revitalization Planning and Sustainable Living Coordination Mechanism", the International Cooperation Open Project of State Key Laboratory of Subtropical Building Science, South China University of Technology [grant numbers: 2019ZA02].

Institutional Review Board Statement: The study was conducted in accordance with the Declaration of Helsinki, and approved by the Institutional Review Board of The Department of Tourism Management, South China University of Technology.

Informed Consent Statement: Informed consent was obtained from all subjects involved in the study.

Data Availability Statement: The data for this study can be accessed through the uploaded affiliated files or by contacting the relevant author.

Acknowledgments: We would especially like to thank Jian Xu and Muchun Li for their assistance in proofreading the manuscript and the International Cooperation Open Project of State Key Laboratory of Subtropical Building Science, South China University of Technology. Furthermore, we are grateful to the 406 park exercisers for their participation in our questionnaire survey.

Conflicts of Interest: The authors declare no conflict of interest.

References

1. Sun, W.T. Planning and layout optimization of urban forest park considering ecological environment protection. *Environ. Sci. Manag.* **2021**, *46*, 43–47. (In Chinese)
2. Liao, Z.-N.; Xu, H.-J.; Ma, J.; Li, M.; He, C.; Zhang, Q.; Xu, S. Seasonal and vegetational variations of culturable bacteria concentrations in air from urban forest parks: A case study in Hunan, China. *Environ. Sci. Pollut. Res.* **2022**, *29*, 28933–28945. [CrossRef] [PubMed]
3. Saninah, T.N.; Rushayati, S.B.; Hermawan, R. Iop in urban forest development at landside of Hang Nadim Batam Airport based on the microclimate and noise study. In Proceedings of the 2nd International Conference on Environment and Forest Conservation (ICEFC), Bogor, Indonesia, 1–3 October 2020; Mindanao State University: Bogor, Indonesia, 2019.
4. Liu, P.; Liu, M.; Xia, T.; Wang, Y.; Guo, P. The relationship between landscape metrics and facial expressions in 18 Urban Forest Parks of Northern China. *Forests* **2021**, *12*, 1619. [CrossRef]
5. Cai, M.; Cui, C.; Lin, L.; Di, S.; Zhao, Z.; Wang, Y. Residents' spatial preference for urban forest park route during physical activities. *Int. J. Environ. Res. Public Health* **2021**, *18*, 11756. [CrossRef]
6. Wang, Y.A.; Chang, Q.; Fan, P.L. A framework to integrate multifunctionality analyses into green infrastructure planning. *Landsc. Ecol.* **2021**, *36*, 1951–1969. [CrossRef]
7. Geng, D.H.; Innes, J.; Wu, W.L.; Wang, G.Y. Impacts of COVID-19 pandemic on urban park visitation: A global analysis. *J. For. Res.* **2021**, *32*, 553–567. [CrossRef]
8. Ugolini, F.; Massetti, L.; Calaza-Martinez, P.; Carinanos, P.; Dobbs, C.; Ostoic, S.K.; Marin, A.M.; Pearlmutter, D.; Saaroni, H.; Sauliene, I.; et al. Effects of the COVID-19 pandemic on the use and perceptions of urban green space: An international exploratory study. *Urban For. Urban Green.* **2020**, *56*, 126888. [CrossRef]
9. Toselli, S.; Bragonzoni, L.; Grigoletto, A.; Masini, A.; Marini, S.; Barone, G.; Pinelli, E.; Zinno, R.; Mauro, M.; Pilone, P.L.; et al. Effect of a park-based physical activity intervention on psychological wellbeing at the time of COVID-19. *Int. J. Environ. Res. Public Health* **2022**, *19*, 6028. [CrossRef]
10. Bamwesigye, D.; Fialova, J.; Kupec, P.; Lukaszkiewicz, J.; Fortuna-Antoszkiewicz, B. Forest recreational services in the face of COVID-19 pandemic stress. *Land* **2021**, *10*, 1347. [CrossRef]
11. Li, F.Y.; Zhu, Z.M.; Zhou, L.; Hu, G.M.; Zhong, L.L. Attraction of green exercise space in suburban forest park based on multiple regression model. *J. Northwest For. Univ.* **2020**, *35*, 288–295. (In Chinese)
12. McCormack, G.R.; Rock, M.; Toohey, A.M.; Hignell, D. Characteristics of urban parks associated with park use and physical activity: A review of qualitative research. *Health Place* **2010**, *16*, 712–726. [CrossRef] [PubMed]
13. Kim, H.; Woo, E.; Uysal, M. Tourism experience and quality of life among elderly tourists. *Tour. Manag.* **2015**, *46*, 465–476. [CrossRef]
14. Han, H.; Hyun, S.S. Customer retention in the medical tourism industry: Impact of quality, satisfaction, trust, and price reasonableness. *Tour. Manag.* **2015**, *46*, 20–29. [CrossRef]
15. Song, Y.; Wu, D.R.; Lv, Q.Z. A study of the application of visual identity system (VI) of rural areas in the landscape design of Banshancun Village. *J. Zhejiang Univ. Technol. (Soc. Sci.)* **2017**, *16*, 266–271. (In Chinese)
16. Cottet, M.; Piegay, H.; Bornette, G. Does human perception of wetland aesthetics and healthiness relate to ecological functioning? *J. Environ. Manag.* **2013**, *128*, 1012–1022. [CrossRef]
17. Ma, B.Q.; Xu, C.Y.; Liu, J.; Chang, C.; Zhao, K.; Kong, X.Q.; Long, J.Y. Visual heterogeneity and visual landscape quality of urban forests as an architectural backdrop. *J. Zhejiang A&F Univ.* **2019**, *36*, 366–374. (In Chinese)
18. Zhang, Y.Q.; Zhu, X.; Gao, M. A study on visual preference of walking path in urban wetland park based on eye tracking. In Proceedings of the 2020 Annual Meeting of Chinese Society of Landscape Architecture, Chengdu, China, 21–23 November 2020; pp. 450–454. (In Chinese).
19. Cassina, L.; Fredianelli, L.; Menichini, I.; Chiari, C.; Licitra, G. Audio-visual preferences and tranquillity ratings in urban areas. *Environments* **2018**, *5*, 1. [CrossRef]
20. Romero, V.P.; Maffei, L.; Brambilla, G.; Ciaburro, G. Acoustic, visual and spatial indicators for the description of the soundscape of waterfront areas with and without road traffic flow. *Int. J. Environ. Res. Public Health* **2016**, *13*, 934.
21. Puyana-Romero, V.; Ciaburro, G.; Brambilla, G.; Garzon, C.; Maffei, L. Representation of the soundscape quality in urban areas through colours. *Noise Mapp.* **2019**, *6*, 8–21. [CrossRef]
22. Preis, A.; Hafke-Dys, H.; Szychowska, M.; Kocinski, J.; Felcyn, J. Audio-visual interaction of environmental noise. *Noise Control Eng. J.* **2016**, *64*, 34–43. [CrossRef]
23. Van Heezik, Y.; Freeman, C.; Falloon, A.; Buttery, Y.; Heyzer, A. Relationships between childhood experience of nature and green/blue space use, landscape preferences, connection with nature and pro-environmental behavior. *Landsc. Urban Plan.* **2021**, *213*, 104135. [CrossRef]
24. Huang, A.S.H.; Lin, Y.J. The effect of landscape colour, complexity and preference on viewing behaviour. *Landsc. Res.* **2020**, *45*, 214–227. [CrossRef]
25. Acoustics-Soundscape—Part 1: Definition and Conceptual Framework. *ISO 12913-1:2014*; ISO: Geneva, Switzerland, 2014. Available online: https://www.iso.org/standard/52161.html (accessed on 20 January 2022).
26. Aletta, F.; Oberman, T.; Kang, J. Associations between positive health-related effects and soundscapes perceptual constructs: A systematic review. *Int. J. Environ. Res. Public Health* **2018**, *15*, 2392. [CrossRef] [PubMed]

27. Hong, X.C.; Zhu, Z.P.; Liu, J.; Geng, D.H.; Wang, G.Y.; Lan, S.R. Perceived occurrences of soundscape influencing pleasantness in urban forests: A comparison of broad-leaved and coniferous forests. *Sustainability* **2019**, *11*, 4789. [CrossRef]
28. Van Renterghem, T.; Vanhecke, K.; Filipan, K.; Sun, K.; De Pessemier, T.; De Coensel, B.; Joseph, W.; Botteldooren, D. Interactive soundscape augmentation by natural sounds in a noise polluted urban park. *Landsc. Urban Plan.* **2020**, *194*, 103705. [CrossRef]
29. Hong, J.Y.; Ong, Z.T.; Lam, B.; Ooi, K.; Gan, W.S.; Kang, J.; Feng, J.; Tan, S.T. Effects of adding natural sounds to urban noises on the perceived loudness of noise and soundscape quality. *Sci. Total Environ.* **2020**, *711*, 134571. [CrossRef]
30. Jian, K. From dBA to soundscape indices: Managing our sound environment. *Front. Eng. Manag.* **2017**, *4*, 184–192.
31. Kang, J.; Aletta, F.; Gjestland, T.T.; Brown, L.A.; Botteldooren, D.; Schulte-Fortkamp, B.; Lercher, P.; van Kamp, I.; Genuit, K.; Fiebig, A.; et al. Ten questions on the soundscapes of the built environment. *Build Environ.* **2016**, *108*, 284–294. [CrossRef]
32. Liu, F.F.; Kang, J. A grounded theory approach to the subjective understanding of urban soundscape in Sheffield. *Cities* **2016**, *50*, 28–39. [CrossRef]
33. Xingyue, F.; Tian, G.; Marcus, H.; Naisheng, X.; Yi, X.; Mengyao, H.; Yuxuan, C.; Ling, Q. Soundscape perceptions and preferences for different groups of users in urban recreational forest parks. *Forests* **2021**, *12*, 468.
34. Hong, X.-C.; Wang, X.; Duan, R.; Zhang, H.; Chi, M.-W.; Lan, S.-R. Evaluation of soundscape preference in forest park based on soundwalk approach. *Tech. Acoust.* **2018**, *37*, 584–588.
35. Li, J.; Burroughs, K.; Halim, M.F.; Penbrooke, T.L.; Seekamp, E.; Smith, J.W. Assessing soundscape preferences and the impact of specific sounds on outdoor recreation activities using qualitative data analysis and immersive virtual environment technology. *J. Outdoor Recreat. Tour.—Res. Plan. Manag.* **2018**, *24*, 66–73. [CrossRef]
36. Liu, J.; Xu, J.; Wu, Z.; Cheng, Y.; Gou, Y.; Ridolfo, J. Soundscape preference of urban residents in China in the post-pandemic era. *Front. Psychol.* **2021**, *12*, 750421. [CrossRef] [PubMed]
37. Ma, K.W.; Mak, C.M.; Wong, H.M. Effects of environmental sound quality on soundscape preference in a public urban space. *Appl. Acoust.* **2021**, *171*, 107570. [CrossRef]
38. The National Physical Fitness Testing Center Issued the Bulletin on the Investigation of National Fitness Activities in 2020. Available online: https://www.ciss.cn/tzgg/info/2021/32029.html (accessed on 20 January 2022). (In Chinese)
39. Yu, Y.; Jia, X.B. Application of "Questionnaire Star" in information technology classroom test. *Ind. Sci. Trib.* **2015**, *14*, 66–67. (In Chinese)
40. Krause, B. Bioacoustics, Habitat ambience in ecological balance. *Whole Earth Rev.* **1987**, *57*, 14–18.
41. Zhao, Z.; Wang, Y.; Hou, Y. Residents' spatial perceptions of urban gardens based on soundscape and landscape differences. *Sustainability* **2020**, *12*, 6809. [CrossRef]
42. Zhao, W.; Li, H.; Zhu, X.; Ge, T. Effect of birdsong soundscape on perceived restorativeness in an urban park. *Int. J. Environ. Res. Public Health* **2020**, *17*, 5659. [CrossRef]
43. Chitra, B.; Jain, M.; Chundelli, F.A. Understanding the soundscape environment of an urban park through landscape elements. *Environ. Technol. Innov.* **2020**, *19*, 100998. [CrossRef]
44. Brown, A.L.; Kang, J.; Gjestland, T. Towards standardization in soundscape preference assessment. *Appl. Acoust.* **2011**, *72*, 387–392. [CrossRef]
45. Calogiuri, G.; Patil, G.G.; Aamodt, G. Is green exercise for all? A descriptive study of green exercise habits and promoting factors in adult norwegians. *Int. J. Environ. Res. Public Health* **2016**, *13*, 1165. [CrossRef] [PubMed]
46. Song, Y.; Wei, W.Z. Research on the Quality Improvement of MICE Tourism in Resource-based Tourism Cities Based on IPA—Take Huangshan City as an example. *J. Chongqing Univ. Sci. Technol. (Soc. Sci. Ed.)* **2013**, *1*, 108–111.
47. Li, H.; Xie, H.; Woodward, G. Soundscape components, perceptions, and EEG reactions in typical mountainous urban parks. *Urban For. Urban Green.* **2021**, *64*, 127269. [CrossRef]
48. Jöreskog, K.G.; Goldberger, A.S. Estimation of a model with multiple indicators and multiple causes of a single latent variable. *J. Am. Stat. Assoc.* **2012**, *70*, 631–639.
49. Hair, J.; Anderson, R.; Tatham, R.; Black, W. *Multivariate Data Analysis*, 5th ed.; Prentice Hall PTR: Hoboken, NJ, USA, 1998.
50. Wang, L.N.; Li, S.S. Misunderstanding and Countermeasures of structural equation model in correction and intermediary analysis. *Chin. J. Health Stat.* **2017**, *34*, 361–363. (In Chinese)
51. Jeon, J.Y.; Jo, H.I. Effects of audio-visual interactions on soundscape and landscape perception and their influence on satisfaction with the urban environment. *Build Environ.* **2020**, *169*, 106544. [CrossRef]
52. Liu, J.; Wang, Y.; Zimmer, C.; Kang, J.; Yu, T. Factors associated with soundscape experiences in urban green spaces: A case study in Rostock, Germany. *Urban For. Urban Green.* **2019**, *37*, 135–146. [CrossRef]
53. Jahani, A.; Kalantary, S.; Alitavoli, A. An application of artificial intelligence techniques in prediction of birds soundscape impact on tourists? mental restoration in natural urban areas. *Urban For. Urban Green.* **2021**, *61*, 127088. [CrossRef]
54. Chen, W.C.; Zheng, W.F. Correlation analysis between landscape preference and active health behavior of tourists in mountain parks—Take Fuzhou Feifengshan Olympic Sports Park as an example. *Shandong For. Sci. Technol.* **2021**, *51*, 7–15. (In Chinese) [CrossRef]
55. Zhai, Y.J.; Baran, P.K.; Wu, C.Z. Spatial distributions and use patterns of user groups in urban forest parks: An examination utilizing GPS tracker. *Urban For. Urban Green.* **2018**, *35*, 32–44. [CrossRef]
56. Hedblom, M.; Heyman, E.; Antonsson, H.; Gunnarsson, B. Bird song diversity influences young people's appreciation of urban landscapes. *Urban For. Urban Green.* **2014**, *13*, 469–474. [CrossRef]

57. O'Brien, E.A. Publics* and woodlands in England: Well-being, local identity, social learning, conflict and management. *Forestry* **2005**, *78*, 321–336. [CrossRef]
58. Rey Gozalo, G.; Barrigon Morillas, J.M.; Montes Gonzalez, D.; Atanasio Moraga, P. Relationships among satisfaction, noise perception, and use of urban green spaces. *Sci. Total Environ.* **2018**, *624*, 438–450. [CrossRef] [PubMed]
59. Zhou, Z.Y.; Kang, J.; Jin, H. Study on sound preference evaluation and its influencing factors in urban historical areas. *Build. Sci.* **2012**, *28*, 40–45, 88. (In Chinese)
60. Chen, Z.S.; Yu, L.M. The present situation and development trend of elderly hearing in China. *Chin. Sci. J. Hear. Speech Rehabil.* **2020**, *18*, 245–249. (In Chinese)
61. Paneerchelvam, P.T.; Maruthaveeran, S.; Maulan, S.; Abd Shukor, S.F. The use and associated constraints of urban greenway from a socioecological perspective: A systematic review. *Urban For. Urban Green.* **2020**, *47*, 126508. [CrossRef]
62. Dong, W.; Zhu, X.; Zhao, X.L. Research on correlation between built environment characteristics of community greenway and intensity of physical activities: A case study of Shenzhen. *Landsc. Archit.* **2021**, *28*, 93–99. (In Chinese)
63. Stigsdotter, U.K.; Corazon, S.S.; Sidenius, U.; Refshauge, A.D.; Grahn, P. Forest design for mental health promotion—Using perceived sensory dimensions to elicit restorative responses. *Landsc. Urban Plan.* **2017**, *160*, 1–15. [CrossRef]
64. Watts, G.R.; Pheasant, R.J. Tranquillity in the Scottish Highlands and Dartmoor National Park—The importance of soundscapes and emotional factors. *Appl. Acoust.* **2015**, *89*, 297–305. [CrossRef]

 *forests*

Article

Multi-Sensory Experience and Preferences for Children in an Urban Forest Park: A Case Study of Maofeng Mountain Forest Park in Guangzhou, China

Jian Xu [1,2,3,4], Lingyi Chen [1], Tingru Liu [1], Tao Wang [5], Muchun Li [1,*] and Zhicai Wu [1,3,4]

[1]   Department of Tourism Management, South China University of Technology, Guangzhou 510006, China
[2]   State Key Laboratory of Subtropical Building Science, Guangzhou 510640, China
[3]   Key Laboratory of Digital Village and Sustainable Development of Culture and Tourism, Guangzhou 510006, China
[4]   Guangdong Tourism Strategy and Policy Research Center, Guangzhou 510006, China
[5]   School of Architecture, Zhijiang College, Zhejiang University of Technology, Shaoxing 312030, China
*   Correspondence: limch@scut.edu.cn; Tel.: +86-181-4288-6885

Abstract: This study developed an analysis based on children's multi-sensory experiences and preferences in urban forest park to make practical suggestions for the design of children's activity areas. Taking Maofeng Mountain Forest Park in Guangzhou, Guangdong province as a case study, based on a face-to-face survey and online questionnaire survey, this study analyzed children's multi-sensory landscape preferences in the park and explored the influence of multi-sensory experiences on children's behavioral experience by establishing a structural equation model. The results reveal that visual, auditory, tactile and olfactory sensations were significantly correlated with children's behavioral experiences. In terms of landscape preference, children preferred landscapes in blue-green tones, original building materials and challenging entertainment programs. Based on these analysis results, the design recommendations for children's activity areas in urban forest parks are discussed.

Keywords: urban forest park; children; multi-sensory experience; preference research

Citation: Xu, J.; Chen, L.; Liu, T.; Wang, T.; Li, M.; Wu, Z. Multi-Sensory Experience and Preferences for Children in an Urban Forest Park: A Case Study of Maofeng Mountain Forest Park in Guangzhou, China. *Forests* **2022**, *13*, 1435. https://doi.org/10.3390/f13091435

Academic Editors: Guang-Yu Wang, Xin-Chen Hong and Jiang Liu

Received: 17 June 2022
Accepted: 28 July 2022
Published: 8 September 2022

Publisher's Note: MDPI stays neutral with regard to jurisdictional claims in published maps and institutional affiliations.

1. Introduction

The urban forest park is based on the ecological landscape of the forest near the city, equipped with appropriate artificial facilities and landscape buildings to meet the demand of urban residents for outdoor activities to return to natural ecology and have leisure vacations [1]. As an important part of the urban green space system, urban forest parks are built following the basic principles of the forest ecosystem and have the functions of recreation, science and education, culture, etc., bringing ecological, economic, social and educational benefits to the city and its residents [2]. Compared with urban parks, urban forest parks focus on forest ecosystems and showcase ecological diversity, presenting a kind of natural and rustic look that is different from artificial urban parks [3]. With the recent concern for urban ecological environment protection, research on the new concept of urban forest parks has increasingly shown its value and potential.

To better realize the various functions of urban forest parks, the multi-sensory experience of park visitors is particularly significant. According to Edward T. Hall, an American anthropologist, human receptors are divided into distance receptors and direct receptors, which form a variety of perceptual sensations such as vision, hearing, smell, touch and taste [4]. These receptors also have different tasks and ranges. The interconnection and cooperation between multiple receptors can realize the complete perception of things and the world [5], and then obtain a more comprehensive and enjoyable personal experience. As a new perspective of experience and cognition, certain theoretical and practical achievements have been developed in the field of multi-sensory research. On the theoretical side, international research focuses on three main directions: health care, multi-sensory landscape construction and sensory experience tourism [6]. There is also an increasing number of studies on audiovisual perceptions such as the "soundscape" [7]. On the practical side, sensory parks designed for specific groups of people or medical purposes are emerging [8], with world-renowned examples of multi-sensory applications such as the Osaka Garden in Japan and the Frosinone Sensory Park in Italy. With the continuous development of theories and practices, multi-sensory ideas continue to permeate innovations in landscape design [9]. As the key point of future urban green infrastructure construction, urban forest parks should be designed and innovated around people's multi-sensory experience to bring all-around green experiences and enjoyment to urban residents.

Increasing attention has been paid to natural activities in child development during the past decade, since children's lack of activities has become a global public health problem [10]. Urban forest parks are some of the most accessible public places for children in terms of natural activities. They play an important role in effectively improving the amount of children's extracurricular activities and promoting children's all-round growth [11]. The positive effects of activities in forest parks on human health have been well documented in the past [12]. However, the best way of maximizing the function of urban forest parks to serve children and improve their multi-sensory experience remains an open question [13]. In 2018, the proportion of domestic tourists aged 18 and below in China reached 12.80% [14], indicating that underage tourists have become more prominent in the total tourist population [15]. At present, the research objects in the field of visitor perception in urban forest parks are more concentrated on adults, not children [16,17]. Moreover, some urban forest parks have no activity areas designed for children, and other forest parks with children's amusement places also make fewer considerations for problems such as children's physical and psychological characteristics [10]. A children's activity area is a special area in an urban outdoor space for children to play, rest, interact, etc. [18]. In some Western cities, children are moving away from outdoor public areas [19], which leads to many problems, such as some children becoming avoidant, timid and lacking in concentration [20,21], which has drawn increasing attention to the importance of children's activity areas for their healthy development [22]. Some scholars have pointed out the need for exclusive children's activity areas in urban forest parks [20]. Based on this, many scholars have used gender differences, differences in household income levels, and ethnic differences as variables to explore children's behaviors and preferences within children's activity areas in urban forest parks [23,24]. Simultaneously, some studies focus on children's health needs and the design of activity areas. Solomon suggested that the design of activity areas should take into account children's physical and mental characteristics, rather than a "McDonald's" type of simple stacking [25]. Cumbo et al. found that children's rich imagination allows them to make full use of the resources in the forest park. A suitable environment for children's fantasies in the forest park is conducive to the dynamics of communication and the realization of imagination, allowing children to have pleasant sensory experiences [26]. To date, there is a paucity of research that explores the needs of child visitors in urban forest parks from a multi-sensory experience perspective. Montessori, a renowned educator, has stated that children grow up experiencing the natural environment primarily through the five senses, which lead to memory, imagination, and thinking [27]. Sensory experiences are a basic way for children to know the external world and play an important role in their growth and development [18]. Compared with adults, children are more sensitive to external sensory stimuli. In addition to seeing and listening, the aroma and touch of urban forest parks can also have a huge impact on children's senses [28]. However, many existing studies mainly focus on the visual and auditory fields, and other sensory effects are often ignored [7]. Previous research results show that tourists have less demand for taste experiences [29], and the olfactory experience is mainly reflected in tourists' eating activities [30], which has a low correlation with forest parks. Accordingly, this study will mainly examine four aspects: vision, hearing, smell and touch. As an exclusive place for children in urban forest parks, the development of children's activity areas should pay more attention to children's sensory experience and participation.

Therefore, under the current situation, in which the lack of movement in the process of children's growth is becoming increasingly prominent and the proportion of child tourists in urban forest parks is gradually increasing, the research of this article has practical and theoretical significance. On one hand, by investigating children's landscape preferences in urban forest parks, this paper puts forward some suggestions for the design of children's activity areas in urban forest parks, hoping to meet children's entertainment needs, so that children can actively participate in extracurricular activities with a greater probability and grow up healthily. On the other hand, at present, there are few studies in the field of urban forest parks that focus on children's groups and multi-sensory experiences. With the gradual liberalization of China's fertility policy and the improvement in people's attention to the growth of children, children's groups are becoming an important topic that cannot be ignored. Based on the summary and utilization of existing research results, this article can make up for the lack of children's sensory experience research to a certain extent and further enrich the research on children's visitor groups in urban forest parks. It also has certain characteristics that can help urban forest park managers to understand the needs of children customers and the common problems existing in the operation of the park, and finally provide a reference for the future strategic development of urban forest parks.

2. Methods

2.1. Case Selection and Data Collection

As shown in Figure 1, Maofeng mountain forest park is located in the northeast of Baiyun District, Guangzhou, Guangdong Province, China, with a total area of 69.72 square kilometers. This park covers the central city of Guangzhou and the new Baiyun International Airport within 20 km, providing an accessible urban green space for families with superior locational conditions [31]. It has a humid southern subtropical climate, with temperatures generally 2–5 °C lower than in the city center. It is dominated by broadleaf and coniferous forests, with a forest cover of 77.63%, while there are several reservoirs and mountain ponds within the park, with a total water area of 2.80 km^2. This park has good forest ecological landscape resources and superior conditions for developing forest tourism, and it was included in the list of provincial forest parks in 2001, being represented as an urban forest park with universal significance for studying children's visiting behavior [32].

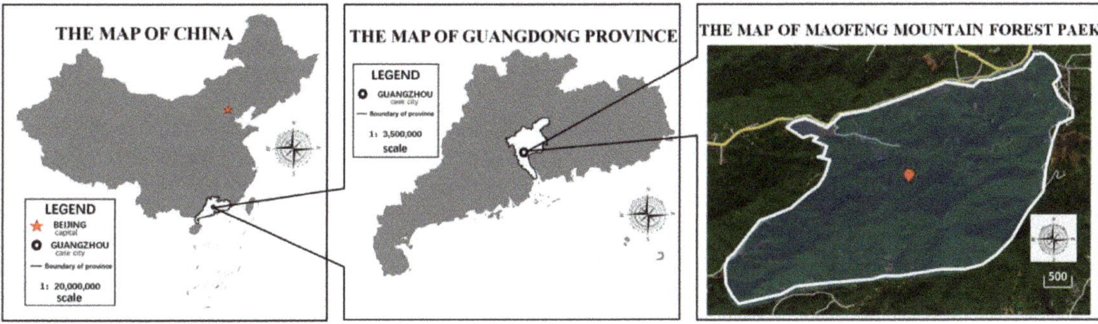

Figure 1. Geographical location of Maofeng Mountain Forest Park in Guangzhou.

Prior to the start of the large-scale questionnaire, this study conducted a pre-study with 10 school-age child volunteers of different ages to ensure that the children in the selected age group could understand the questionnaire questions in a basic way, and could complete the questionnaire in an average of 4 min 53 s (min = 3′28″, max = 9′28″). Moreover, investigators tested the children's intelligence level with simple general knowledge questions prior to the interview. Meanwhile, to ensure that the children fully understood the questionnaire, the survey team members and the parents of children assisted as support staff throughout the interview. However, they were not allowed to interfere with or intentionally guide the children's true thoughts. After completing the questionnaire, each child who successfully completed the whole task received a gift, which attracted more children and stimulated sharing among the children's group.

This study conducted a multi-sensory experience questionnaire survey of child visitors aged 6–14 years in Maofeng Mountain Forest Park during clear daylight hours (9 a.m.–5 p.m.) from January 25 to 21 February 2022, a time when the weather in Guangzhou was more comfortable than average (5–24 °C); therefore, residents were more willing to travel. Furthermore, it was the Chinese New Year period with a high probability of parent–child travel, which was suitable for this study. In terms of research methods, the research respondents were children aged 6–14 years old, who received elementary school education and had a certain level of literacy [33]. Moreover, the sample size required for this study was large, such that the data collection of the questionnaire survey was divided into two modes: an on-site face-to-face survey and an online survey. The on-site survey used a random sampling method by observing and questioning children of appropriate age at four different locations (Jinguo Forest, Tianhu Lake, Maofeng Qinyuan and Ancient Temple) in Maofeng Mountain Forest Park. As for the online survey, the online questionnaire was distributed to age-appropriate children who had been to Maofeng Mountain Forest Park after obtaining the permission of their guardians. Furthermore, an online video was used to assist children in real time in order to avoid excessive parental intervention in the process of children's texts.

In the end, the team members interviewed 157 children at the park site, excluding respondents who were not of the right age and were unwilling to cooperate, finally collecting a total of 126 on-site questionnaires. The online survey placed more emphasis on the authenticity of the subjects, and a total of 224 online questionnaires were collected. Excluding inconsistencies and incomplete answers, there were 336 valid questionnaires left online and offline, with an effective rate of 96%.

2.2. Questionnaire Design

This study used the definition of children in the relevant Chinese laws [34], and defined the interviewed group as children aged 6–14 years. The study first summarized the role of urban forest parks in children's growth and the importance of multi-sensory design in children's activity areas. Then, the article investigated the multi-sensory experience and preferences of children tourists and analyzed the collected questionnaire data quantitatively, followed by suggestions for the design of children's activity areas based on the analysis results. Finally, limitations and directions for future research were given at the end of the article.

This study applied a questionnaire survey method and collected 336 valid questionnaires. These questionnaires used a five-point Likert scale to examine the relationship between multi-sensory perceptions and children's multi-sensory behavioral experiences. The Likert scale can quickly and intuitively obtain the degree of recognition of a subject and quantify it, and has the characteristics of strong operability and analysis [35]. In this study, children's multi-sensory perceptual situation, the sense of behavioral experience and sensory attention were categorized into five levels: very good (2), better (1), neutral (0), bad (−1) and very bad (−2). Due to the limited cognitive level of the children group, the questionnaire was designed to have simple and friendly textual expression, so that children could better understand the content. There were also colorful pictures associated with questions in the questionnaire to give children a more visualized and vivid presentation. In addition, the questionnaire was divided into several parts. In every questionnaire distributed, each section was randomly arranged and combined in order to avoid order effects. The following dimensions were adopted to design the questionnaire in this study.

2.2.1. Multi-Sensory Perception and Behavioral Experience Dimensions

The multi-sensory perception section of the questionnaire drew on Zhang et al.'s scale on the influence of multi-sensory perception on the restoration effect of urban green spaces [8], and the behavioral experience section drew on Cui et al.'s study on user experience evaluation methods [36], finally designed to investigate children's multi-sensory behavioral experiences along four dimensions (visual sensation, auditory sensation, tactile sensation and olfactory sensation). Additionally, this study investigated children's perceptual preferences for the above four senses, collecting the senses that children paid the most and least attention to.

2.2.2. Landscape Preferences and Design Dimensions for Children's Activity Areas

Children's visitation environment settings are divided into three aspects: contact with nature, aesthetic appearance and opportunities for recreation and play [37]. Accordingly, this study also investigated children's design preferences regarding children's activity areas in terms of five aspects: (1) color; (2) building types; (3) vegetation types; (4) landscape types; and (5) activity types. Specifically, regarding the fifth item, "activity types", in addition to the conventional amusement programs and quality-development programs, this study also set up outdoor learning activities about knowledge of the natural environment with reference to the findings of Labintah and Shinozaki [38]. Through the setting of the above questionnaire questions, this study can, to a certain extent, find what children expect of the children's activity areas in urban forest parks.

2.2.3. Individual Relevance Dimension

Socio-demographic characteristics and visiting habits were investigated as relevant factors, specifically: gender, age, frequency and duration of visits and periods of visits [8]. This study investigated the potential influence of these individual factors on children's multi-sensory experiences and preferences in forest parks.

2.3. Technology Roadmap

As shown in Figure 2, this paper assumes that there is a dependency between children's multi-sensory perceptions, landscape preferences, and children's behavior in the forest park, and applies it to our questionnaire design. Visual sensation, auditory sensation, tactile sensation and olfactory sensation together influence children's landscape preferences and behavioral experience, and children's multi-sensory perception and landscape preferences together influence their behavior. There is a close relationship between multi-sensory perception, behavioral experience and landscape preference, which makes it possible to formulate guidelines on how to design children's activity areas. After completing the questionnaire design, this paper validated the conjectures through a series of data analyses. The first was a reliability and validity analysis to establish that the questionnaire data collected were reliable and internally consistent. Secondly, the results obtained from the collected questionnaire data were briefly described through descriptive analysis, which was used to analyze children's overall preference for urban forest park landscapes. Finally, by building structural equation models, this paper explored the correlation between children's multi-sensory perceptions and children's behavioral experiences in urban forest parks.

2.4. Reliability and Validity Analysis of Respondents Sample

The data samples were tested for reliability and validity using the SPSS reliability test with IBM SPSS Statistics 26. As shown in Table 1, the Cronbach's alpha value of the total variables of the scale was 0.804, and the Cronbach's alpha values of the five latent variables were all in the range of 0.637 to 0.903. The combined reliability (CR) was also higher than 0.60, indicating that the observed variables of each latent variable of the scale were reasonably designed, and the internal stability of the sample data was good.

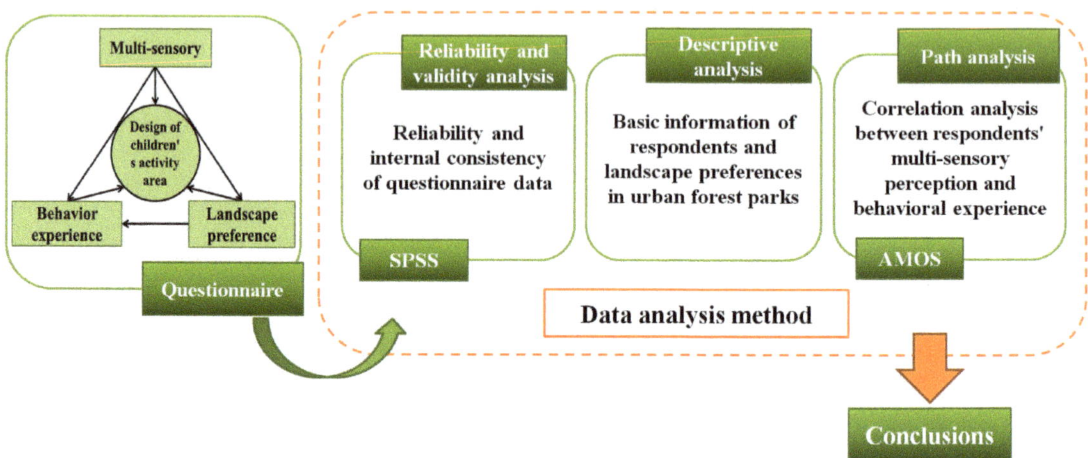

Figure 2. Research framework.

Table 1. Reliability and validity analysis.

Latent Variable	Observation Variable	α	KMO	p Value
Visual sensation	8	0.796		
Auditory sensation	6	0.738		
Tactile sensation	7	0.903		
Olfactory sensation	5	0.637		
General behavior experience	5	0.682		
Total	31	0.804	0.822	0.000

In terms of validity, the KMO value was 0.822, which was greater than 0.6. The p value (0.000) was less than 0.001, indicating that the validity of the measurement model was high and the selected observed variables had a certain correlation and validity.

2.5. Path Analysis of Respondents Sample

This study used IBM SPSS Amos 26 Graphics to construct a structural equation model for the data obtained from the questionnaire. In this study, the specific questions about multi-sensory perception in the questionnaire are listed as observation variables, and the children's multi-sensory perception experience and general behavioral experience, which are the variable that the research wants to explore, are listed as potential variables, as shown in Figure 3.

The fit indices of the structural equation models for all observed variables are shown in Table 2. The fitting index test is an important link to evaluate the effectiveness of the structural equation model [39]. The fitting index is the consistency between the hypothetical theoretical model and the actual data. The higher the value of the fitting index, the better the consistency between the theoretical model and the actual data. From the first index, TLI, proposed by Tucker and Lewis [40] in 1973, to NTLI, proposed by Marsh and Balla [41] in 1996, there are more than 40 named fitting indexes officially published in the literature. At present, the RMSEA (root mean square error of approval), AGFI (adjusted good fit index), TLI (Tucker–Lewis index), CFI (comparative fit index), SRMR (standardized root mean square residual) and other popular fitting indexes are recommended in academic circles [42]. This paper referred to the commonly used fitting indexes in academia and the selection of fitting indexes in the study of multi sensory restoration effect of urban green space by Zhang et al. [8], and finally chose Chi-square/df, RMSEA, GFI, AGFI, TLI and CFI as the fitting indexes for the study. The chi-square/df and RMSEA in the model reached the desired values. The values of GFI, AGFI, TLI, and CFI in the model were slightly lower than the ideal values, but the overall fitting degree is acceptable for these data [8,30,43]. The results suggest that all scales used in this study formed adequate measurement models, and the construct validity of the measures was confirmed.

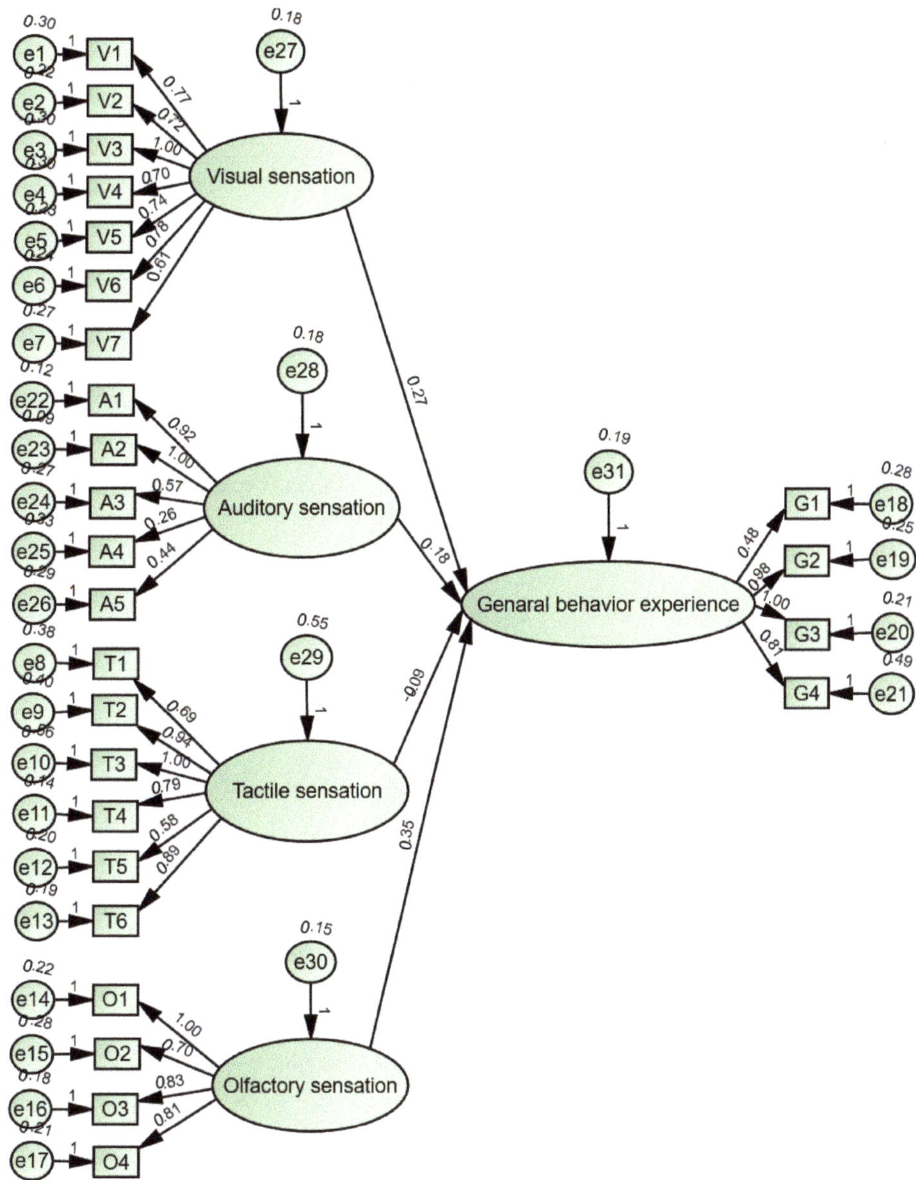

Figure 3. Structural equation model regression coefficients.

The standardized regulation weights of the structural equation model are used to describe the relationship between elements in the model. When the weight is greater than zero, it indicates that the two elements are positively correlated. When the weight is less than zero, it indicates that the two elements are negatively correlated. The higher the value, the greater the correlation.

Table 2. Tests of structural equation model fit indices.

Fitting Index	Ideal Value	Acceptable Value	Model Predictive Value	Whether Passes Test
Chi-square/df	1–2	1–3	1.839	Pass
GFI	>0.9	>0.7	0.892	Pass
AGFI	>0.9	>0.7	0.871	Pass
RMSEA	<0.08	<0.09	0.050	Pass
TLI	>0.9	>0.7	0.874	Pass
CFI	>0.9	>0.7	0.885	Pass

3. Results

3.1. Descriptive Analysis

The descriptive statistics of the sample are presented in Figure 4. In terms of socio-demographic characteristics, regarding gender structure, the proportion of male and female respondents was relatively balanced (Figure 4a). In terms of age, the largest proportion of respondents (57.13%), more than half, were between 10 and 12 years old (Figure 4b). Regarding the average frequency of park visits, the majority of children (45.50%) visited the forest park once a month, with slightly fewer choosing "Once a week or more" and "Semiannual or less" (Figure 4c). In terms of the personal duration of park visits, more than half of the children (56.25%) stayed between one hour and three hours, while nearly one-third (28.57%) stayed between half an hour and one hour (Figure 4d). As for the period of visits, nearly half of the respondents (43.75%) chose to visit the park in the morning, and more than one-third (35.71%) chose to visit in the afternoon (Figure 4e).

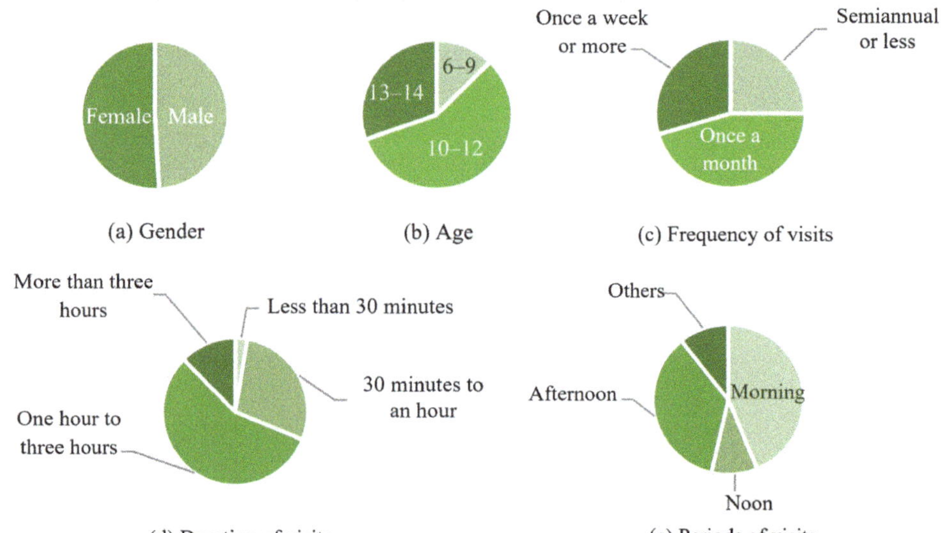

Figure 4. Description of surveyed children's basic characteristics and activities in the urban forest park. (**a**) Gender; (**b**) age; (**c**) frequency of visits; (**d**) duration of visits; and (**e**) periods of visits. All items were set as single-choice questions.

As shown in Figure 5, in terms of personal color preference, nearly two-thirds of the respondents preferred green (69.64%) and blue (16.96%) (Figure 5a). As for the landscape color combination preference, more than half of the respondents preferred a combination of blue and green water and wood landscapes (65.18%), 42.86% of the respondents preferred planted landscapes with various colors, and only 9.82% of the respondents preferred large-scale planted landscapes with high contrast (Figure 5b).

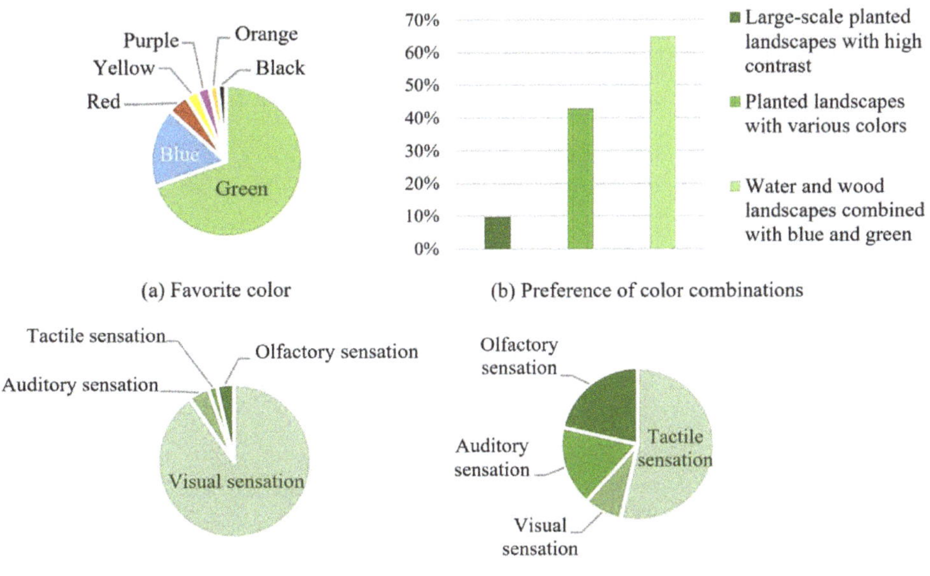

Figure 5. Description of surveyed children's perceived preferences in the urban forest park. (**a**) Favorite color; (**b**) preference of color combination; (**c**) sensation that concerned them the most; and (**d**) sensation that concerned them the least. Items (**a**,**c**,**d**) were set as single-choice questions, while item (**b**) was set as a multiple-choice question.

For the overall perception of children's sensations in the forest park, visual sensations (90.18%) were those that children paid the most attention to in the forest park, and tactile sensations (53.57%) were those that the children were least concerned with (Figure 5c,d).

Regarding the children's landscape preferences in urban forest parks, as shown in Figure 6, in terms of the choice of building materials, original ecological materials such as wood (61.61%) were the most popular ones. Elastic materials such as rubber (42.86%) and soft materials such as cotton (41.96%) followed. Rough-texture materials such as hemp rope (25.00%) and light and durable materials such as plastics (16.96%) were chosen by fewer people, while only 9.82% of the respondents chose smooth and hard materials such as steel (Figure 6a).

In terms of the type of vegetation designed, there was no significant difference in preference between different vegetation types. The largest number of children preferred lawn (62.50%), followed by flowering plants (56.25%). In contrast, the preference for forest vegetation was relatively low (41.96%–44.64%), but the differences presented between the three different types of tree species were not significant (Figure 6b). It could be seen that children expressed a certain preference for the rich and diverse types of forest park vegetation. The comparison of people choosing lawn vegetation versus forest vegetation also showed that children's preference for taller forest vegetation was lower than for lawn vegetation.

The same small difference was also seen in the preference for landscape types, where the largest number of children chose comfortable rest areas made with gallery frames (66.07%) and feature amusement facilities (60.71%); wide squares or open spaces (62.50%) and beautiful natural landscapes (50.00%) were next in number. Novel artificial ornaments were chosen by the least children (43.75%) (Figure 6c). It was clear that feature amusement facilities, plazas or open spaces for free movement, and beautiful natural landscapes were preferred, being endorsed by half or more of the children surveyed. In addition, comfortable rest areas were the most popular form of landscape for the children surveyed, which emphasizes the need for children to rest after activities and sightseeing, showing that the construction of rest areas for children should not be neglected in forest parks.

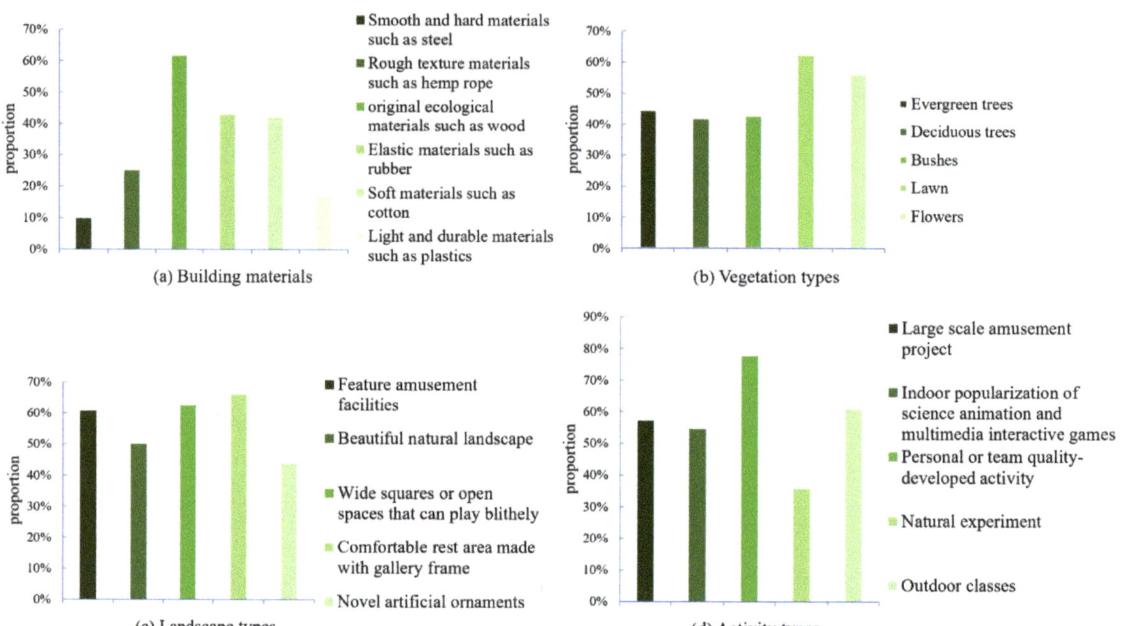

Figure 6. Children's preference for children's activity areas in urban forest parks. (**a**) Building materials; (**b**) vegetation types; (**c**) landscape types; and (**d**) activity types. All items were set as multiple-choice questions.

Focusing on the types of activities in the children's area, the results show that the highest number of people chose personal or team quality-developed activities (77.68%), more than half of the respondents also chose outdoor classes (60.71%), large-scale amusement projects (57.14%), and the indoor use of science animation and multimedia interactive games (54.46%), and the smallest number of respondents chose nature experiments (35.71%) (Figure 6d). It could be seen that personal or team quality-developed programs, such as rock climbing and canoeing, which were hands-on and challenging, were the most interesting activities for children. Amusement programs, on the other hand, still received majority support as traditional children's recreational activities.

In terms of visual perception, the mean values of "V2 Lush vegetation" "V7 Rich colors" were all at a high level, being greater than 1.4 (1.43–1.57), and only the value for "V1 Regular landscape" (1.38) was slightly lower than the other six items, indicating that the children interviewed had a high perception of most of the visual experience variables in the forest park. In terms of auditory perception, the mean values of all potential variables were higher than 1 point, indicating that children showed strong agreement in terms of the perception of each auditory experience. In particular, the children perceived beautiful natural sounds such as "A1 Natural sounds" and "A2 Animal sounds" more favorably than comfortable artificial sounds such as "A3 Background music", "A4 No traffic noise" and "A5 Others' sounds of activities". As for tactile perception, the differences between latent variables were larger. As can be seen from Figure 7, the means of "T1 Suitable apparent temperature",

"T2 Suitable somatosensory wind speed", and "T3 Paved road" were all below 1 (0.89–0.93), which were between the levels of general to relatively agreeable attitudes. The means of the last three, "T4 Natural landscapes touch", "T5 Artificial landscapes touch", and "T6 Safety perception of artificial landscape", were relatively high (1.08–1.38), indicating that children paid slightly less attention to the dimensions related to physical sensation and road leveling. However, for natural or artificial specific landscapes, the degree of tactile perception was higher. In terms of olfactory perception, the children showed a stronger agreement (1.33–1.61), ranging from agreeing to strongly agreeing, in all four experience perception subdivisions, especially for "O3 Fragrant natural beings" (1.61). Regarding the overall sensory experience of the forest park, "G1 mood" received the highest perceptual agreement (1.74), while "G4 Content attraction" presented a slightly lower perception (1.15) compared to the other three. However, in general, the mean scores of "G1 Mood"–"G4 Content attraction" dimensions were higher than 1, showing positive attitudes ranging from agreeing to strongly agreeing, indicating that respondents had a better overall behavioral experience in the forest park and tended to have a pleasant experience of activity in the park (Table 3 and Figure 7).

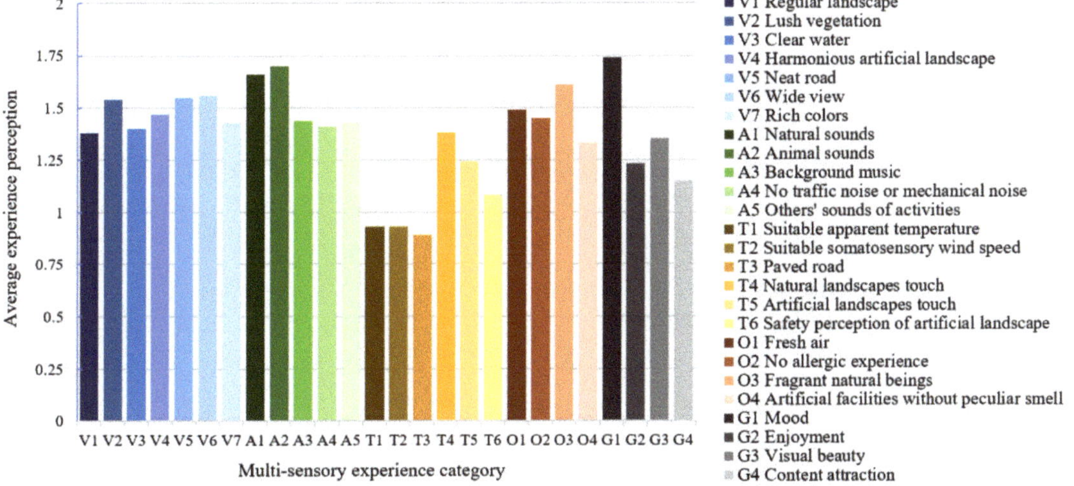

Figure 7. Bar chart of perceived mean values of children's multi-sensory experiences in an urban forest park (more details of items are shown in Table 3).

In this study, an in-depth investigation of the "O1 Fresh air" and "O3 Fragrant natural beings" was conducted. In "O1 Fresh air", the reasons that children thought the air in urban forest parks was not fresh were presented. A few respondents who expressed bad and completely bad attitudes toward the air cleanliness considered that the air quality in the park had decreased due to dust or garbage (Figure 8). In "O3 Fragrant natural beings", the study also investigated children's perceptual descriptions of flowers, trees and soil in urban forest parks. Most respondents described this item with words such as "fresh" and "fragrant", indicating that children's olfactory perception of air, natural plants and trees was relatively good (Figure 9).

3.2. Path Analysis of Multi-Sensory Perception

The standardized regression weights of the structural equation model are shown in Table 4. In terms of visual sensation, the most influential indicator was V3 clear water (0.613), followed by V6 wide view (0.564), V5 neat road (0.554) and V2 lush vegetation (0.542). From this conclusion, it could be seen that children paid more attention to simple and accessible visual experiences than to the more complex criteria such as landscape regularity and colorfulness. Therefore, the design of children's activity areas needed to focus on the positive influence of natural scenery on their willingness to experience things. For auditory sensation, the standardized regression weights were larger for A1 natural sounds (0.752) and A2 animal sounds (0.821), and smaller for A4 no traffic noise (0.186) and A5 other sounds of activities (0.327). This meant that children were more sensitive to and were fond of natural sounds and animal sounds, and more repulsed by traffic noise and noisy human sounds. Accordingly, the design of children's activity areas should avoid areas in the park that are close to downtown areas and factories, and should be close to areas with more vegetation and animals.

Table 3. Descriptive analysis of the perception of multi-sensory experiences in the urban forest park ($n = 336$).

Code	Items	Mean	Standard Deviation
V	**Visual sensation**		
V1	Regular landscape	1.38	0.64
V2	Lush vegetation	1.54	0.56
V3	Clear water	1.40	0.69
V4	Harmonious artificial landscape	1.47	0.63
V5	Neat road	1.55	0.57
V6	Wide view	1.56	0.59
V7	Rich colors	1.43	0.58
A	**Auditory sensation**		
A1	Natural sounds (the sounds of water flow, wind blowing leaves, etc.)	1.66	0.52
A2	Animal sounds (the sounds of birds and insect, etc.)	1.70	0.52
A3	Background music (broadcast music, live music, etc.)	1.44	0.57
A4	No traffic noise (car whistle, etc.) or mechanical noise (sounds from factories and lawn mowers, etc.)	1.41	0.59
A5	Others' sounds of activities (the sounds of conversation, whistle, footsteps, singing, large-scale activities, etc.)	1.43	0.57
T	**Tactile sensation**		
T1	Suitable apparent temperature	0.93	0.80
T2	Suitable somatosensory wind speed	0.93	0.94
T3	Paved road	0.89	1.06
T4	Natural landscapes touch (water flow, plants, land, etc.)	1.38	0.70
T5	Artificial landscapes touch (amusement facilities, rest seats, etc.)	1.24	0.63
T6	Safety perception of artificial landscape	1.08	0.80
O	**Olfactory sensation**		
O1	Fresh air	1.49	0.61
O2	No allergic experience	1.45	0.60
O3	Fragrant natural beings (trees, flowers, soil, etc.)	1.61	0.53
O4	Artificial facilities without peculiar smell (buildings, amusement facilities, etc.)	1.33	0.55
G	**General behavior experience**		
G1	Mood	1.74	0.58
G2	Enjoyment	1.23	0.69
G3	Visual beauty	1.35	0.67
G4	Content attraction	1.15	0.81

In terms of tactile sensation, T4 natural landscape touch (0.842) and T6 safety perception of artificial landscape (0.833) had high correlations, indicating that children had clear perceptions of the touch of natural objects that they could directly come into contact with and the safety of the artificial landscape, while they were not sensitive to the perception of external temperature and wind speed. In terms of olfactory sensation, its correlation with O1 fresh air (0.632) was stronger than other variables, indicating that children paid more attention to air odor in terms of olfactory sensation compared with the odor of natural and artificial landscapes, and it was especially important to keep the air in the children's activity area clean and fresh. In terms of general behavioral experience, the standardized regression weights for mood, enjoyment, visual beauty, and content attraction were 0.402, 0.689, 0.720, and 0.483, indicating that children paid more attention to G3 visual beauty (0.720) and G2 enjoyment (0.689) in their general behavioral experience.

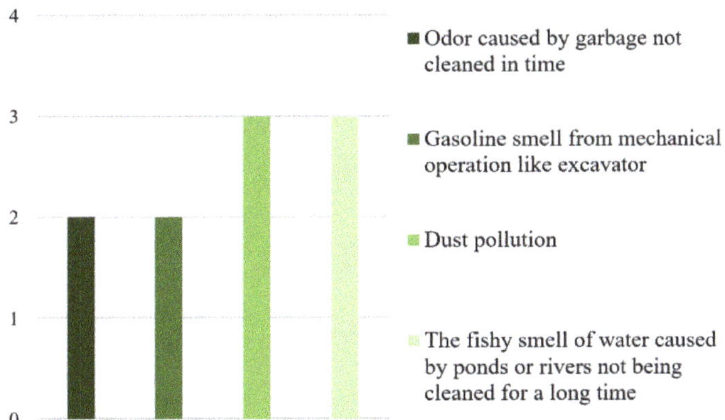

Figure 8. The reasons that children thought the air in urban forest parks was not fresh ($n = 4$).

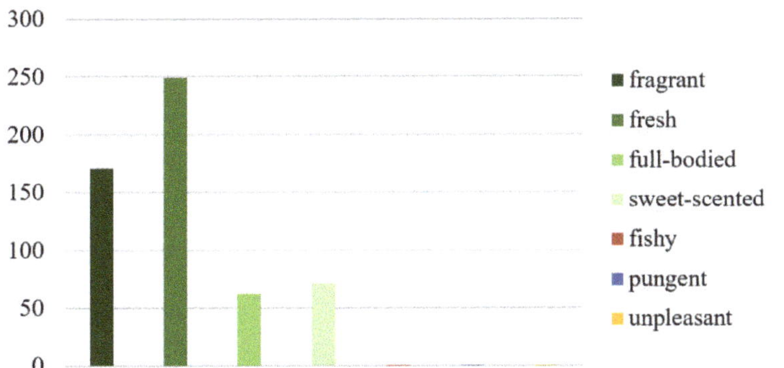

Figure 9. Children's perceptual descriptions of flowers, trees and soil in the urban forest parks ($n = 334$).

Standardized weights and their statistical significance obtained from the structural equation model are presented in Table 5. Visual, auditory, tactile and olfactory sensations all had p-values < 0.05 and were positively and significantly associated with general behavioral experience [8]. In addition, the standardized weights for visual, auditory, tactile, and olfactory sensations were 0.268, 0.176, -0.085, and 0.348, which implied that olfactory sensation had the greatest influence on children's behavioral experiences, followed by visual and auditory sensation, and tactile sensation had the least influence on children's behavioral experiences. Meanwhile, visual, auditory and olfactory sensations were positively correlated with behavioral experience, and tactile sensation was negatively correlated with behavioral experience.

Table 4. Structural equation model standardized regression weights of observed variables.

Items	Standardized Regression Weights
Visual sensation	
V1 Regular landscape	0.510
V2 Lush vegetation	0.542
V3 Clear water	0.613
V4 Harmonious artificial landscape	0.476
V5 Neat road	0.554
V6 Wide view	0.564
V7 Rich colors	0.450
Auditory sensation	
A1 Natural sounds (the sounds of water flow, wind blowing leaves, etc.)	0.752
A2 Animal sounds (the sounds of birds and insect, etc.)	0.821
A3 Background music (broadcast music, live music, etc.)	0.424
A4 No traffic noise (car whistle, etc.) or mechanical noise (sounds from factories and lawn mowers, etc.)	0.186
A5 Others' sounds of activities (the sounds of conversation, whistle, footsteps, singing, large-scale activities, etc.)	0.327
Tactile sensation	
T1 Suitable apparent temperature	0.639
T2 Suitable somatosensory wind speed	0.741
T3 Paved road	0.707
T4 Natural landscapes touch (water flow, plants, land, etc.)	0.842
T5 Artificial landscapes touch (amusement facilities, rest seats, etc.)	0.696
T6 Safety perception of artificial landscape	0.833
Olfactory sensation	
O1 Fresh air	0.632
O2 No allergic experience	0.451
O3 Fragrant natural beings (trees, flowers, soil, etc.)	0.605
O4 Artificial facilities without peculiar smell (buildings, amusement facilities, etc.)	0.560
General behavior experience	
G1 Mood	0.402
G2 Enjoyment	0.689
G3 Visual beauty	0.720
G4 Content attraction	0.483

Table 5. Results of path analysis: standardized estimates (n = 336).

Hypothesis	Estimate	S.E.	C.R.	p
General behavior experience←Visual sensation	0.268	0.086	3.121	0.002 *
General behavior experience←Auditory sensation	0.176	0.080	2.186	0.029 *
General behavior experience←Tactile sensation	−0.085	0.043	−1.998	0.046 *
General behavior experience←Olfactory sensation	0.348	0.103	3.381	***

Note: * means $p < 0.05$, *** means $p < 0.001$, S.E. means standard error, C. R. equivalent to the value of T in statistics.

4. Discussion

Although the previous literature has attempted to explore the relationship between sensory perceptions and behavioral experiences of crowds in urban parks [16], few scholars have analyzed this from children's perspective by combining four different senses. In this study, the subjective evaluation method was used to explore the relationships among multi-sensory perception, children's behavioral experiences and landscape preferences based on the questionnaire data of children in Maofeng Mountain Forest Park in Guangzhou, and the following conclusions were obtained. Specifically, this study confirmed the direct link between visual, auditory, tactile and olfactory multi-sensory perceptions on children's behavioral experiences, indirectly revealed the relationship between multi-sensory perception and children's landscape preferences, and finally integrated multi-sensory perception, behavioral experiences, and landscape preferences into a whole, providing a comprehensive exploration of children's behavior and psychology in urban forest parks from a multi-sensory perspective. Therefore, this study could enrich the evidence of the multi-sensory perception and behavioral psychology of children to some extent.

4.1. General Behavioral Experience Findings of Children in Urban Forest Parks

4.1.1. The Sensory Perception with the Highest Level of Children's Self-Attention Was Visual and with the Lowest Level of Attention Was Tactile

In terms of general sensory perception in the forest park, vision was the sensation that the interviewed children paid the most attention to in the forest park (90.18%) and touch was the sensation that the interviewed children paid the least attention to (53.57%).

This finding further validates the conclusion of Lyu and Zhang (2021), that eyes are the most basic tools for children to perceive external information, and that children understand and perceive objects through the world as seen by the eyes, which makes children pay more attention to and favor visual inputs in the development of their senses. In terms of tactile sensations, this study presents further exploration of and reflection on Lyu and Zhang's (2021) findings through an empirical approach. It has been argued that the sense of touch is the most intuitive process that children experience [44]; however, this study found that vision is the sensation that the interviewed children paid the least attention to in the forest park of all four sensations, which suggests that the extent of children's tactile perception needs to be assessed from the children's own perspective and based on the actual situation.

4.1.2. Visual, Auditory, Tactile and Olfactory Sensations Were Significantly Correlated with Children's Behavioral Experiences

The results of the operational tests of the structural equation model show that visual, auditory, tactile and olfactory sensations were all significantly correlated with children's general behavioral experience in the urban forest park, indicating that visual, auditory, tactile and olfactory sensations all influence children's overall sense of experience in the forest park. This finding is consistent with existing research by Yaswinda (2016), which suggest that children's direct and authentic experiences require multiple senses to deepen their perception and understanding of what they see, hear, smell and touch [45].

4.1.3. Olfactory Sensation Has the Greatest Influence on Children's Behavioral Experiences, Followed by Visual and Auditory Sensations, and Tactile Sensation Has the Least Influence on Behavioral Experiences

Among children's behavioral experiences in urban forest parks, the most influential is the need for olfactory sensation (0.348), which indicates that good olfactory sensation can improve the satisfaction of potential child visitors to a greater extent, with the odors of O1 fresh air (0.632) and O3 fragrant natural beings (0.605) having the greatest impact on children's overall olfactory sensation. This conclusion is different from previous scholars' view that vision is the most influential of all senses on behavioral experience [46], probably because the improvement of the overall urban environment and landscape in recent years has led to better visual sensation in forest parks in general. Coupled with the lack of social experiences of children compared to adults, which makes the amount of park contrast insufficient, the impact of visual sensation on experience satisfaction is relatively lower. More empirical research is needed to support the effects of olfactory sensation on children's behavioral experiences and their causes. In addition, unpleasant odors directly stimulate children's respiratory tracts, causing children to produce behavioral feedback such as using body language and voice language to express resistance, while visual sensations do not have such obvious feedback. The effects of visual and auditory sensations on behavioral experience were 0.268 and 0.176, and the visual sensation of air freshness and natural landscape in urban forest parks had positive effects in enhancing children's behavioral experience dimensions. Comfortable auditory sensations generated in urban forest parks, including natural soundscapes and pleasant artificial soundscapes, can enhance children's experience in forest parks and regulate their behavior. Traffic noise and noisy people have a low impact on enhancing the experience. The absolute value of the standardized coefficient of tactile sensation is 0.085, which has a low impact on children's behavioral experience.

4.1.4. Olfactory Sensation Was Negatively Correlated with Children's Behavioral Experience

Olfactory sensation was negatively correlated with children's behavioral experience (−0.085), which shows that lower somatosensory, landscape tactile sensations result in a higher general experience. The reasons for this may be due to some error in the results due to less consideration of tactile factors in the setting of children's general behavioral experience variables, which needs further research and investigation. Moreover, it may be due to the individual differences in perceptual preferences during childhood. Previous studies have confirmed that children enjoy activities and challenges involving risk, speed, excitement, stimulation, and uncertainty [47]. To some extent, uneven surfaces, lower physical sensations and landscape tactile sensations can stimulate children's sense of adventure and challenge, and enhance their experience satisfaction.

4.2. General Landscape Preferences for Children in Urban Forest Parks

4.2.1. Children Prefer Blue and Green Tones in Urban Forest Parks

In terms of visual color preference, this study consistently ascertains that blue and green received higher support from the children surveyed in both single color preference and color combination preference statistics. This is in line with the findings of Child et al. (1968), who found that children and adolescents from 6 to 18 years of age preferred cooler colors and that blue and green were favorite colors for all age groups. In addition, children's preference for highly saturated colors generally diminishes with age [48]. Furthermore, blue and green landscapes (water and wood landscapes, etc.), as the main components and basic elements of forest parks, fit better with children's perceived impressions of forest parks; thus, these two colors in forest parks give children a better sense of connection and intimacy [49].

4.2.2. Children Prefer Soft and Original Ecological Building Materials in Urban Forest Parks

In terms of building tactile preferences, children's more delicate skin than that of adults makes them prefer soft and comfortable tactile sensations. Wood, cotton and rubber are the preferred building materials for children, while the rough, cold and hard touches of materials such as twine, steel and plastic are resisted by them. This finding echoes the established knowledge by Friso et al. (2015) that natural materials such as wood and fiber are usually associated with comfort. Additionally, the degree of elasticity and softness that these natural materials have usually bring better comfort and pleasure to children and offer them the sense of security [50].

4.2.3. There Is Little Difference between Children's Perceived Preferences for Vegetation and Landscape Types in Urban Forest Parks

Focusing on the vegetation types in urban forest parks, lawns, flowers, and trees are all preferred by children. However, in general, vegetation types such as lawn and flowers with a low field of view are preferred over vegetation types such as trees with a higher field of view, which may be due to children's high demand for social activities in the forest park and the fact that children's perception of the size, positive recreational use and shade capacity of forest vegetation is not as strong as that of adults [51]. In contrast, the situation regarding floristic plants, which were chosen by more than half of the children, is consistent with the findings of Talal and Santelmann (2021), indicating that most people of all ages have some preference for floristic plants with color [51].

Regarding the landscape types, most children show preferences for recreatilal facilities, squares, natural landscapes, and rest areas. Among them, feature amusement facilities, plazas or open spaces for free movement, and beautiful natural landscapes were strongly endorsed by the children surveyed, which reflects the same finding as previous studies, that the most important influences that attract children to parks are active recreational facilities and natural landscapes [52]. In addition, comfortable rest areas were the most popular form of landscape for the children surveyed, which emphasizes the need for children to rest after activities and sightseeing, showing that the construction of rest areas for children should not be neglected in urban forest parks. The importance of rest areas in relation to children's outdoor activities deserves further research and attention in the future.

4.2.4. Children Prefer Quality-Development Programs That Are Somewhat Hands-On and Challenging in Urban Forest Parks

Childhood is a time in life when people are more imaginative, curious, and adventurous than in other periods [47]. In terms of activity programs, children generally show more interest in and a preference for personal or team quality-development programs that are hands-on and challenging, such as rock climbing and canoeing. This result is also consistent with the effectiveness of the "Adventure Obstacle Course" derived from the study by Labintah and Shinozaki. When children participate in this kind of adventure course, such as going through the mudflats, they have the opportunity to use their sensory and tactile skills and gain more immersion and enjoyment [38].

Children also prefer science animations and interactive games that reveal the mysteries of nature, etc., indicating that the introduction of activity programs about nature science knowledge in forest parks is equally attractive and relevant.

4.3. Design Recommendations of Children's Activity Areas in Urban Forest Parks

Children are the participants of their own activity areas. However, at present or in the past, designers have tended to ignore children's own feelings and needs, and to be too subjective during the design process [53]. Thus, based on the above findings on children's multi-sensory experiences and landscape preferences in forest parks of this study, the following recommendations are made for the design of children's activity areas in urban forest parks.

4.3.1. Overall Design

Many previous studies have investigated and designed children's learning and therapeutic environments from a multi-sensory perspective to better facilitate children's interaction and the learning experience [54,55]. As a place for children to rest, play, learn and interact, children's activity areas should pay more attention to children's perceptions and preferences for landscapes, and meet the needs of children's unique group experiences. As the basic window for children to perceive and understand the world, multi-sensory perception is an important reference for the design of children's activity areas. It is suggested that the design and construction of children's activity areas in urban forest parks should fully consider the multi-sensory experience of children, and bring better experiences and natural perceptions to children through the interplay of sight, sound, touch and smell. Secondly, it is also important to optimize the functional zoning and activity design of children's activity areas to better meet children's preferences and enhance their experience and interest.

4.3.2. Visual Design

Vision is the sensory perception that receives the most attention from children in forest parks (Figure 5), accounting for approximately 80% of all human perceptual activities, and is critical to the visual landscape design of children's activity areas. In terms of color, influenced by the main color palette of the park, the interviewed children, aged 6–14 years old, had significantly more favorable perceptions and stronger preferences for green and blue in the forest park. In terms of landscape color combinations, more than half of the respondents expressed a preference for water and wood landscapes with a combination of blue and green colors, while less than half of the children chose planted landscapes with richer and brighter colors. This is mainly because the color perception of children aged 6–14 years old is gradually becoming similar to adults', and the high purity of colors is likely to cause visual fatigue, so they begin to pay attention to and prefer less pure colors [28]. Therefore, in the color arrangement of children's activity areas, it is recommended to focus on the perceptions and preferences of children of different ages for colors, and to reasonably match colors of different purity to reduce the visual fatigue of children in environments with high-purity colors for a long time. Moreover, the color characteristics of the forest park should be appropriately integrated to make the color of the activity area more harmonious with the surrounding landscape, which can also effectively reduce children's cognitive dissonance with landscape colors in urban forest parks. In terms of vegetation, the interviewed children had the highest preference for lawn vegetation, followed by floral plants, and a lower preference for forest trees (evergreen trees, deciduous trees and shrubs). The main reason is that forest vegetation is less ornamental and less palpable, and most of the trees are tall, such that children are affected by their height and easily feel alienated from these large-scale landscapes. Consequently, when designing children's activity areas, it is recommended that designers focus on the scale of the landscape, so that the spatial distances and the size of objects in the activity areas are closer to the perceptual habits and standards of children, which is in line with "the accessibility and safety principles" of Ding et al. (2019), suggesting that children's line of sight and height should be specially considered when designing the equipment of urban residential areas [53].

4.3.3. Auditory Design

Hearing is a perceptual pathway second only to vision, and a good auditory environment can make children happy and relaxed and has certain physiological health functions [28]. In this study, it was found that the interviewed children had some positive emotions toward natural sounds, animal sounds, and beautiful background music, while they had negative emotions toward traffic and mechanical noises, and the sounds of other people's activities. Accordingly, it is suggested that the auditory design of children's activity areas focuses on creating a natural sound environment where natural sounds (water, wind, etc.) and animal sounds (insects, birds, etc.) are coordinated with each other, such as musical fountains, small zoos, etc., and supplemented with beautiful radio music or live music performances. This approach can create a comfortable and harmonious sound atmosphere to enrich children's aesthetic perception of the activity area. This suggestion echoes the established theory of positive soundscape design, and can bring people a better experience of the sound environment through the organic superposition of various sound elements [56]. Simultaneously, paying attention to attenuating artificial noises such as traffic and machinery to soothe the mood also allows children in the activity area to better immerse themselves in the natural atmosphere.

4.3.4. Tactile Design

Tactile sensation refers to direct contact, and is the most intuitive sensory activity for children to perceive the world. However, this study found that more than half of the respondents felt that touch was the least relevant to their sensory experience in forest parks (Figure 5), and that there is an urgent need to improve tactile perception. For the choice of construction materials, the interviewed children preferred natural, soft, and textured materials such as wood, rubber, and cotton, and had poorer tactile experiences with hard and rough materials such as steel and twine. Therefore, this study suggests that the tactile design of urban forest parks should strengthen the materials and textures of landscapes, as well as the facilities and equipment in activity areas. The landscape quality of natural objects such as water flow and trees in the activity area should be improved through timely pruning and cleaning, and using soft and smooth construction materials to build facilities.

A tactile experience walking path is also an infrastructure worth considering, by integrating a variety of different materials. As Ding's (2019) research showed, children can better understand the different shapes and textures through different ground materials and then enrich their tactile experience and intuitive sensory feelings, finally creating psychological resonance [53].

4.3.5. Olfactory Design

Past studies have shown that human olfactory ability decreases with age, and children's perception of smell is more sensitive than that of adults [28]. In this study, children's olfactory perception of natural landscapes such as air, natural plants and trees was relatively good. However, a very small number of respondents reported that dust and garbage influenced their perception of air quality in the park. As for artificial buildings and amusement facilities, the perceptions varied widely among subjects, with nearly one-quarter of children perceiving artificial buildings in the park to have a certain pungent odor. Consequently, it is recommended to increase the odor cooperation between different natural objects in the process of activity area olfactory design, reduce the discordant odor factors in the olfactory environment, and also strengthen the environmental management and odor inspection to create a fresh and natural olfactory environment. In turn, this can deepen children's experience and impression of spatial places and the landscape environment, and a good olfactory environment can promote their physical and mental health [57].

4.3.6. Functional Partitioning and Activity Design

With regard to the functional partitioning and activities of children's activity areas, the survey data of this study showed that the interviewed children had some recognition of areas with specific functions such as resting, playing and viewing. Among these areas, children pay most attention to the comfortable corridor rest areas, while they are relatively less interested in the novel artificial decorations. In terms of activities, this study also provided the interviewed children with options for activities such as amusement facilities, quality-development activities, and outdoor classes with reference to the contents of nature experience and education [58]. The results show that all of the nature activity programs, except for the nature experiments, were supported and preferred by more than half of the children, with the largest number choosing individual or team quality-development programs, nearly 4/5 of the total number. Accordingly, this study suggests that children's activity areas can also be enhanced by strengthening the construction of various functional zones and introducing various kinds of rich nature activities to improve children's overall experience in the activity areas, so as to promote children's physical and mental health while enhancing their perception of nature and cognitive ability.

5. Conclusions

This study explored children's behavioral experiences and landscapes preference from a multi-sensory perspective and provided some suggestions for the construction of children's activity areas in urban forest parks. Focusing on children's behavioral experiences, a direct association was ascertained between visual, auditory, tactile, and olfactory sensations and children's behavioral experiences. In comparison, it was found that olfactory sensation had the greatest influence and tactile sensation had the least on children's behavioral experiences. In addition, olfactory sensation was negatively correlated with children's behavioral experiences, calling for further explanation in future studies. Children's landscape preferences in urban forest parks were investigated and summarized under a multi-sensory perspective, which further strengthened the interconnection between multi-sensory perception and children's landscape preferences. This study took a more comprehensive look at children's landscape preferences from five main perspectives: color tones, building materials, vegetation types, landscape types and activity types. Operationally, multi-sensory perceptions and preferences should be seriously considered when designing children's activity areas in urban forest parks, particularly the integration of the above findings regarding children's behavioral experiences and landscape preferences.

This study reveals a potential link between multi-sensory perception, children's behavioral experiences and landscape preferences in urban forest parks, which provides a new way of thinking about how to create appropriate activity areas for children. Nevertheless, the limitations of this study should be admitted: the taste variable is more difficult to consider in urban forest parks; thus, the interaction and influence of children's five senses (visual, auditory, tactile, olfactory and taste) in forest parks cannot be completely analyzed. Additionally, this study specifically found that olfactory sensation had a greater impact on children's behavioral experiences than the other three senses, and that there was a negative correlation between tactile sensation and children's behavioral experiences, all of which need to be further explored and explained in detail in future studies. Children's multi-sensory perceptions in forest parks are an area worth exploring, and more in-depth additional research is needed, achieved through increased research time, more case sites, and greater sample sizes.

Author Contributions: Conceptualization, J.X., M.L., L.C. and T.L.; methodology, J.X., M.L., L.C. and T.L.; software, T.L. and L.C.; validation, J.X. and M.L.; formal analysis, T.L. and L.C.; investigation, L.C. and T.L.; resources, J.X., L.C. and T.L.; data curation, T.L. and L.C.; writing—original draft preparation, L.C. and T.L.; writing—review and editing, J.X., L.C., T.L. and T.W.; visualization, L.C. and T.L.; supervision, M.L. and J.X.; project administration, J.X. and M.L.; funding acquisition, Z.W., J.X. and M.L. All authors have read and agreed to the published version of the manuscript.

Funding: This research was funded by State Key Laboratory of Subtropical Building Science, South China University of Technology (No. 2019ZA02); Youth fund of national natural science foundation projects 'Research on visual landscape evaluation method of traditional villages in Guangdong based on deep learning' (52208057). Ministry of Education Humanities Social Sciences Research Project (20YJC760101); Zhejiang Provincial Natural Science Foundation of China (LQ20E080021); Guangzhou Philosophy & Social Science Planning Project (2022GZGJ131); Research on major basic theory of Social Science Planning in Guangdong Province (GD21ZDZGL01); Guangzhou Philosophy and Social Science Development "14th five-year plan" project (2022GZYB33).

Data Availability Statement: The data presented in this study are available upon request from the corresponding author.

Acknowledgments: We are very thankful to the International Cooperation Open Project of State Key Laboratory of Subtropical Building Science, South China University of Technology. We also appreciate the valuable comments of editors and anonymous reviewers.

Conflicts of Interest: The authors declare no conflict of interest.

References

1. Cai, M.; Cui, C.; Lin, L.; Di, S.; Zhao, Z.; Wang, Y. Residents' Spatial Preference for Urban Forest Park Route during Physical Activities. *Int. J. Environ. Res. Public Health* **2021**, *18*, 11756. [CrossRef] [PubMed]
2. Nowak, D.J.; Hoehn, R.E.; Bodine, A.R.; Greenfield, E.J.; O'Neil-Dunne, J. Urban forest structure, ecosystem services and change in Syracuse, NY. *Urban Ecosyst.* **2016**, *19*, 1455–1477. [CrossRef]
3. Feng, Z.L. Discussion on the Development of Urban Forest Parks under the Background of Ecological Civilization Construction. *Mod. Gard.* **2022**, *45*, 114–115. (In Chinese)
4. Hall, E.T. *The Hidden Dimension*, 1st ed.; Doubleday: New York, NY, USA, 1966; Volume 6.
5. Sun, Y.Y.; Piao, Y.J.; Zhu, W.J. Research on five senses design of landscape. *Inf. Agric. Sci. Technol. (Mod. Gard.)* **2010**, *12*, 37–40. (In Chinese)
6. Xi, L.; Qiu, E.F.; Zhang, Z.Y.; Zhang, C.; Cao, L.W. Research status and trend analysis of five senses landscape at home and abroad. *World For. Res.* **2020**, *33*, 31–36. (In Chinese)
7. Hong, X.; Wang, G.; Liu, J.F.; Dang, E. Perceived Loudness Sensitivity Influenced by Brightness in Urban Forests: A Comparison When Eyes Were Opened and Closed. *Forests* **2020**, *11*, 1242. [CrossRef]
8. Zhang, T.; Liu, J.; Li, H. Restorative effects of multi-sensory perception in urban green space: A case study of urban park in Guangzhou, China. *Int. J. Environ. Res. Public Health* **2019**, *16*, 4943. [CrossRef]
9. Tang, Z.; Liu, B.Y. Research progress of visual landscape assessment. *Landsc. Archit.* **2015**, *9*, 113–120. (In Chinese)

10. Ma, Y.C.; Chen, Y.; Gao, Z.M. Research on Children Friendly Park Based on the connection between children and nature. *Land Green.* **2021**, *6*, 53–55. (In Chinese)

11. Fang, X.; Gao, T.; Hedblom, M.; Xu, N.; Xiang, Y.; Hu, M.; Chen, Y.; Qiu, L. Soundscape Perceptions and Preferences for Different Groups of Users in Urban Recreational Forest Parks. *Forests* **2021**, *12*, 468. [CrossRef]

12. Li, J.; Wang, G.; Wang, Z.; Wang, W.J.; Chen, H.; He, M. Comparative study of the physiological and psychological effects of forest and urban auditory stimulus on humans. *Int. J. Geoheritage Parks* **2021**, *9*, 363–373. [CrossRef]

13. Shen, Y.; Jin, R.R.; Liao, Y.H.; Xie, C.; Li, M. Mechanism for Building Child-Friendly Communities Based on Multi-Party Co-Construction: A Case Study on Community Practice in Changsha. *China City Plan. Rev.* **2021**, *30*, 43–54.

14. Deng, S.Y.; Qiu, Q.H. Analysis and development forecast of domestic tourism. *Tour. Overv.* **2018**, *16*, 21. (In Chinese)

15. Maynard, E.; Barton, S.J.; Rivett, K.S.J.; Maynard, O.; Davies, W. Because 'grown-ups don't always get it right': Allyship with children in research—From research question to authorship. *Qual. Res. Psychol.* **2020**, *18*, 518–536. [CrossRef]

16. Mediastika, C.E.; Sudarsono, A.S.; Kristanto, L.; Tanuwidjaja, G.; Sunaryo, R.G.; Damayanti, R. Appraising the sonic environment of urban parks using the soundscape dimension of visually impaired people. *Int. J. Urban Sci.* **2020**, *24*, 216–241. [CrossRef]

17. Chen, C.; Yuan, Z.J.; Zhu, H. Playing, parenting and family leisure in parks: Exploring emotional geographies of families in Guangzhou Children's Park, China. *Child. Geogr.* **2019**, *18*, 463–476. [CrossRef]

18. Ding, J.; Cai, L.L. Design and Research on ecological landscape of children's activity area in Wetland Park. *Agric. Technol.* **2021**, *41*, 135–137. (In Chinese)

19. Valentine, G. *Public Space and the Culture of Childhood*; Routledge: Oxford, UK, 2004.

20. Bourke, J. Children's experiences of their everyday walks through a complex urban landscape of belonging. *Child. Geogr.* **2017**, *15*, 93–106. [CrossRef]

21. Louv, R. *Last Child in the Woods: Saving Our Children from Nature-Deficit Disorder*; Education Review; Algonquin Books: Chapel Hill, NC, USA, 2008.

22. Liu, L.; Shi, N.; He, Y.L.; Yang, H.; Pan, Y.J.; Yu, W.K.; Shen, Y.; Yu, Y.F.; Yang, Y.F.; Wei, L.H. Construction practice of Children Friendly City. *Urban Plan.* **2022**, *46*, 44–52. (In Chinese)

23. Marquet, O.; Hipp, J.A.; Alberico, C.O.; Huang, J.-H.; Mazak, E.; Fry, D.; Lovasi, G.S.; Floyd, M.F. How Does Park Use and Physical Activity Differ between Childhood and Adolescence? A Focus on Gender and Race-Ethnicity. *J. Urban Health* **2019**, *96*, 692–702. [CrossRef]

24. Marquet, O.; Aaron Hipp, J.; Alberico, C.O.; Huang, J.H.; Fry, D.; Mazak, E.; Lovasi, G.S.; Floyd, M.F. Park use preferences and physical activity among ethnic minority children in low-income neighborhoods in New York City. *Urban For. Urban Green.* **2019**, *38*, 346–353. [CrossRef]

25. Maxwell, L.E. Review of The Science of Play: How to Build Playgrounds That Enhance Children's Development, by Susan G. Solomon. *Child. Youth Environ.* **2015**, *25*, 301–303. [CrossRef]

26. Cumbo, B.J.; Leong, T.W. Wearable audio-video recorders as a tool for investigating child play experiences in nature. In Proceedings of the Annual Meeting of the Australian Special Interest Group for Computer Human Interaction, Association for Computing Machinery, Parkville, VIC, Australia, 7–10 December 2015; pp. 618–622.

27. Moretti, E. *The Best Weapon for Peace: Maria Montessori, Education, and Children's Rights*; University of Wisconsin Press: Madison, WI, USA, 2021.

28. Dong, Z.G.; Zhang, X.Y. Analysis of urban children's activity space design based on five senses. *Archit. Cult.* **2021**, *7*, 152–153. (In Chinese)

29. Hao, L.; Lin, T.; He, K.; Sui, M.F.; Zhang, L.Y.; Ding, G.C. Analysis on the Experience Needs of Five Senses and the Willingness of Potential Tourists to Experience Forest Health Trail. *Taiwan Agric. Res.* **2020**, *1*, 62–68. (In Chinese)

30. Wu, X.Y.; Liang, X.C.; Zhao, Y. An Exploratory Study on Tourists' Tri-dimensional Experience from the Perspective of Embodiment: The See Again Series as an Example. *Tour. Hosp. Prospect.* **2021**, *5*, 66–83. (In Chinese)

31. Fei, N.C.; Chen, B.F.; Shi, X. Monitoring of animal and plant diversity resources in Maofeng Mountain Forest Park, Guangzhou. *Ecol. Sci.* **2018**, *37*, 191–194. (In Chinese)

32. Tang, X.; Chen, Z.R.; Yu, W.H.; Ning, Y.M.; Wu, Y.; Zou, F.S. Investigation on bird species diversity in Maofeng Mountain Forest Park, Guangzhou. *J. Guangzhou Univ. Nat. Sci. Ed.* **2018**, *17*, 52–57. (In Chinese)

33. Kodan, H. Determination of Reading Levels of Primary School Students. *Univers. J. Educ. Res.* **2017**, *5*, 1962–1969. [CrossRef]

34. Yang, Y.J. Research on the testimony of child victims in child abuse and sexual assault cases. *J. Lanzhou Inst. Educ.* **2017**, *33*, 164–166. (In Chinese)

35. Han, G.H.; Fan, B. The Effect of Semantic Differences of Likert-type Questionaires on Scientific Measurements. *Sci. Technol. Prog. Policy* **2017**, *34*, 1–6. (In Chinese)

36. Cui, Y.; Hou, W.J.; Li, X.M. Research on quantitative method of UX based on core experience goals. *Exp. Technol. Manag.* **2012**, *29*, 353–356. (In Chinese)

37. Mahidin, A.M.M.; Maulan, S.B. Understanding Children Preferences of Natural Environment as a Start for Environmental Sustainability. *Procedia Soc. Behav. Sci.* **2012**, *38*, 324–333. [CrossRef]

38. Labintah, S.; Shinozaki, M. Children Drawing: Interpreting School-group Student's Learning and Preferences in Environmental Education Program at TanjungPiai National Park, Johor Malaysia. *Procedia Soc. Behav. Sci.* **2014**, *116*, 3765–3770. [CrossRef]

39. Wen, Z.L.; Hou, J.T.; Marsh, H. Structural equation model testing: Fit index and chi-square criterion. *Acta Psychol. Sin.* **2004**, *2*, 186–194. (In Chinese)
40. Tucker, L.R.; Lewis, C. Reliability coefficient for maximum likelihood factor analysis. *Psychometrika* **1973**, *38*, 1–10. [CrossRef]
41. Marsh, H.W.; Balla, J.; Hau, K.T. An evaluation of incremental fit indices: A clarification of mathematical and empirical properties. In *Advanced Structural Equation Modeling: Issues and Techniques*; Psychology Press: London, UK, 1996.
42. Wen, H.; Liang, Y.S. The essence of the commonly fitted exponential test for structural equation models. *J. Psychol. Sci.* **2015**, *38*, 987–994. (In Chinese)
43. Ye, C.X.; Wang, J.J. Research on satisfaction path of mathematical modeling flipping teaching based on structural equation model. *High. Educ. Sci.* **2016**, *6*, 76–83+51. (In Chinese)
44. Lyu, H.; Zhang, F.F. Research on the landscape application of five senses theory in children's Nature Education Park. *J. Chifeng Univ. Nat. Sci. Ed.* **2021**, *37*, 33–36. (In Chinese)
45. Yaswinda, Y. In Science Learning Model Based on Multisensory-Ecology in Early Childhood Education: A Conceptual Model. In Proceedings of the 3rd International Conference on Early Childhood Education (ICECE 2016), Bandung, Indonesia, 11–12 November 2016.
46. Zhao, J.W.; Hu, Q.; Zhang, L.; Zhu, X.J. A research overview of soundscape and visual aesthetics. *J. Southeast Univ. Philos. Soc. Sci.* **2015**, *17*, 119–123.
47. Stephenson, A.M. Physical Risk-taking: Dangerous or endangered? *Early Years* **2003**, *23*, 35–43. [CrossRef]
48. Child, I.L.; Hansen, J.A.; Hornbeck, F.W. Age and sex differences in children's color preferences. *Child Dev.* **1968**, *39*, 237–247. [CrossRef] [PubMed]
49. Heezik, Y.; Freeman, C.; Falloon, A.; Buttery, Y.; Heyzer, A. Relationships between childhood experience of nature and green/blue space use, landscape preferences, connection with nature and pro-environmental behavior. *Landsc. Urban Plan.* **2021**, *213*, 104135. [CrossRef]
50. Friso, V.R.; da Silva, J.C.R.P.; Landim, P.d.C.; Paschoarelli, L.C. Ergonomic Analysis of Visual and Tactile Information of Materials Used in the Manufacture of Toys. *Procedia Manuf.* **2015**, *3*, 6161–6168. [CrossRef]
51. Talal, M.L.; Santelmann, M.V. Visitor access, use, and desired improvements in urban parks. *Urban For. Urban Green.* **2021**, *63*, 127216. [CrossRef]
52. Loukaitou-Sideris, A.; Sideris, A. What Brings Children to the Park? Analysis and Measurement of the Variables Affecting Children's Use of Parks. *J. Am. Plan. Assoc.* **2010**, *76*, 89–107. [CrossRef]
53. Ding, Y.; Lei, L.; Du, J.; He, Q.; Ming, X.; Feng, Y.; Yu, X. Analysis on the Space Design of Children's Outdoor Activities in Urban Residential Areas in China. In Proceedings of the 4th International Conference on Contemporary Education, Social Sciences and Humanities (ICCESSH 2019), Moscow, Russia, 17–19 May 2019.
54. Cosentino, G. Exploring Multi-Sensory Interaction to Enhance Children' Learning Experience. In *Interaction Design and Children*; Association for Computing Machinery: New York, NY, USA, 2021; pp. 644–647.
55. Shan, J.; Mei, H. A Theoretical and Practical Review on Multi-Sensory Interactive Space Design for Autistic Children. In Proceedings of the 2020 the 4th International Conference on E-Education, E-Business and E-Technology, Shanghai, China, 5–7 June 2020; pp. 61–65.
56. Weng, M. Aural landscape design. *Chin. Landsc. Archit.* **2007**, *12*, 46–51. (In Chinese)
57. Dong, Y.P.; Zhang, W. Children's health garden sensory experience design. *Xiandai Hortic.* **2016**, *4*, 110–112. (In Chinese)
58. Lieflander, A.K.; Frohlich, G.; Bogner, F.X.; Schultz, P.W. Promoting connectedness with nature through environmental education. *Environ. Educ. Res.* **2013**, *19*, 370–384. [CrossRef]

 *forests*

Article

Network Text Analysis of Visitors' Perception of Multi-Sensory Interactive Experience in Urban Forest Parks in China

Jian Xu [1,2,3,4], Jingling Xu [1], Ziyang Gu [1], Guangwei Chen [1], Muchun Li [1,*] and Zhicai Wu [1,3,4]

[1] Department of Tourism Management, South China University of Technology, Guangzhou 510006, China
[2] State Key Laboratory of Subtropical Building Science, Guangzhou 510641, China
[3] Key Laboratory of Digital Village and Sustainable Development of Culture and Tourism, Guangzhou 510641, China
[4] Guangdong Tourism Strategy and Policy Research Center, Guangzhou 510641, China
[*] Correspondence: limch@scut.edu.cn; Tel.:+86-18142886885

Abstract: Urban forest parks play a crucial role in contributing to the urban environment, residential well-being, and social welfare. Visitors' perception of multi-sensory interactive experiences in urban forest parks is an important source of information for landscape planning. Whilst data elicited from visitors via questionnaires are temporally and spatially restricted, online media provide a public platform for the direct and comprehensive expression of park experiences beyond such restrictions. To look into visitors' multi-sensory interactive experiences in an urban forest park in China, a total of 7447 reviews of such were collected from four authoritative online platforms using Python, and the ROSTCM tool was used to generate semantic and social networks out of the data set. The results showed that urban forest park visitors' sensory experiences are dominated by visual and olfactory perceptions, followed by audio-visual and visual-tactile interactions. Among them, visual perception displays the highest degree of specificity and diversity, while tactile and gustatory perceptions are relatively infrequent and singular. The landscapes that affect visitors' perceptual preferences mainly include floriculture, green vegetation, soundscapes, and sanitation utilities. Moreover, both the fresh air and the agreeable environment have a significant positive impact on visitors' perceptions. The above findings not only have practical implications for the landscape planning and design of urban forest parks, but also provide theoretical insights into the evaluation of natural landscapes in urban forest parks from the perspective of tourists' multi-sensory experiences.

Keywords: network text analysis; multi-sensory interaction; perception; urban forest parks; China

Citation: Xu, J.; Xu, J.; Gu, Z.; Chen, G.; Li, M.; Wu, Z. Network Text Analysis of Visitors' Perception of Multi-Sensory Interactive Experience in Urban Forest Parks in China. *Forests* **2022**, *13*, 1451. https://doi.org/10.3390/f13091451

Academic Editors: Xin-Chen Hong, Jiang Liu and Guang-Yu Wang

Received: 21 July 2022
Accepted: 6 September 2022
Published: 9 September 2022

Publisher's Note: MDPI stays neutral with regard to jurisdictional claims in published maps and institutional affiliations.

1. Introduction

Urban forest parks, as important ecological service systems in residential areas, are often called the "Heart of the City" due to their essential role in maintaining the overall ecological balance [1,2]. A good urban forest park can purify urban air, improve health and well-being, provide space for exercising, recreation, entertainment, and ecotourism [3], and help prevent obesity, increase immunity, and reduce the incidence of chronic diseases [4]. In addition, studies have demonstrated that the therapeutic landscape design and implementation in accordance with multi-sensory theory in urban forest parks play an active role in providing mental relaxation [5,6]. Multi-sensory interaction means engagement with the environment through all sensory organs, including the eyes, ears, nose, tongue, and skin [5]. Accordingly, identifying the specific elements that are characteristically appealing to forest park visitors can not only help optimize the visitors' interaction with the park but also improve the experiential benefits of the park [7]. With the aging of the population and the expansion of sub-health problems, the relationship between urban forest parks and human health has received increasing attention [8,9]. Currently, the global public health crisis (COVID-19) has engendered an increasing awareness of the health-promoting and recreational functions of urban forest parks [10,11]. International treaties and guidelines such as the European Landscape Convention (ELC) also point out the importance of public participation in landscape assessment [12]. Government agencies and research designers have long relied on traditional techniques of data collection, such as subjective evaluation methods and questionnaires, to study tourists' perceptual preferences [13].

However, traditional data collection techniques are time-consuming and painstaking, and can only reach a limited number of respondents. Moreover, they may suffer from social desirability bias, and questionnaires often under-represent respondents' views due to the restriction of predetermined items [14]. As a consequence, policymakers and research designers are unable to obtain a wide range of public insights. With technological developments, live presentation simulation methods [15] and assistive technologies, such as eye-tracking technology and GIS, are widely used to examine visitors' evaluation of landscapes [16,17]. Nevertheless, these emergent methods are often constrained by various research conditions, including environment, funding, and sample size, and hence the conclusions may not be generalizable to wider contexts.

The emergence of interactive computer-mediated technology has facilitated the collection of research data, and reviews posted on social media platforms have started to be used as an alternative source of research data in recent studies [18]. Currently, there is a growing trend in China to share daily experiences and personal feelings on media platforms such as MicroBlog and Meituan [19]. Review data on social media platforms, also known as crowd-sourced data, refer to user-generated content shared on social networking platforms [20]. This type of data is provided by non-professional organizations and citizens, rather than professional experts and scientists. crowd-sourced data are available in a variety of formats, including textual materials, images, and videos [21], and are highly participatory, spontaneous, extensive, and representative [22]. Compared with other data collection methods, crowd-sourcing is less time-consuming, less costly, and less restricted by space [23]. Numerous studies have revealed that crowd-sourced data play a more important role in advancing landscape perception and preference research [24]. To date, the method has been widely used by academic workers as well as commercial practitioners to investigate tourists' perceived tourism experiences, architectural landscape design, social public marketing, etc. [25].

In addition, crowd-sourced data can be applied for network text analysis which has a number of advantages over traditional data analytic methods. Firstly, network text analysis can deal with a massive amount of data with high processing efficiency [26]. Secondly, it can reveal the key concerns and sentiments manifested by different user groups [27]. Thirdly, network text analysis can integrate crowd-sourced data from multiple sources to improve the accuracy of results [28]. Therefore, this paper aims to examine the multi-sensory interactive perceptions of urban forest park visitors through network text analysis.

The significance and originality of this study can be highlighted as follows:

(1) It is hoped that this study will contribute to the construction of a theoretical system of urban forest natural landscape evaluation from the perspective of visitors' sensory experiences, and offer practical implications for landscape planning and design of urban forest parks.

(2) As multi-sensory interactive experience is a relatively important area of ecotourism research, the result of this study can help enrich the knowledge domains and the theoretical development of accessibility in recreation and tourism, sensory experience [24].

(3) This study makes an important contribution to extending the application scope of online text analysis methods from the field of consumer behavior to the field of multi-sensory interaction.

2. Literature Review

2.1. Network Text Analysis

Network text analysis is a method of analyzing text data collected from the Internet with the help of the ROSTCM semantic network and social network generating tool [24].

Text mining is a relatively new research field, and the main method is to collect a large amount of text data from the Internet (social media, newspapers, blogs, websites, forums, etc.) by computers, and after a series of pre-processing, a comprehensive analysis of the valid texts is performed by means of management information system and other technologies to find new knowledge connections and construct new models (Figure 1) [22]. ROSTCM is a data analysis application developed by Professor Shen Yang of Wuhan University with such functions as word segmentation, word frequency statistics, and NetDraw (Visual analysis) [24]. Feature words relating to the research topic are extracted from the text data for quantitative analysis and association mining, which can generate new knowledge domains through the reorganization of useful information [24].

Figure 1. Text mining process.

Text mining is mainly comprised of node analysis and cluster analysis. Specifically, the network text consists of connected nodes, where points represent specific words and feature values, and edges represent the degree of connectivity between individual words or feature values [22]. The number of edges owned by a node is the degree of connectivity between points, and the higher the degree of centrality of a node, the greater the influence it plays on the whole network [23]. Cluster analysis is the process of grouping objects into different clusters by their intrinsic associations [24]. Zhu J explored the emotional changes of urban forest park visitors during the outbreak of COVID-19 by mining and analyzing Microblog users' review data [19]; Wan C investigated visitors' preferences and values inspired by urban parks based on social media data in HK [20]. This technique has been widely used to explore users' emotional responses and affective preferences. But there are few studies using the method to examine visitors' multi-sensory interaction experience reviews.

2.2. Multi-Sensory Interaction

Multi-sensory interaction refers to one's integrated engagement with the environment through all sensory organs, including the eyes, ears, nose, tongue, and skin [5]. Currently, the theory of multi-sensory interactive experience has attracted increasing attention from landscape designers and scholars with an interest in designing healing landscapes and promoting human multi-sensory interactive experience [29]. Mei H conducted a case study of Stanley Park in Canada to analyze the therapeutic functions of multi-sensory-based plant landscapes [5]. With the normalization of epidemic prevention and control, medical research on COVID-19 is constantly updated, and recent studies have shown that the senses of COVID-19 patients are severely affected, with 5.3% inflicted with taste disorders and 59.69% with smell disorders [10]. The high contagiousness of COVID-19 has led the public to pay more attention to health exercises and rehabilitation training [10,11]. This trend foreshadows another new construction goal for urban forest parks and underscores the design and improvement of landscapes with multi-sensory interactive experiences.

However, previous studies on multi-sensory interaction are mostly conducted through traditional methods, such as on-site questionnaires and the odor walking method [25,30]. Some scholars devised on-site questionnaires to access individuals' soundscape preferences and color perception in urban recreational forest parks [30,31]. Other scholars used odor walking and semantic difference methods to induce a subjective evaluation of aroma perception and explored the relationship between the scent landscape and olfactory perception of phyto-communities in parks [30]. Such studies, however, due to inherent methodological limitations, only dealt with single sensory dimensions, such as visual, auditory, olfactory, and so on [31,32], and thus had little to offer concerning the overall landscape perception of urban forest parks.

2.3. Visitor Preferences for Urban Forests

A plethora of research has been conducted concerning people's preferences for green spaces. It has been revealed that visitor preferences can be detected by visitors' destination preferences and their attention to certain park facilities [29], and correlation analysis and text selection are the common methods used to measure such preferences [33]. Despite the bulk of relevant research, most studies are focused on people's preferences for specific amenities or recreational facilities, such as trails in US urban and suburban forest parks [13], or different ways of green space maintenance, such as the sustainable management of green spaces of an urban forest park in the capital of Austria [34], rather than visitors' overall preferences for the parks as a whole.

Tourists' preferences for urban forests are found to be closely related to such properties of forests that can help relieve physical and mental stress and bring a sense of security [13,34]. The stronger the visitors' preference is for a particular attraction, the richer their perception of its features will be [35]. In other words, visitors' preferences for and their perceptions of forest park features are intertwined. While previous studies have identified characteristics of visitors' preferences for green spaces [36], few studies have used web-based text analysis to explore visitors' perceptions of urban forest parks.

In summary, the evaluation of multi-sensory interaction is an important part of urban green space planning and design, and the use of crowd-sourced data can provide a faster and more comprehensive understanding of visitors' preferences and values in urban forest parks, thus promoting urban forest park landscape planning [37]. In this spirit, this paper intends to conduct a case study of the Baiyun Mountain Urban Forest Park, Guangzhou, China, by visualizing the keywords in visitors' network reviews through the NetDraw tool and then exploring visitors' perception of multi-sensory interactive experience through connectivity factor analysis of interconnected nodes between different keywords within the matrix and systematic clustering methods [24]. This is an endeavor that aims to reveal the types of park landscapes favored by most visitors by answering the following questions:

(1) What preferences are characteristic of visitors' multi-sensory interaction with urban forest parks?
(2) What are the landscapes that influence visitors' perceptual preferences?
(3) What aspects are most noteworthy in the landscape planning and design of urban forest parks in China according to tourists' sensory preferences?

3. Methods

3.1. Research Scope

Baiyun Mountain Urban Forest Park in Guangzhou City, Guangdong Province, China, is located in the southern subtropical monsoonal maritime climate zone. Its geographical scope is sketched in Figure 2. Thanks to its unique climatic conditions and abundant natural resources, seven sightseeing areas have been developed for public access, including the Mingchun Valley and the Yuntai Garden, namely the largest natural-style birdcage and the largest garden-style garden in China. The Park boasts 876 species of precious plants, and the air is filled with birdsong and floral scent in all seasons [38]. Delicious snacks of Cantonese style served in the Park are even more memorable for visitors. The special scenic resources and comprehensive dining and touring infrastructure bring visitors a uniquely fulfilling multi-sensory interactive experience [39]. In addition, located in the center of Guangzhou City, the Park is easily accessible and is the most frequently visited urban forest park in the city. The park covers an area of 28 square kilometers and comprises more than 30 linked hilly peaks. There are four entrances to the park, each of which is close to a traffic lane and can be easily accessed by visitors. The park provides undifferentiated access to all visitors, either local or non-local, from home or abroad. The entrance fee to the park is RMB 5 (CHI 0.7) per visit and some of the internal attractions are available for a separate fee, but none of the tickets cost more than RMB 20 (CHI 2.8). Most areas of the park are free for visitors. Therefore, it can be utilized as a typical case to shed light on the landscape planning and design of urban forest parks.

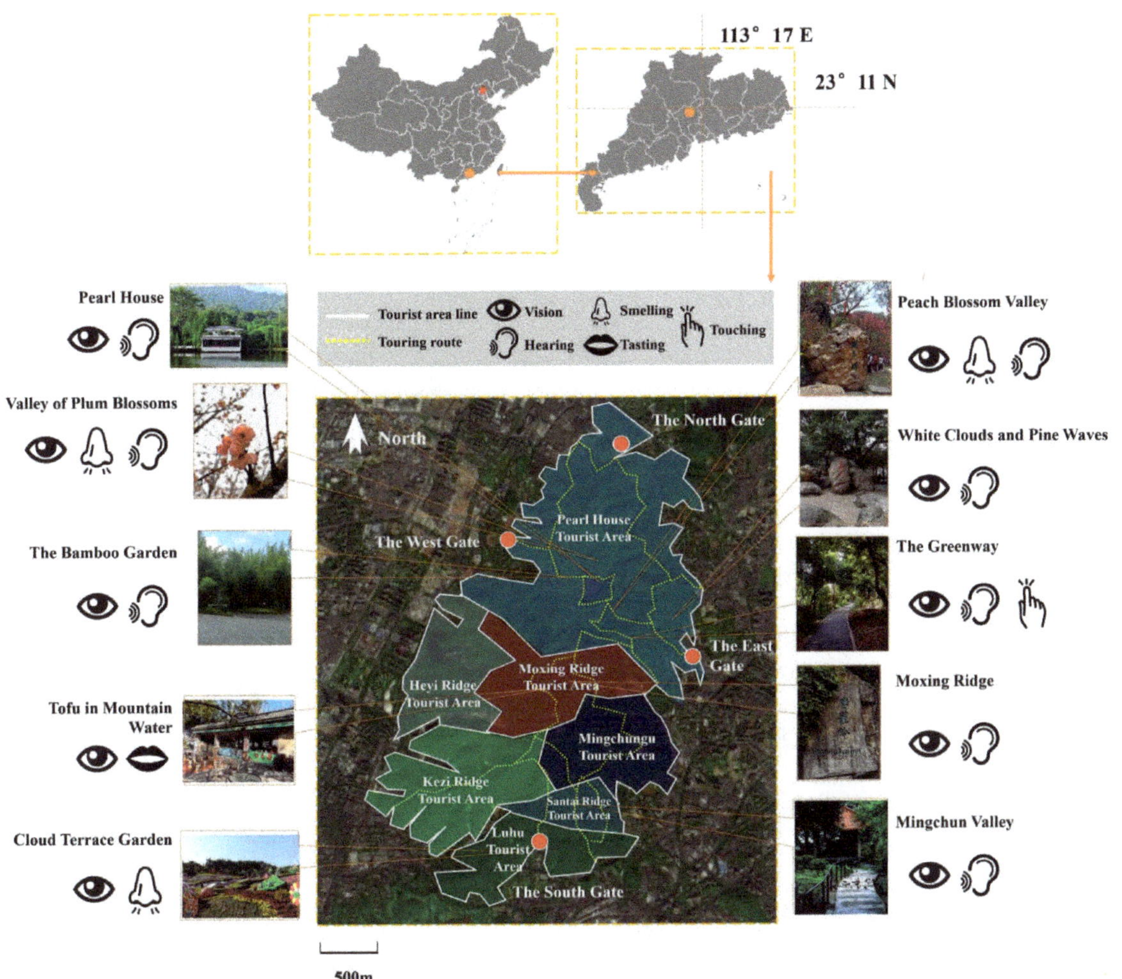

Figure 2. The map of Guangzhou Baiyun Mountain Urban Forest Park.

3.2. Research Framework

The network texts extracted in this study are narrations and reviews of visitors' multi-sensory perceptions of the Park. To guarantee the representativeness of the sample, we collected visitors' reviews posted online throughout an entire year spanning from January 2017 to January 2022. We used Python to retrieve and process our initial set of candidate texts from four social networking platforms, namely Mafengwo, Ctrip, Meituan, and MicroBlog. The GET request was used to procure review data from the servers of the four aforementioned platforms, and the POST request was to send the returned data to the servers for parsing. Considering the usage rate of each website, the proportion of positive reviews, and the comprehensiveness of sampling, we finally gathered a total of 10,593 reviews, the composition of which is shown in Figure 3.

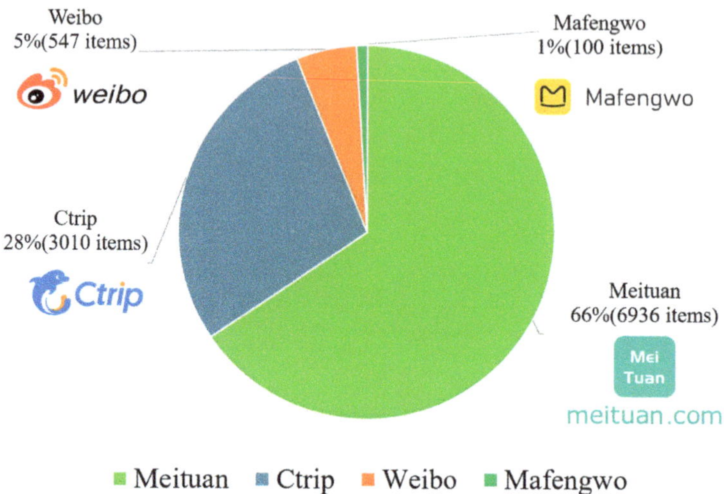

Figure 3. The number of network reviews of Baiyun Mountain, China.

The usefulness of the data crowd-sourced from the four platforms has been testified by other studies. Mafengwo is a popular travel site in China with over 100 million registered users and information covering more than 60,000 destinations worldwide [40]. Ctrip is a leading travel agency in China that provides a one-stop service for travelers from hotel reservations to attraction tickets and has the largest market share of approximately 60%. Nowadays, Ctrip has become a data source widely used in studies on the tourism industry [41]. Meituan is the biggest on-demand services platform in China and its users can share reviews on it regarding all aspects of life [42]. Sharing daily emotions and states on social media has become part of everyday life for MicroBlog users. As one of the most representative online social media in China, the Sina MicroBlog had 523 million active users in 2020 [43]. We chose these four platforms because large and reputable platforms have mature business models and ample clients who could offer opinions with adequate variance [44]. The four platforms are all free for registration and can be logged into with no age requirements. Similar studies that have used data from these platforms are showcased in Table 1.

Table 1. The application of the four platforms in related research.

Platform	Study Area	Data Type	Key Findings	Reference
Meituan meituan.com	Accommodation Sharing, Tourist attraction	User comments	A 15.4% increase in an attraction's online popularity after the entry of accommodation sharing.	[45]
	Text segmentation, Keyword extraction	User comments	Platforms offering fast, low-cost services can increase user satisfaction and dependability.	[46]
	Text mining, Tourist Attractions	User comments	The results demonstrate that the characteristics of scenic spots, service attitude, and tourist facilities are the focuses of tourist evaluation.	[47]

Table 1. *Cont.*

Platform	Study Area	Data Type	Key Findings	Reference
Ctrip	Content analysis, Tourist Preferences	Travel notes	It turned out that most Chinese honeymoon tourists prefer Asian and European countries, especially island countries for honeymoon tourism	[48]
	Emotional Analysis	Tourist reviews	The findings determined the words that most closely represent the demands and emotions of this customer base	[49]
	Tourism Destination Image	Tourist reviews	The tourist image of Guangzhou is mainly composed of cognitive, emotional, and conative image	[50]
MicroBlog	Geospatial Semantics Analysis	Short Texts	The cities can be classified into three groups according to their geospatial semantic components, i.e., tourism-focused, life-focused, and religion-focused cities.	[51]
	Temporal and spatial analysis	Microblog text	Changes in visitor sentiment are influenced by the epidemic, the level of the economy, and geographical location.	[52]
Mafengwo	Destination Image, Emotion Analysis	Travel notes	The tourists' perception of the destination image, cognitive theme, and emotional experience has different effects on the tourist experience.	[53]
	Tourism Experience	User comments	The changes and analysis characteristics of the tourism experience index under the three-time dimensions.	[54]

As mentioned earlier, visitors to the park are of various geographical locations, including local residents, and tourists from home and abroad. As the study aims to analyze visitors' sensory interaction with the park, there is no need to draw a distinction between local and non-local tourists; nor is there any need to distinguish between frequent and infrequent visitors. Anyone who visited the park and left their comments on any of the four platforms was included in the sample. The majority of comments in our dataset were given by young and middle-aged users, probably because children and older people use the online social platform less, which virtually constitutes an unavoidable defect for all research using web-based data. Nonetheless, we did remove duplicate texts and comments made by the same account to ensure that no bias would be germinated by identical data. We didn't collect any other demographic information about the respondents and we kept them anonymous throughout the study.

Despite the bulky size of the data, some of the texts were duplicates, some deviated from the topic, and some described other sensory experiences unrelated to the park. Therefore, prior to the network text analysis, initial filtering was carried out. When the scraping of textual content from URLs was all set up, 7447 valid data were retained for further analysis.

Subsequently, the NetDraw function in the ROSTCM tool was used for content analysis. The top 50 high-frequency feature words related to multi-sensory interaction were extracted by the pre-processing steps: merging and replacing synonyms. Next, CONCOR analysis was conducted to construct nodal blocks to identify the relationship between blocks with the help of the Pearson correlation of co-occurrence matrix between keywords, so as to analyze the multi-sensory interactive experience characteristic of visitors to the park and to provide implications for future multi-sensory landscape planning and construction of urban forest in China (Figure 4).

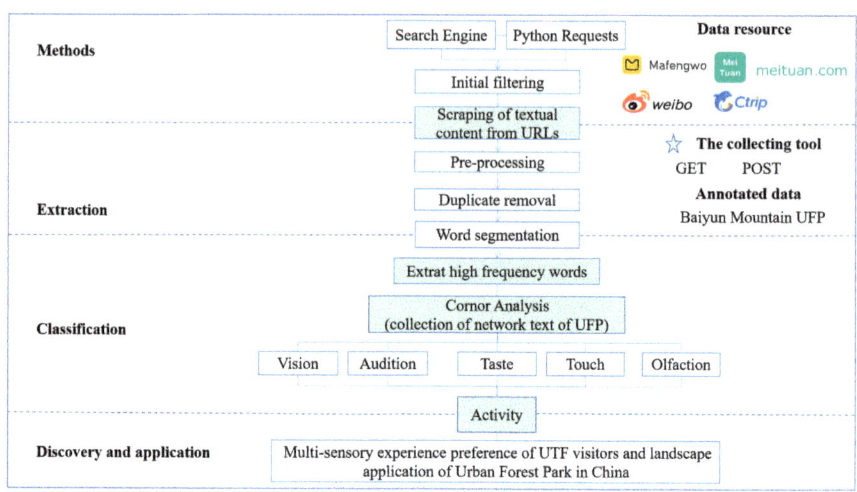

Figure 4. Research framework.

4. Results

4.1. Content Analysis of High-Frequency Words

Table 2 shows the top 50 high-frequency words in the reviews of Baiyun Mountain Urban Forest Park. Previous studies have demonstrated that the frequency of words can be an important proxy for preference rating and can reveal the degree of importance of features and values [23]. The higher the frequency, the more important the characteristics and values are. The lexical analysis of keywords shows that the visitors predominantly use objective nouns to describe their cognitive experience of multi-sensory interaction.

Table 2. High-frequency characteristic word list.

High-Frequency Words (Frequency)	High-Frequency Words (Frequency)
Landscape (3297)	Panorama (63)
Environment (1532)	Creek (62)
Air (1319)	Sunrise (61)
Air refreshing (1013)	Birds' twitter and fragrance of flowers (58)
Aesthetics (876)	Broad (57)
Facilities (490)	Mountains and waters (55)
Beautiful (424)	Taste (54)
Sightseeing (274)	Vegetation (54)
Respiration (237)	Sunset (53)
Peach blossom (195)	Plants (53)
Forest (161)	Delicacies (51)
Nightscape (140)	Restaurants (49)
Overlook (139)	Plank road (46)
Sanitary (123)	Tea-drinking (45)
Greening (122)	Dine (43)
Fresh (116)	Plum blossom (43)
Bean curd jelly (113)	Neatness (42)
Quietness (110)	Sunshine (40)
Luxuriance (79)	Visual field (40)
Green mountains and waters (78)	Flesh flowers (39)
Umbrage (72)	Soughing of the wind in the pines (37)
Catering (70)	Cool (36)
Delicious (69)	Lawn (36)
Green (67)	Gardens (34)
Mountain road (65)	Crowded conditions (33)

The high frequency of 'air' and 'fresh' is consistent with the results of Wang's research which states that good air quality is what people like most about urban forest parks, and that access to fresh air is a driving force for park visits [55]. The number of reviews containing '(fresh) air' totaled 2332, accounting for about 31% of the whole dataset. The results indicate that the olfactory interactive experience is of significant prominence to visitors in the park. Next to the number of positive affective expressions in response to the natural environment is that of the aesthetic expressions in response to the landscape environment [56]. The frequent use of such terms as 'Beautiful' confirms that visually driven aesthetic perception receives significant attention in landscape evaluation [56], thus reaffirming visitors' clear preference for natural elements in urban forest parks [8]. Flowers, wildlife, and water sound, are the natural landscape elements most favored by visitors. In respect of visitors' preferred non-natural landscape features, infrastructure environments, such as sanitation (123) and facilities (490), are more popular than lawns (36) and gardens (34). This may be closely related to the behaviors of the park visitors, such as picnicking and exercising in the park.

Word cloud image, an information text visualization technique that visually reflects the differences in word group importance by using a layout algorithm to represent word frequency in terms of text size, supplemented by a variety of color displays, shows that visual dimensions such as scenery, facilities, environment, and aesthetics are richly perceived, and olfactory dimensions such as air and breathing are saliently perceived (Figure 5).

Figure 5. Word Cloud image of the high-frequency characteristic word.

4.2. CONCOR Analysis of Keywords

To further explore the association between keywords and multi-sensory interactive experiences, CONCOR analysis was performed to visually derive the connectivity patterns within the network and to cluster keywords with similarities [24]. As shown in Figure 6, the filtered high-frequency words cluster into seven groups in order of feature vector scores.

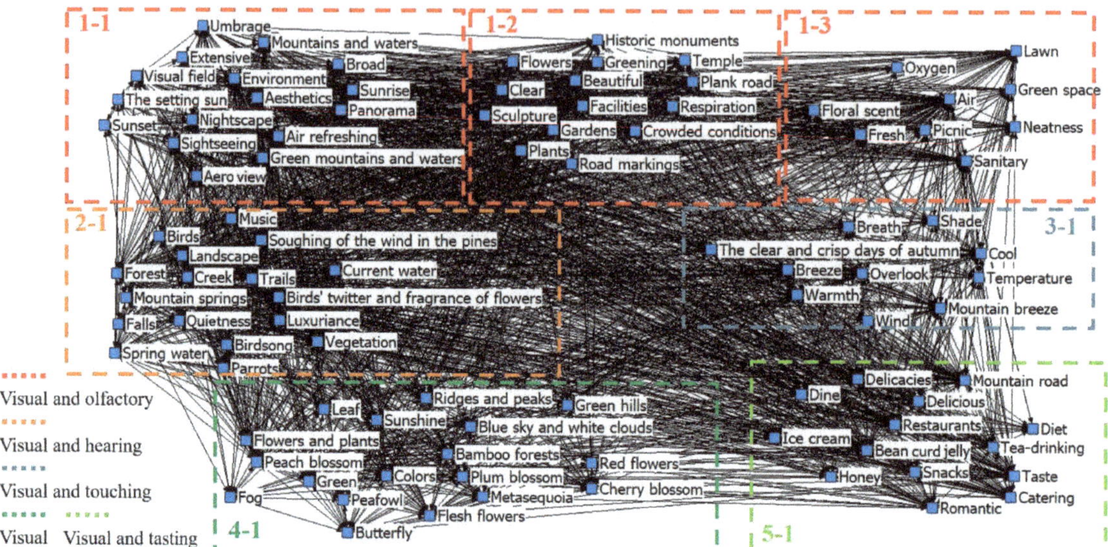

Figure 6. Visualization results of CONCOR analysis.

The high-frequency words subsumed in each cluster are listed as follows.

1-1. Natural landscape based on visual and olfactory interactive perception: Environment, Air refreshing, Aesthetics, Sightseeing Green mountains and waters Umbrage, Nightscape, Visual field, Panorama, Sunrise, Extensive, Mountains and waters, Broad, Sunset, The setting sun, Aero view

1-2. Humanistic landscape based on visual and olfactory interactive perception: Facilities, Beautiful, Respiration, Greening, Plants, Temple, Plank road, Gardens, Historic monuments, Flowers, Clear, Sculpture, Road markings, Crowded conditions

1-3. Visual and olfactory interactive experience during outdoor activities: Air, Fresh, Sanitary, Neatness, Picnic, Lawn, Flower scent, Oxygen, Greenspace

2-1. Urban forest park elements based on audio-visual interactive perception: Landscape, Forest, Creek, Quietness, Birdsong, and fragrance of flowers, Luxuriance, Soughing of the wind in the pines, Vegetation, Springwater, Current water, Parrots, Trails, Mountain springs, Music, Birdsong, Birds, Falls

3-1. Natural scenery based on visual and tactile interactive perception: Overlook, Cool, Mountain breeze, Breath, Warmth, The clear and crisp days of autumn, Temperature, Breeze, Wind, Shade

4-1. Natural landscapes based on the visual interactive perception: Peach blossom, Cherry blossom, Flesh flowers, Bamboo forests, Plum blossom, Metasequoia, Colors, Sunshine, Peafowl, Butterfly, Ridges and peaks, Blue sky and white clouds, Red flowers, Green, Leaf, Flowers and plants, Fog, Green hills

5-1. Gustatory perception and experience activities: Bean curd jelly, Delicacies, Dine, Mountain road, Restaurants, Diet, Taste, Tea-drinking, Delicious, Snacks, Honey, Ice cream, Romantic, Catering

What the CONCOR analysis revealed, as shown in Figures 7–13, is the significant association between urban forest park visitors' experience and multi-sensory interaction preference theory [57].

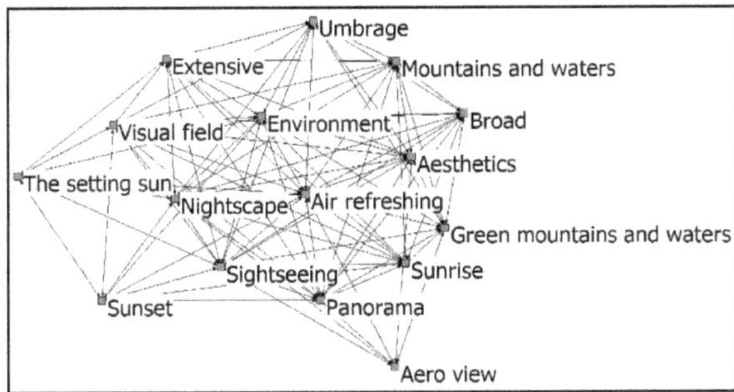

Figure 7. 1-1. Natural landscape based on visual and olfactory interactive perception.

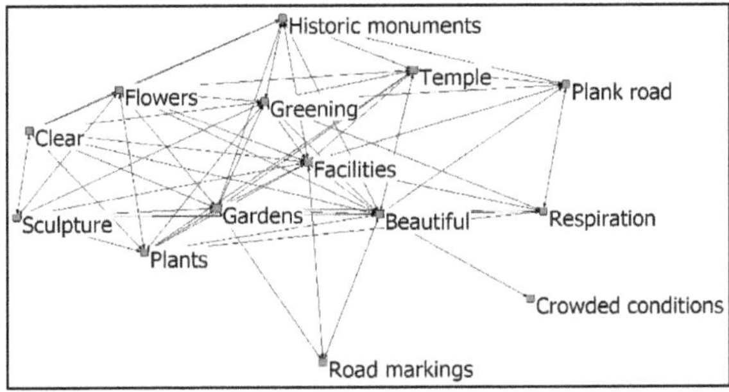

Figure 8. 1-2. Humanistic landscape based on visual and olfactory interactive perception.

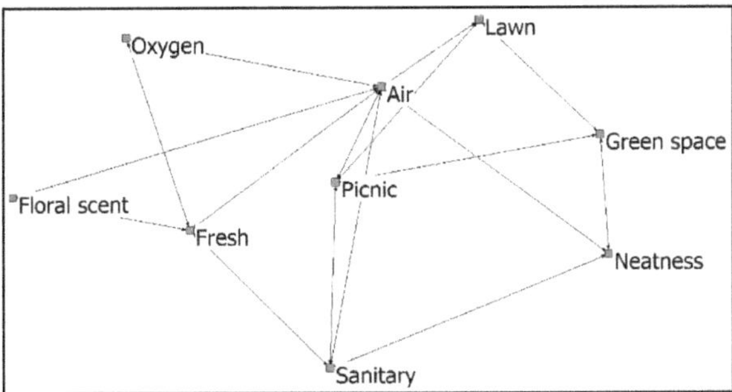

Figure 9. 1-3. Visual and olfactory interactive experience during outdoor activities.

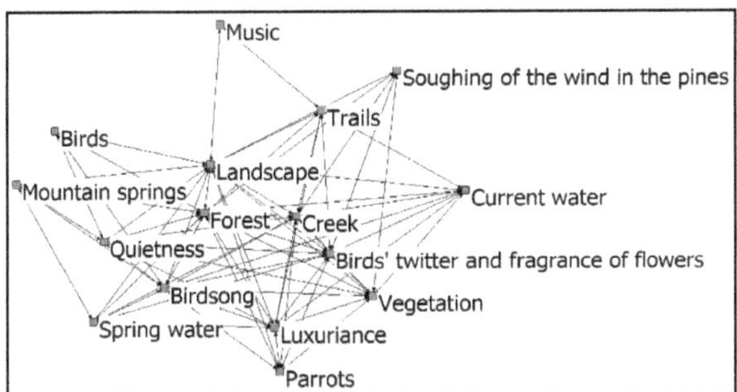

Figure 10. 2-1. Urban forest park elements based on audio-visual interactive perception.

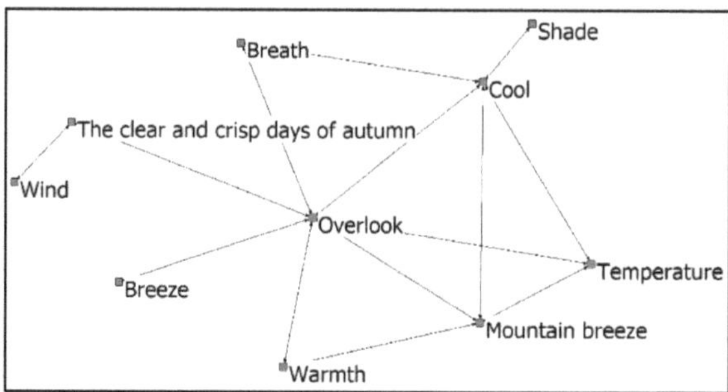

Figure 11. 3-1. Natural scenery based on visual and tactile interactive perception.

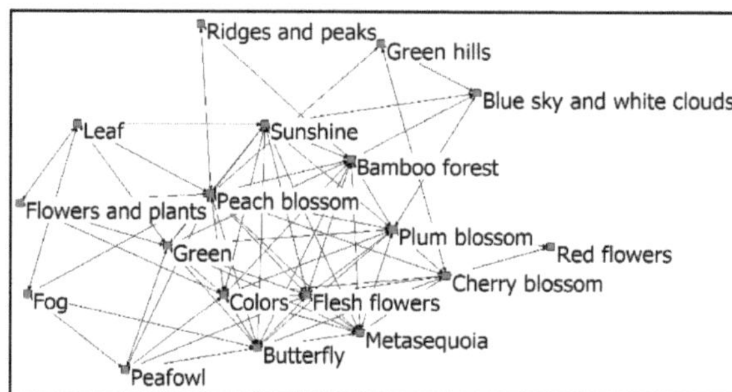

Figure 12. 4-1. Natural landscapes based on the visual interactive perception.

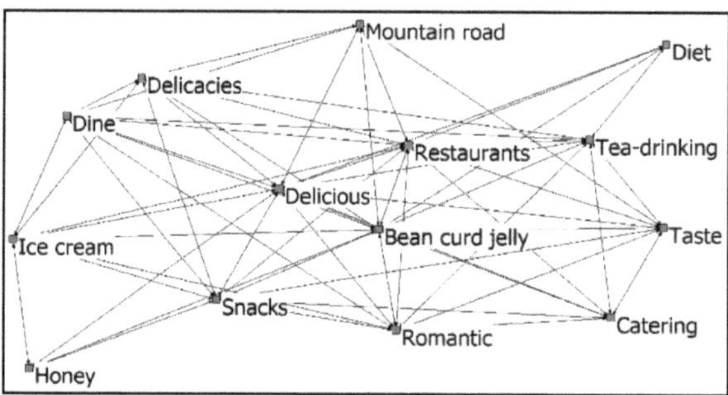

Figure 13. 5-1. Gustatory perception and experience activities.

Vision, the major means of awareness and cognition, is a subjective image of an object produced by stimulating the visual organs [58], and olfactory sensation is the main way of fresh air perception [32]. Clusters 1-1, 1-2, and 1-3 mainly present the clusters of tourists' perceptions based on visual-olfactory interaction. Cluster 1-1 represents the objectivity of tourists' descriptions of the natural environment. The cluster image shows that tourists not only have a higher perception of 'Environment', 'Nightscape', and 'Panorama' but can also give aesthetics evaluation to the Urban Forest Park. The high centrality of 'Air refreshing' indicates that tourists' perception of air is significant in their sightseeing behaviors. Cluster 1-2 describes the human landscape of the park. The close connection between facilities and breathing indicates that tourists pay high attention to the cleanness of the environment and the freshness of the air when they enter into the humanistic landscape tour area [59].

The tourists' focus on the integration of natural experience and humanistic landscape confirms the feasibility of the principle that urban forest park planning should take humanistic connotations into account [60].

Cluster 1-3 is a collection of outdoor activities carried out by tourists in Baiyun Mountain Urban Forest Park. The clustering image shows that 'Picnic' is closely connected with green space, lawn, and air, which indicates that tourists pay more attention to the service level of infrastructure of the park when they have picnics. The frequent occurrence of 'Flower fragrance', 'Oxygen', and 'Fresh' bears out that visitors' olfactory interactive perception activities are significant and specific. Clusters 1-1,1-2 and 1-3 show that tourists pay critical attention to air perception when using the park.

In terms of multi-sensory interactive experience, although the visual experience dominates, the sense of hearing is necessary for people to obtain information and fully perceive their surroundings [14–16]. A bulk of academic research has been conducted from the perspective of audio-visual interaction [57]. Cluster 2-1 is a group of words relating to the soundscape, such as 'Soughing of the wind in the pines', 'Birdsong', 'Music', and other soundscape sources such as 'Stream', 'Current water', 'Mountain spring' and 'Parrot'. It shows that visitors be as a significant preference for natural and humanistic soundscapes, a moderate preference for the natural soundscape, including the sound of water, pines, birdsong, and so on, and a lower preference for humanistic soundscapes [58]. This can be attributed to the fact that the regional planning of Baiyun Mountain Urban Forest Park is mainly oriented toward ecological conservation [38].

People are constantly in direct contact with the external environment and thus resonate with the natural environment. 'Temperature', 'Cool', and 'Autumn' in Cluster 3-1 are evidence of visitors' tactile perception of the outside world. The cluster being centered around the verb 'Overlooking' indicates that the tactile perceptual interaction is more pronounced when they are viewing from high places.

Cluster 4-1 displays visually dominant sensory elements, including floral categories such as 'Cherry blossom' and 'Red flowers'; forest landscapes such as 'Green mountains' and 'Bamboo forests'; 'Butterfly', animals such as 'Peafowl' and 'Butterfly'; high-level landscapes such as 'Blue sky and white clouds', 'Fog', and 'Sunshine'. The convergence of many types of visual landscapes illustrates that human visual perception is characteristic of motion and variation. Moreover, the cluster being centered around flowers and sunlight suggests that tourists have higher perceptual preferences for flower landscapes and sunlight, followed by animals. The preference for flowers can be accounted for by the location of the park as the city of Guangzhou is known as a city of flowers in China, and the preference for sunshine can be ascribed to the hot and humid weather. Cluster 5-1 involves the catering services in the park, with 'Bean curd jelly', 'Tea-drinking', 'Snacks' and 'Restaurant' pointing to the products or services therein and 'Delicious' and 'Romantic' relating to tourists' sensory experience of the cuisine and the atmosphere. These two clusters being obviously apart indicates that visitors involved in different types of activities have different multi-sensory interaction preferences for the landscapes in the urban forest park. While sightseers have a stronger perceptual preference for visual and natural landscapes, picnickers give more weight to the perceptual experience of food and sanitation.

As shown in Figure 14, the overall share of clusters is presented in a tree diagram.

The cluster analysis identified the following characteristics of tourist's multi-sensory interactive perception preferences in urban forest parks:

Visual perception dominates the preferences for multi-sensory interactive experiences, which is consistent with the result of relevant medical studies showing that people rely on their eyes for 87% of the information coming from the outside world and that 75%-90% of human activities are caused by vision [59]. As shown by the six cluster images related to the visual experience, tourists have the strongest preference for flowers, followed by green areas, and lastly, the infrastructure.

Air freshness is the hotspot of olfactory perception interaction. The three cluster images of visual and olfactory interactions show that tourists' sensory experience of urban forest park landscapes is mostly focused on visual and olfactory perceptions. The factors that attract visitors with different travel purposes include the fresh air, the green plants, and the diverse floral fragrances, among which the fresh air is the most significant.

Figure 14. Tree diagram of the share of each cluster.

A natural soundscape is preferred over a humanistic soundscape. Visitors' salient perception of natural soundscapes in the park stems from the park's long history of soundscape relics development. The combination of motion and stillness is characteristic of urban forest park landscapes as visitors rely more on auditory-visual senses to gain interactive perception. The study reveals that tourists have a stronger preference for perceptual preferences for water sounds and animal sounds.

The tactile experience is focused on the interaction with the natural landscape. Tactile interactions with the natural vegetation landscape are least reported in the reviews, and the few contents being reviewed point to the weather in the upland areas of the park after climbing activities.

The gustatory experience is closely related to humanistic activities. People's gustatory experience in the park takes place mainly during picnics and tea-drinking activities. The results suggest that tourists tend to have a singular taste in the natural landscape in the park.

5. Discussion

By summarizing the major findings of the current study and drawing on relevant views of other scholars, this section further explores the perceptual characteristics of visitors' multi-sensory interactive experience in urban forest parks and offers some significant implications for the landscape design of urban forest parks.

5.1. Perceptual Characteristics of Visitors' Multi-Sensory Interactive Experience in Urban Forest Parks

This study found that there was obvious multi-sensory interaction in visitors' perceived experience in urban forest parks, which is consistent with the findings of most studies with a focus on single-dimensional sensory interactions [31,32]. Among the five dimensions of sensory interaction, the most prominent were the visual and olfactory perceptions in this study, which corroborates the influence of natural and human soundscapes on visitors' audio-visual interactions as demonstrated by visitors' comments, and fills the gap in existing studies [57]. Consistent with other studies on visitors' multi-sensory perception of urban forest park, this study also clarifies the dominance of visual perception in multi-sensory interaction [60]. What is noteworthy is that, this study found that when visitors are exercising, picnicking, sightseeing or doing other activities in urban forest parks, their perceptual interaction with the environment is predominantly visual and olfactory, as they are easily attracted to specific flowers such as peach blossom and cherry blossoms, which. is a new finding in multi-sensory interactive experience research.

5.2. Effect of Multi-Sensory Interactive Experience on Visitors' Perception of Activities

This study found that the neatness of the natural landscape and environment can affect visitors' ability to perceive ongoing activities, which is consistent with the findings of Song X and Xiao J [31,32]. Visitors are keenly concerned about the neatness of public infrastructures, such as trails and restrooms, and during their visits to urban forest parks whether the environment is neat or not is easily noticeable to them. This study also found that visitors' gustatory perception of the park under study is closely related to the restaurant in the park and the culture of Cantonese cuisine. The gustatory perception is mainly manifested in the evaluation of dining tastes and the atmosphere of the restaurant, which indicates that the gustatory perception should be further explored in future research on visitors' interactive experience in urban forest parks [61]. Another new finding worthy of mentioning is that visitors' tactile experience was mainly focused on indirect contact with the natural landscape, especially their exposure to the wind in the forests and the weather conditions after climbing up to high grounds [5]. The shortage of review texts concerning direct tactile perception in this study reveals the absence of tactile sensory interaction facilities for visitors in urban forest parks and the necessity to provide design recommendations in subsequent research.

5.3. Implications of Multi-Sensory Interactive Experience for Landscape Planning

This study can also offer practical implications for the future landscape design of urban forest parks from the perspective of multi-sensory interactive experience.

(1)   Improving the neatness and richness of the visual landscape of urban forest parks:

As discussed earlier, the neatness and richness of the landscape are important factors affecting visitors' audiovisual interaction. Therefore, on one hand, the parks should improve the neatness of overall infrastructure and the park environment through regular maintenance, such as cleaning the trails, restrooms, and the restaurants. On the other hand, based on visitors' high perception of floral landscape, urban forest park managers can take advantage of the climate features and regularly hold flower-themed exhibitions to enrich visitors' visual experience.

(2)   Constructing tour areas where the olfactory landscape can be visualized:

A new finding of this study is visitors' rich use of visual-olfactory perception of urban forest parks. However, up to date, there are relatively few studies on the construction of olfactory sensory landscapes in China's urban forest parks. It is thus suggested that Chinese urban forest park researchers and designers refer to the principle applied in the building of an olfactory sensory garden in Sheffield, UK. Brightly colored flowers can be planted on either side of the entrance to the garden to provide visual guidance, trees and shrubs should be planted around the exhibition area to facilitate fragrance concentration, and exhibition junctions should be designed to face the prevailing wind so that the wind can partially pass through the garden, carrying the fragrance to every corner of the exhibition area [62]. The core of the olfactory landscape lies in the visitors' subjective awareness, behavior, and perceptual range, and subjective factors are more difficult to control in the design [63]. Therefore, we can refer to Belgian olfactory artist Peter de Cupere's "Scent Scenario" installation, which uses "fog" to attract visitors to stop and feel [64].

(3)   Strengthening the protection of human soundscape relics while developing natural soundscape resources in a scientific and diversified manner:

This study found that the human and natural soundscapes were significantly perceived by visitors to the urban forest park [65]. Unfortunately, it has been shown in earlier research that the soundscape remains of Baiyun Mountain urban forest park are at the risk of disappearing [38]. It has been revealed that bird twittering accompanied by insect chirping and water flowing or light music accompanied by temple bells could enhance participants' sense of involvement and significantly reduces stress [66]. Therefore, in respect of humanistic soundscape conservation, park managers can conserve valuable sound as a resource by preserving historical tapes. On top of that, park managers can develop natural sound landscape resources in many ways, for example, by creating sound sequences by using the sound of water falling on different building materials or the sound of wind blowing through different buildings, to encourage visitors to identify the soundscape of different materials and buildings to enhance their interactive experience.

(4)　Enriching tactile perception through directly accessible infrastructure:

The study found that visitors to urban forest parks have a weaker tactile perception of water bodies, plants, and park landscape facilities, except for the wind. Therefore, various forms of water landscape, such as over-water wooden boardwalks and pools, can be built in the future to provide people with a hydrophilic tactile experience. More than that, paths can be paved with cobblestones, bark, and wood chips, and signposts can be made with carved designs to extend visitors' visual and olfactory interactive experience with their hands and feet as they are the most directly sensitive organs for tactile perception [5]. In addition, to ensure that visitors can fully enjoy the beauty and novelty of the park design, the plants and flowers on both sides of the road in the urban forest park should be regularly pruned and maintained; toxic, thorny, hard, and untouchable plants should be avoided.

(5)　Enriching gustatory perception by providing various culinary services:

The study found that the gustatory experience of visitors to urban forest parks is relatively homogeneous. To enrich people's gustatory perception and to spread local culinary culture, managers can improve catering services in the park by setting up more restaurants of different styles or enriching the variety of food and beverage served. Managers can also integrate into the food services "natural oxygen bars", "flower foods" and other theme-based restaurants to create a brand that is unique to the urban forest park.

6. Conclusions and Further Work

This study reflects the perceptions of different stakeholders about their urban forest park experiences. A distinct advantage of this study is that it reveals the multi-sensory interactive experience of visitors to an urban forest park. The data collected are highly authentic and representative as they are tourists' self-reports of direct experiences. In regard to visitors' overall multi-sensory preferences, it is found that the perception of visual and olfactory landscapes was the most prominent, followed by the audio-visual perception of natural and humanistic soundscapes, while the tactile and gustatory perceptions were the least reported. These findings demonstrate that visitors' evaluative texts should be taken into account in landscape design.

The innovative use of crowd-sourced data ensures that visitors to the urban forest park are maximally represented, though some groups, such as those who do not like to share their experiences on social networking platforms, may have been under-represented. Besides, our dataset could lend itself to a wider range of qualitative and quantitative analyses, including sentiment analysis with a view to identifying visitors' positive and negative emotions towards urban forest parks.

However, the generalizability of this study is also limited by the use of web-based data. First, the demographic information of may be ambiguous as users may have concealed their age, gender, place of residence, etc. In addition, sentiment analysis is one of the important elements of network text analysis [19], but this study only focuses on the multi-sensory interactive experience without further exploring emotional or attitudinal issues, such as the sentiment polarity of the visitors' review texts, which could help urban forest park designers make finer-grained decisions. Therefore, in-depth interviews and fieldwork combined with qualitative and quantitative analysis can be conducted to detect the emotional attitudes behind visitors' sensory experiences.

Author Contributions: Conceptualization, J.X. (Jian Xu), M.L., J.X. (Jingling Xu) and Z.G.; methodology, Z.G. and J.X. (Jingling Xu); validation, J.X. (Jian Xu) and M.L.; data analysis, Z.G. and J.X. (Jingling Xu); resources, G.C.; writing-original draft preparation, J.X. (Jingling Xu); writing-review and editing, Z.G., J.X. (Jian Xu) and M.L.; visualization, J.X.; supervision, M.L.; project administration, J.X. (Jian Xu) and M.L.; funding acquisition, Z.W. and M.L. All authors have read and agreed to the published version of the manuscript.

Funding: This study is part of the project called 'The Study on Rural Tourism Revitalization Planning and Sustainable Living Coordination Mechanism' and funded by the International Cooperation open Project of State Key Laboratory of Subtropical Building Science, South China University of Technology (No. 2019ZA02); China youth fund of national natural science foundation projects 'Research on visual landscape evaluation method of traditional villages in Guangdong based on deep learning'(No. 52208057); Guangzhou Philosophy & Social Science Planning Project (No. 2022GZGJ131).

Data Availability Statement: Data are available upon the enclosure in the ZIP folder.

Acknowledgments: We would especially like to thank Jian Xu and Muchun Li for their assistance in proofreading the manuscript and the International Cooperation open Project of State Key Laboratory of Subtropical Building Science, South China University of Technology. Furthermore, we are grateful to Guangwei Chen for his contribution to Python programming.

Conflicts of Interest: The authors declare no conflict of interest.

References

1. Zhou, C.; Yan, L.; Yu, L.; Wei, H.; Guan, H.; Shang, C.; Chen, F.; Bao, J. Effect of Short-term Forest Bathing in Urban Parks on Perceived Anxiety of Young-adults: A Pilot Study in Guiyang, Southwest China. *Chin. Geogr. Sci.* **2018**, *29*, 139–150. [CrossRef]
2. Han, X.; Sun, T.; Cao, T. Study on landscape quality assessment of urban forest parks: Take Nanjing Zijinshan National forest Park as an example. *Ecol. Indic.* **2020**, *120*, 106902. [CrossRef]
3. Ferretti-Gallon, K.; Griggs, E.; Shrestha, A.; Wang, G. National parks best practices: Lessons from a century's worth of national parks management. *Int. J. Geoheritage Parks* **2021**, *9*, 335–346. [CrossRef]
4. Li, X.; Chen, C.; Wang, W.; Yang, J.; Innes, J.L.; Ferretti-Gallon, K.; Wang, G. The contribution of national parks to human health and well-being: Visitors' perceived benefits of Wuyishan National Park. *Int. J. Geoheritage Park.* **2021**, *9*, 1–12. [CrossRef]
5. He, M.; Wang, Y.; Wang, W.J.; Xie, Z. Therapeutic plant landscape design of urban forest parks based on the Five Senses Theory: A case study of Stanley Park in Canada. *Int. J. Geoheritage Park.* **2022**, *10*, 97–112. [CrossRef]
6. Liu, H.; Li, F.; Li, J.; Zhang, Y. The relationships between urban parks, residents' physical activity, and mental health benefits: A case study from Beijing, China. *J. Environ. Manag.* **2017**, *190*, 223–230. [CrossRef]
7. Kil, N.; Stein, T.V.; Holland, S.M.; Kim, J.J.; Kim, J.; Petitte, S. The role of place attachment in recreation experience and outcome preferences among forest bathers. *J. Outdoor Recreat. Tour.* **2021**, *35*, 100410. [CrossRef]
8. Chen, B.; Qi, X. Protest response and contingent valuation of an urban forest park in Fuzhou City, China. *Urban For. Urban Green.* **2018**, *29*, 68–76. [CrossRef]
9. Yu, W.; Jianyu, H.; Jin, Q. Five Senses Experiential Therapeutic Landscape: A Case of Qinhuangdao Community Park. *J. Hebei Univ. Environ. Eng.* **2022**, *32*, 6. (In Chinese)
10. Kumar, L.; Kahlon, N.; Jain, A.; Kaur, J.; Singh, M.; Pandey, A. Loss of smell and taste in COVID-19 infection in adolescents. *Int. J. Pediatr. Otorhinolaryngol.* **2021**, *142*, 110626. [CrossRef]
11. Qian, K.; Yahara, T. Mentality and behavior in COVID-19 emergency status in Japan: Influence of personality, morality and ideology. *PLoS ONE* **2020**, *15*, e0235883. [CrossRef] [PubMed]
12. Eiter, S.; Vik, M.L. Public participation in landscape planning: Effective methods for implementing the European Landscape Convention in Norway. *Land Use Policy* **2015**, *44*, 44–53. [CrossRef]
13. Wang, X.; Zhang, J.; Wu, C. Users' recreation choices and setting preferences for trails in urban forests in Nanjing, China. *Urban For. Urban Green.* **2022**, *73*, 127602. [CrossRef]
14. Kai, C.; Xinchen, H.; Zhouyu, L.; Liying, Z.; Siren, L. The Construction of Evaluation System of Healing Landscape in Forest Park Based on GST and AHP. *Acta Agric. Univ. Jiangxiensis* **2017**, *39*, 118–126. (In Chinese)
15. Sacchelli, S.; Favaro, M. A Virtual-Reality and Soundscape-Based Approach for Assessment and Management of Cultural Ecosystem Services in Urban Forest. *Forests* **2019**, *10*, 731. [CrossRef]
16. Liu, Y.; Hu, M.; Zhao, B. Audio-visual interactive evaluation of the forest landscape based on eye-tracking experiments. *Urban For. Urban Green.* **2019**, *46*, 126476. [CrossRef]

17. Zheng, Y.; Lan, S.; Chen, W.Y.; Chen, X.; Xu, X.; Chen, Y.; Dong, J. Visual sensitivity versus ecological sensitivity: An application of GIS in urban forest park planning. *Urban For. Urban Green.* **2019**, *41*, 139–149. [CrossRef]
18. Ho, C.-I.; Chen, M.-C.; Shih, Y.-W. Customer engagement behaviours in a social media context revisited: Using both the formative measurement model and text mining techniques. *J. Mark. Manag.* **2021**, *38*, 740–770. [CrossRef]
19. Zhu, J.; Xu, C. Sina microblog sentiment in Beijing city parks as measure of demand for urban green space during the COVID-19. *Urban For. Urban Green.* **2020**, *58*, 126913. [CrossRef]
20. Wan, C.; Shen, G.Q.; Choi, S. Eliciting users' preferences and values in urban parks: Evidence from analyzing social media data from Hong Kong. *Urban For. Urban Green.* **2021**, *62*, 127172. [CrossRef]
21. Yun, E.; Park, Y. Extraction of scientific semantic networks from science textbooks and comparison with science teachers' spoken language by text network analysis. *Int. J. Sci. Educ.* **2018**, *40*, 2118–2136. [CrossRef]
22. Drieger, P. Semantic Network Analysis as a Method for Visual Text Analytics. *Procedia-Soc. Behav. Sci.* **2013**, *79*, 4–17. [CrossRef]
23. Godnov, U.; Redek, T. Application of text mining in tourism: Case of Croatia. *Ann. Tour. Res.* **2016**, *58*, 162–166. [CrossRef]
24. Li, Y.; Chen, X.; Liu, P.; Guan, H. *Recommending Heritage Tourist Destinations, Involves Constructing Positive Evaluation Semantic Network Based on ROSTCM Semantic Network and Social Network Generating Tool, and Determining Negative Evaluation Theme of Heritage Tourism Field*; China National Intellectual Property Administration: Beijing, China, 2021; p. 607816.
25. Schrammeijer, E.A.; Van Zanten, B.T.; Verburg, P.H. Whose park? crowd-sourcing citizen's urban green space preferences to inform needs-based management decisions. *Sustain. Cities Soc.* **2021**, *74*, 103249. [CrossRef]
26. Li, J.; Xu, L.; Tang, L.; Wang, S.; Li, L. Big data in tourism research: A literature review. *Tour. Manag.* **2018**, *68*, 301–323. [CrossRef]
27. Lyu, J.; Khan, A.; Bibi, S.; Chan, J.H.; Qi, X.G. Big data in action: An overview of big data studies in tourism and hospitality literature. *J. Hosp. Tour Manag.* **2022**, *51*, 346–360. [CrossRef]
28. Hu, M.; Li, H.; Song, H.; Li, X.; Law, R. Tourism demand forecasting using tourist-generated online review data. *Tour. Manag.* **2022**, *90*, 104490. [CrossRef]
29. Shuting, W.; Xinchen, H. A Study of the Relationship between Color Characteristics of Urban Parks and Tourists' Perception Psychology: A Case Study of South Part of Minjiang Park, Fuzhou City. *J. Environ. Manag.* **2019**, *17*, 37–41+72. (In Chinese)
30. Bae, E.J.; Jang, A.R.; Park, H.; Yoon, J.Y. Investigating knowledge structure and research trends in child and adolescent health literacy research through network text analysis. *J. Pediatr. Nurs.* **2022**, *67*, 57–63. [CrossRef]
31. Song, X.; Wu, Q. Study on smellscape perception and landscape application of fragrant plants. *Urban For. Urban Green.* **2021**, *67*, 127429. [CrossRef]
32. Xiao, J.; Tait, M.; Kang, J. Understanding smellscapes: Sense-making of smell-triggered emotions in place. *Emot. Space Soc.* **2020**, *37*, 100710. [CrossRef]
33. Arnberger, A.; Eder, R. Are urban visitors' general preferences for green-spaces similar to their preferences when seeking stress relief? *Urban For. Urban Green.* **2015**, *14*, 872–882. [CrossRef]
34. Keith, S.J.; Larson, L.R.; Shafer, C.S.; Hallo, J.C.; Fernandez, M. Greenway use and preferences in diverse urban com-munities: Implications for trail design and management. *Landsc. Urban Plan.* **2018**, *172*, 47–59. [CrossRef]
35. Ebenberger, M.; Arnberger, A. Exploring visual preferences for structural attributes of urban forest stands for restoration and heat relief. *Urban For. Urban Green.* **2019**, *41*, 272–282. [CrossRef]
36. Campagnaro, T.; Vecchiato, D.; Arnberger, A.; Celegato, R.; Da Re, R.; Rizzetto, R.; Semenzato, P.; Sitzia, T.; Tempesta, T.; Cattaneo, D. General, stress relief and perceived safety preferences for green spaces in the historic city of Padua (Italy). *Urban For. Urban Green.* **2020**, *52*, 126695. [CrossRef]
37. Jiang, L.; Xinwei, T. Reflections on Restorative Landscape Design Based on Landsenses Ecological Concept. *Landsc. Ar-Chitecture* **2021**, *28*, 107–112. (In Chinese)
38. Qian, W.; XiaoMei, Y. A survey Study on Soundscape Resources of Guangzhou Baiyun Mountain. *South Archit.* **2013**, *3*, 86–89.
39. Liang, H.; Tao, L. Analysis on the Experience Needs of Five Senses and the Willingness of Petential Tourists to Experience Forest Healthcare Trails. *Taiwan Agric. Res.* **2020**, *1*, 62–68. (In Chinese)
40. Mainolfi, G.; Marino, V. Destination beliefs, event satisfaction and post-visit product receptivity in event marketing. Results from a tourism experience. *J. Bus. Res.* **2020**, *116*, 699–710. [CrossRef]
41. Sun, S.; Law, R.; Schuckert, M. Mediating effects of attitude, subjective norms and perceived behavioural control for mobile payment-based hotel reservations. *Int. J. Hosp. Manag.* **2019**, *84*, 102331. [CrossRef]
42. Lu, B.; Yan, L.; Chen, Z. Perceived values, platform attachment and repurchase intention in on-demand service platforms: A cognition-affection-conation perspective. *J. Retail. Consum. Serv.* **2022**, *67*, 103024. [CrossRef]
43. Jin, Y.; Yan, A.; Sun, T.; Zheng, P.; An, J. Microblog data analysis of emotional reactions to COVID-19 in China. *J. Psychosom. Res.* **2022**, *161*, 110976. [CrossRef] [PubMed]
44. Bl, A.; Wang, Z.A.; Zhang, S.C. Platform-based Mechanisms, Institutional Trust, and Continuous Use Intention: The Moderat-ing Role of Perceived Effectiveness of Sharing Economy Institutional Mechanisms. *Inf. Manag.* **2021**, *58*, 103504.
45. Song, H.; Xie, K.; Park, J.; Chen, W. Impact of accommodation sharing on tourist attractions. *Ann. Tour. Res.* **2020**, *80*, 102820. [CrossRef]
46. Yong-juan, W.; Guang-hua, Y.; Li-nan, S.; Peige, L. Text Mining and Analysis of Meituan User Review Text. In *International Conference on Green Communications and Networking*; Springer: Cham, Switzerland, 2020; pp. 151–155.

47. Heled, J.; Ying, Z.; Yuan, A. Research on the Promotion of Word of Mouth in Tourist Scenic Spots Based on Web Text Mining—The Case Study of Wanlu Valley in Guangdong Province. In *MATEC Web of Conferences*; EDP Sciences: Les Ulis, France, 2018; Volume 173, p. 03060.

48. Wu, H.; Chen, Y.L.; Zhang, X.T. Study on Destination Selections of China's Honeymoon Tourists Based on Word Frequency Analysis. In Proceedings of the International Conference on Strategic Management (ICSM 2016), Chengdu, China, 10–11 March 2016; pp. 544–550.

49. Carreon, E.C.A.; Nonaka, H.; Hiraoka, T.; Kumano, M.; Hirota, M.; Ito, T. Emotional Contribution Analysis of Online Reviews. In Proceedings of the International Conference on Artificial Life and Robotics (ICAROB), Beppu, Japan, 1–4 February 2018; pp. 359–362.

50. Sen, Y.; Ling, Z.; Xu, L. Analysis of Tourism City Image Based on Big Data of Network Text-Take Guangzhou as an Example. In *IOP Conference Series: Materials Science and Engineering*; IOP Publishing: Bristol, UK, 2020; Volume 806, p. 012048.

51. Xu, J.; Hu, L. Geospatial Semantics Analysis of the Qinghai–Tibetan Plateau Based on Microblog Short Texts. *ISPRS Int. J. Geo-Inf.* **2021**, *10*, 682. [CrossRef]

52. Kang, Q.; Zhang, J.; Li, C.; Chen, H.; Meng, L. Temporal and spatial analysis of the impact of epidemic situation on tourists' emotional fluc-tuation in China based on text mining. In Proceedings of the International Conference on Algorithms, Microchips and Network Applications, Zhuhai, China, 6 May 2022; SPIE: Bellingham, DC, USA, 2022; Volume 12176, pp. 12–17.

53. Huang, W.; Zhu, S.; Yao, X. Destination Image Recognition And Emotion Analysis: Evidence From User-Generated Con-tent Of Online Travel Communities. *Comput. J.* **2020**, *64*, 296–304. [CrossRef]

54. Hui, H.; Li, P.F.; Zhang, R.J. Research on Time Difference of Tourism Experience Based on Big Data Analysis—Taking the First Batch of National AAAAA Level Tourist Attraction in China as An Example. In Proceedings of the 8th Annual China Marketing International Conference (CMIC), Electr Network, 20–30 June 2020; pp. 1105–1115. Available online: https://www.webofscience.com/wos/alldb/full-record/WOS:000684378100098 (accessed on 21 July 2022).

55. Wang, Y.; Chang, Q.; Fan, P. A framework to integrate multifunctionality analyses into green infrastructure planning. *Landsc. Ecol.* **2020**, *36*, 1951–1969. [CrossRef]

56. Mameno, K.; Kubo, T.; Oguma, H.; Amagai, Y.; Shoji, Y. Decline in the alpine landscape aesthetic value in a national park under climate change. *Clim. Chang.* **2022**, *170*, 1–18. [CrossRef]

57. Chen, H.; Qiu, L.; Gao, T. Application of the eight perceived sensory dimensions as a tool for urban green space assessment and planning in China. *Urban For. Urban Green.* **2018**, *40*, 224–235. [CrossRef]

58. He, J.Y. Online Plant Landscape Evaluation System of Minjiang Riverside Park in Fuzhou City Based on Analytic Hierarchy Process. In Proceedings of the 2021 13th International Conference on Measuring Technology and Mechatronics Automation (ICMTMA), Beihai, China, 16–17 January 2021; pp. 603–606.

59. Dai, T.; Zheng, X. Understanding how multi-sensory spatial experience influences atmosphere, affective city image and be-havioural intention. *Environ. Impact Assess. Rev.* **2021**, *89*, 106595. [CrossRef]

60. Lu, X. Current Situation and Trend Analysis of International and National Five Sense Landscape Research. *World For. Res.* **2020**, *33*, 31–36. (In Chinese)

61. Kim, S.; Park, E.; Fu, Y.; Jiang, F. The cognitive development of food taste perception in a food tourism destination: A gastrophysics approach. *Appetite* **2021**, *165*, 105310. [CrossRef]

62. Si, Y.Q.; Zhu, Q. A preliminary study on the plant landscape design of urban cemetery under the theory of five senses. *Shanxi Archit.* **2022**, *48*, 173–175+194. (In Chinese)

63. Song, C.; Ikei, H.; Miyazaki, Y. Physiological effects of forest-related visual, olfactory, and combined stimuli on humans: An additive combined effect. *Urban For. Urban Green.* **2019**, *44*, 126437. [CrossRef]

64. Xiao, J.; Feng, H.; Xie, H. From "Smell" to "Smellscape": A systematic review of smellscape research and design methods. *J. Hum. Settl. West China* **2021**, *36*, 7–14. (In Chinese)

65. Zhao, J.; Xu, W.; Ye, L. Effects of auditory-visual combinations on perceived restorative potential of urban green space. *Appl. Acoust.* **2018**, *141*, 169–177. [CrossRef]

66. Hong, J.Y.; Ong, Z.-T.; Lam, B.; Ooi, K.; Gan, W.-S.; Kang, J.; Feng, J.; Tan, S.-T. Effects of adding natural sounds to urban noises on the perceived loudness of noise and soundscape quality. *Sci. Total Environ.* **2019**, *711*, 134571. [CrossRef]

 *forests*

Article

Audio-Visual Analysis of Visitors' Landscape Preference for City Parks: A Case Study from Zhangzhou, China

Yonghong Gan * , Yibin Zheng and Lihui Zhang

School of Biological Science and Biotechnology, Minnan Normal University, Zhangzhou 363000, China
* Correspondence: gyhmnnu@gmail.com

Abstract: Soundscape perception is increasingly recognized as an important part of landscape preference and environmental experience. However, few studies have juxtaposed visual landscape preference and soundscape preference to compare their contributions to overall landscape preference. This paper aims to quantify and compare the contribution of audiovisual perception to visitors' overall park landscape preference. The landscape preferences of visitors at seven sample sites in a city park were investigated through field questionnaires in three dimensions: visual landscape, acoustic landscape, and audiovisual landscape. The results showed that visitors' visual landscape preference (VLP = 7.53) was generally higher than soundscape preference (SP = 7.08), while the influence of auditory preference (57%) on overall landscape preference (OLP) was found to be greater than that of visual preference (43%). The ratio of audio/visual contribution to the overall landscape preference decreased as the average sound level of the sample sites increased. Of all the population characteristics, only the educational level (sig = 0.034) could be used as an effective predictor of OLP (Impact coefficient = −0.103). In addition, older visitors rated OLP lower than young visitors, and females rated OLP lower than males. It was found that visual harmony, color richness, color contrast, plant coverage, and plant diversity were the main visual landscape attributes that influenced visitors' visual preferences, while acoustic harmony, quietness, sound vitality, and acoustic richness were the main soundscape attributes that impacts visitors' auditory preference. The results of this study may be useful for park landscape design and regeneration.

Keywords: visitors' landscape experience; contribution of audiovisual perception; soundscape; urban forest; perceived landscape attributes

Citation: Gan, Y.; Zheng, Y.; Zhang, L. Audio-Visual Analysis of Visitors' Landscape Preference for City Parks: A Case Study from Zhangzhou, China. *Forests* **2022**, *13*, 1376. https://doi.org/10.3390/f13091376

Academic Editors: Xin-Chen Hong, Jiang Liu and Guang-Yu Wang

Received: 4 August 2022
Accepted: 25 August 2022
Published: 28 August 2022

Publisher's Note: MDPI stays neutral with regard to jurisdictional claims in published maps and institutional affiliations.

1. Introduction

Studies have shown that urban forest parks are effective in reducing stress [1–3], restoring attention [4], and improving physical and mental health [5–8] through the process of sensory interaction between visitors and nature in the city. Therefore, the planning and design of urban forest parks are generally valued by policy makers and have been built in large numbers in the context of ecological civilization and sustainable development. However, it is worth noting that the health performance of the park is influenced by visitation frequency, which in turn depends on the impact of visitors' sensory experience and satisfaction with urban forest parks [9,10].

Visitors' landscape experience is a multisensory and comprehensive experience [11]. However, visual perception has long been considered to dominate all other human sensory perceptions [12,13], particularly in the field of urban landscape planning and design practices as well as related scientific research. By contrast, nonvisual design elements have been largely overlooked due to their intangible nature [11,14]. Traditionally, the main consideration of auditory perception and sound has been limited to measures to reduce, control and manage noise [15–18]. In this context, the rich information and specific implications of sound [19] and their important impact on human perception and experience [16,20,21] have been mostly ignored. This continued until the late 1960s when the soundscape concept was introduced by R. Murray Schafer [22]. Based on the concept of "soundscape", the acoustic environment can be treated as a kind of resource [23] and the acoustic aspects of urban green space can be regarded as similar to the visual aspects of design [24].

In fact, in terms of visual aspects alone, visual preference research has undergone a paradigm shift from single-attribute to multiple-attribute interaction. Most landscape preference studies in the past have only analyzed the effect of a single visual landscape attribute on landscape preference, and it is often difficult to obtain consistent results between different studies using this research paradigm [25–27], sometimes even obtaining completely opposite results [28,29]. With updated research tools, especially the introduction of psychological research methods, including EEG techniques [30,31] and eye-tracking [32–34], researchers have been able to gain a deeper and more comprehensive understanding of the relationship between visual landscape attributes and landscape preferences. For example, Liu et al. used eye-tracking technology to study the effect of landscape complexity on landscape preferences in different types of environments [34]. Zhang et al. explored the effects of the interaction of multiple visual attributes, including openness, richness, order, and depth on landscape preferences [35].

Following this diversified and integrated research direction, the interactive, multisensory, and holistic nature of landscape perception has begun to receive more attention in recent years [16,36–40]. Some attention has been paid to the effects of audio-visual interaction on landscape perception, including the influence of visual information on auditory perception [15,41,42] and the impact of sound elements on visual perception [20,39,43]. However, most audio-visual interaction studies do not give equal importance to visual and auditory perception, treating visual perception as the main object of study and auditory perception as one of its influencing factors [20,21,43,44]. One of the most important reasons for this approach is that vision is the most important sensory channel through which humans obtain most of their environmental information (approximately 80% of total information) [45,46]. In any case, does this emphasis necessarily imply that visual perception also has an overwhelming advantage over other human sensory perception in people's perception of landscape and experience of the environment?

In response to the above question, an indoor landscape preference experiment with 63 students as subjects was conducted in a previous study [47]. Participants were asked to rate their preference for six visual-only, six audio-only, and thirty-six combined audio-visual recordings of six different types of landscapes, respectively. That research concluded that "under certain conditions, the contribution of auditory preferences to landscape preferences can largely exceed visual preferences". However, the conclusions drawn under laboratory conditions must be verified in real-world scenarios, and the conditions of their application and scope of application should be further clarified. In this context, this paper aims to (1) compare and validate the contributions of visitors' visual and auditory preferences to overall landscape preferences (validating previous research findings), (2) analyze the main influencing factors of visitors' visual and auditory preferences, and (3) establish a model for predicting landscape preferences in urban parks that considers both visual and audio perception. To achieve this goal, an on-site questionnaire survey was performed in a city park with 210 visitors. The relationship between visual preference and overall landscape preference, as well as the relationship between audio preference and overall landscape preference, was first examined. Their respective contributions to landscape preference were then measured, and the main influencing factors of visitors' visual preference and audio preference were analyzed.

2. Method

2.1. Study Area and Sample Sites

Zhongshan Park, formerly known as the 1st Park of Zhangzhou City, covers an area of about 45,000 square meters and is situated at the intersection of Xinhua West Road and Yan'an South Road, Zhangzhou City, Fujian Province. The park is adjacent to the ancient city of Zhangzhou (an important part of the Minnan Cultural and Ecological Reserve, one of the first national Cultural and Ecological Reserve in China). It is an urban park with both historical and cultural heritage and contemporary atmosphere. The landscape around the park consists of residential, commercial neighborhoods, cultural and tourist areas, and the related soundscape elements mainly include the sound of traffic, conversation, and footsteps. The park has a rich variety of visual landscape types with different types of sound elements. To improve the relevance of the research results and to take into account the heterogeneity of the park's internal environment, we chose specific sample sites rather than the whole park as the objects of landscape evaluation. The functional division of the park, the composition of the visual and auditory landscapes, and the distribution of pedestrian flow and the main attractions were all considered when setting up the sample sites. Seven sample sites (SS1–SS7) were set up in this study: the Zhangzhou Liberation Memorial Pavilion, the Raining Corridor, the Promenade, the Dragon Column Pavilion, the Zhongshan Square, the Seven Star Pool, and the Senior Activity Room (Figure 1). Among them, the Liberation Memorial Pavilion and the Zhongshan Square are representatives of the historical and cultural landscape. The Raining corridor, the Promenade, and the Senior Activity Room are the main places for leisure activities. The Dragon Column Pavilion is a place for sports and fitness, and the Seven Star pool is the only water feature in the park.

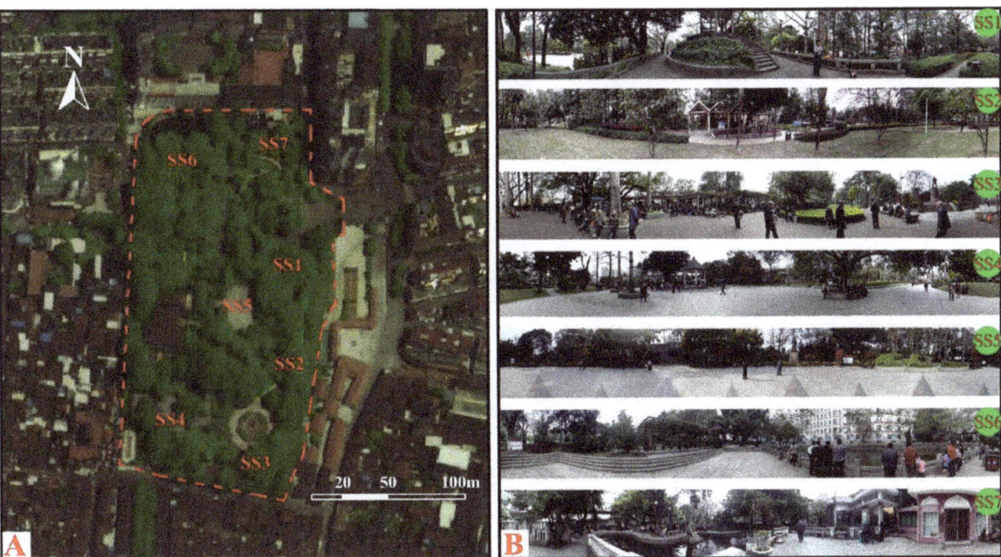

Figure 1. Park image from Amap (source: http://ditu.amap.com; accessed on 20 January 2020) (**A**) and panoramic photos of seven sample sites in Zhongshan Park (**B**).

2.2. Survey of the Composition of Visual Landscape and Soundscape

Survey of the visual landscape included the recording and classification of visual landscape elements and the taking of 360-degree panoramic photos (Figure 1B). Visual landscape elements of the park can be classified into two types: natural landscape elements and artificial landscape elements [20,48]. Specifically, natural landscape elements include lakes, lawns, trees, and rocks, while artificial landscape elements cover sculptures, children's play facilities, trash cans, buildings, and pavilions. On the basis of determining the visual landscape elements of the 360-degree panoramic photos, a 10 mm $\times$ 10 mm grid was superimposed on the panoramic photos, and the proportion of the visual landscape elements (excluding the sky) was calculated. The results showed that, except for SS1 and SS2, the proportion of natural elements was lower than 50% in all other five sample sites, indicating that the naturalness of the park was relatively low (Figure 2).

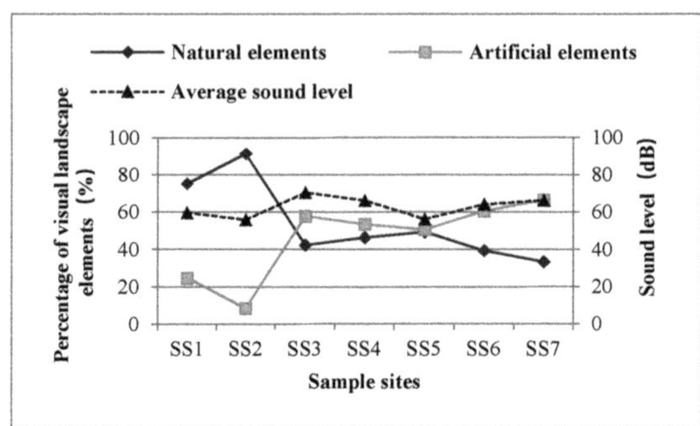

Figure 2. Visual landscape element composition and sound level of the park.

While carrying out the questionnaire survey, the investigators recorded the frequencies of the soundscape elements over a five-minute period and synchronously used the AWA6218A noise meter (1.5 m above the ground) to collect the sound level data at the sample sites. In the survey of soundscape elements, three types of sounds can be distinguished, namely natural sounds, living (man-made) sounds, and artificial sounds. Among them, natural sounds include wind, water and bird songs, living sounds include singing, talking, and playful sounds, and artificial sounds involve broadcasting and music. In general, the soundscape of the park is mainly composed of artificial sounds and living sounds. Specifically, music has the highest frequency among the artificial sounds, while singing and talking dominate the living sounds. In addition, the natural sounds are mainly birdsongs (Figure 3). Since the survey was conducted on a weekend with many visitors in the park, the average sound level in the park was also relatively high (55.9 dB–70.5 dB). By comparison, it was found that the average sound level at the sample sites was positively correlated with the proportion of artificial landscape elements. That is, the larger the proportion of artificial landscape elements, the higher the average sound level.

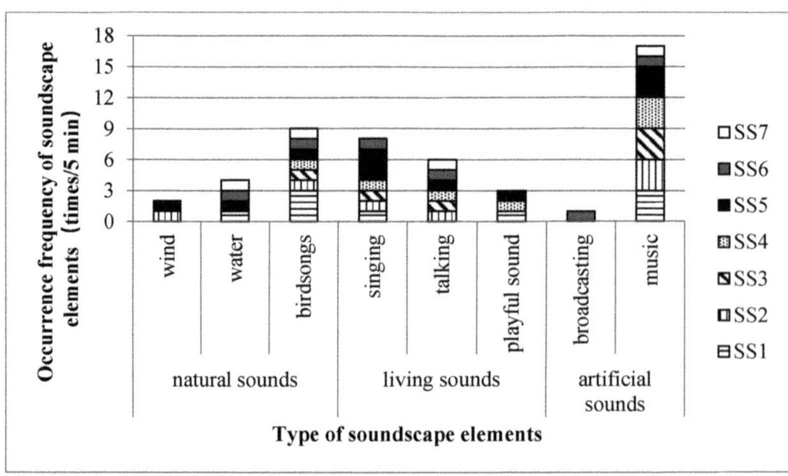

Figure 3. Frequency of soundscape elements.

2.3. Questionnaire Design and Data Collection

The questionnaire consists of two parts: the collection of background information of visitors and the evaluation of park landscape perception (the main part of the questionnaire). The first part includes visitors' gender, age, educational background and occupation, and the second part includes two parts: evaluation of perceived attributes of the audio-visual landscape and evaluation of visitors' landscape preference.

Research had shown that people's subjective evaluation had a certain "objectivity", and the overall visual evaluation of landscape (visual landscape preference) was based on cognitive judgment of certain visual landscape attributes (visual perceptual attribute evaluation) [49]. Based on this, it can be reasonably inferred that the soundscape preference is a comprehensive evaluation based on the evaluation of certain soundscape perceptual attributes. It can be further inferred that the overall audiovisual landscape preference is a comprehensive evaluation based on the judgment of visual landscape preference and soundscape preference. Therefore, the main parts of the questionnaire were specifically designed as follows: (1) Perceived Visual Landscape Attributes (PVLA) and Perceived Soundscape Attributes (PSA); (2) Visual Landscape Preference (VLP) and Soundscape Preference (SP); and (3) Audiovisual Overall Landscape Preference (OLP). VLP refers to the visitor's preference rating of the visual landscape, SP is related to the visitor's preference of the soundscape, and OLP is the visitor's overall rating of the landscape from both visual and audio aspects. In this questionnaire, participants were first asked to evaluate the PVLA and PSA of the sample sites, then to assign values to the VLP and SP of the sample sites respectively, and finally to evaluate the OLP for each sample site. Based on the systematic review of literature [19,34,35,49–53] and the actual situation of Zhongshan Park, a total of 16 audiovisual landscape perception attribute indicators were selected (11 visual landscape perception attribute indicators and five soundscape perception attribute indicators) (Table 1).

Table 1. Indicators of Perceived Visual Landscape Attributes (PVLA) and Perceived Soundscape Attributes (PSA).

PVLA	Level (1–5)	PSA	Level (1–5)
Naturalness	Highly artificial—Highly natural	Quietness	Very noisy—Very peaceful
Openness	Highly closed—Highly open	Sound Vitality	Very boring—Very interesting
Plant Diversity	Highly Monotonous—Highly diverse	Acoustic Richness	Very poor—Very rich
Color Contrast	Very low—Very high	Acoustic Harmony	Highly disharmonious—Highly harmonious
Color Richness	Very poor—Very rich	Acoustic Interference	Very low—Very high
Plant Coverage	Highly sparse—Highly dense		
Plant Hierarchy Change	Highly monotonous—Highly varied		
Number of Landscape Buildings	Very few—Very much		
Relief	Highly flat—Highly fluctuating		
Visual Harmony	Highly disharmonious—Highly harmonious		
Neatness	Very messy—Very tidy		

To quantify participants' perception of the audiovisual landscape, a semantic differential method was used, in which the participants were asked to rate landscape and soundscape attributes using a set of opposite adjective pairs on a scale of 5 (1–5). Taking the naturalness indicator as an example, 1 = highly artificial, 2 = artificial, 3 = neither artificial nor natural, 4 = natural, and 5 = highly natural. The higher the score, the closer the rating is to the meaning of the adjective on the right. In addition, Visual Landscape Preference (VLP), Soundscape Preference (SP), and Overall Audio-visual Landscape Preference (OLP) were assigned on a 10-point scale (1 = least preferred, 10 = most preferred) based on the visitor's sensory experience.

The questionnaire survey was conducted simultaneously at seven sample sites on Sunday, 26 March 2017, from 2:30–5:30 pm, during good weather (breeze). Thirty questionnaires were randomly distributed to park visitors at each sample site, and a total of 210 questionnaires were distributed (all returned), of which 203 were valid questionnaires (85 males and 118 females) (Table 2). Most participants had visited the park several times in a year and were familiar with the visual and acoustic landscape of the park. No sensory-related problems were reported by any of the respondents. Before completing the questionnaires, the researchers emphasized to the participants the importance of distinguishing between visual landscape, soundscape and overall audiovisual landscape when evaluating visitors' landscape preferences.

Table 2. The personal characteristics of the respondents.

Attributes	Numbers of Each Categorization
Age	1. <15(14), 2. 15–24(76), 3. 25–34(47), 4. 35–44(27), 5. 45–59(13), 6. >60(26)
Gender	1. male (85), 2. Female (118)
Educational background	1. primary school (17), 2. secondary school (42),3. high school and trade/technical/vocational college (53), 4. college (91)
Occupation	1. employed (59), 2. retired (24), 3. student (81), 4. other (39)
Visit frequency	1. several times in a year (99), 2. once in a month (15), 3. once in a week (42), 4. twice or thrice in a week (29), 5. everyday (18)
Length of stay	1. <60 min (85), 2. 1 to 3 h (91), 3. 3 to 5 h (19), 4. >5 h (8)

2.4. Statistical Analysis

Statistical analysis was performed using IBM SPSS 20.0 software. First, the reliability and validity of the questionnaire data were tested. Second, correlation analysis was conducted to compare the relationship between VLP, SP and OLP, and stepwise multiple linear regression analysis was conducted to further compare the contribution of VLP and SP to OLP. Finally, correlation analysis was performed to further explore the relationships between VLP and PVLA as well as between SP and PSA.

3. Results

3.1. Reliability and Validity Test

First, the reliability of the questionnaire data was tested by the Cronbach's alpha coefficient. The Cronbach's alpha value was 0.873 (>0.7), indicating high reliability of the data. Then, the Kaiser–Meyer–Olkin (KMO) test was used to test the validity of the questionnaire data. The KMO measure of sampling adequacy was 0.712 (>0.7), so the validity test was acceptable.

3.2. Visitors' Evaluation of VLP, SP, and OLP

There was a large consistency among visitors' visual landscape preference (VLP), soundscape preference (SP) and audiovisual overall landscape preference (OLP) for Zhong-shan Park, with a mean value of about 7.0, indicating that visitors were not very satisfied with the sensory environment of the park (Figure 4). Most of the sample sites showed the following evaluation results: visual landscape preference (VLP = 7.53) > audio-visual overall landscape preference (OLP = 7.38) > soundscape preference (SP = 7.08), which indicated that the auditory experience was relatively weak in visitors' park landscape experience. This result may be related to the fact that the current soundscape environmental design in parks has not received sufficient attention.

3.3. Relationship between VLP, SP, and OLP

The KS (Kolmogorov–Smirnov) test for the normality of the data distribution showed that the two-tailed asymptotic probability of the landscape preference data for most sample sites was $p < 0.05$, so the following correlation analysis was performed using Spearman correlation analysis. The Spearman correlation analysis of VLP, SP, and OLP (Table 3) revealed that the relationships of VLP-SP, VLP-OLP, and SP-OLP were significantly and positively correlated in all sample sites except VLP-SP in SS6, indicating that there was a significant intrinsic correlation between them, especially the correlation of VLP-OLP and SP-OLP. By further comparison, it was found that all sample sites (including the combination of all sample sites) except SS3 showed SP—OLP > VLP—OLP > VLP—SP, indicating that soundscape preference was generally higher than visual landscape preference in terms of correlation with audiovisual overall landscape preference.

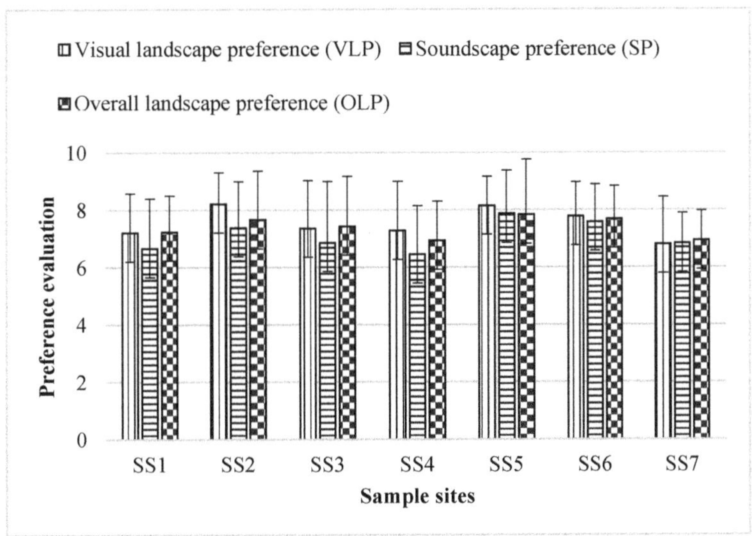

Figure 4. Evaluation on park landscape preference.

Table 3. Correlation analysis between VLP, SP, and OLP (Spearman).

Sample Site (SS)	VLP-SP	VLP-OLP	SP-OLP
SS1 (n = 29)	0.608 **	0.645 **	0.756 **
SS2 (n = 29)	0.491 **	0.581 **	0.764 **
SS3 (n = 28)	0.554 **	0.924 **	0.604 **
SS4 (n = 29)	0.461 *	0.576 **	0.823 **
SS5 (n = 28)	0.674 **	0.771 **	0.866 **
SS6 (n = 30)	0.252	0.374 *	0.600 **
SS7 (n = 30)	0.560 **	0.549 **	0.796 **
Total SS (n = 203)	0.567 **	0.658 **	0.769 **

** $p < 0.01$; * $p < 0.05$.

To further quantify and compare the effects of VLP and SP on OLP, a stepwise multiple linear regression analysis was performed with VLP and SP as independent variables and OLP as the dependent variable. As showed in Table 4, VLP together with SP could effectively predict the value of OLP, and the fitness of the regression equations was higher than 60% for most sample sites (5 out of 7 sample sites, except for SS5 and SS7). When all sample sites (Total SS) were included, the predictive power of SP was higher than that of VLP, as 55.6% variance of OLP could be explained by SP compared to 44.4% by VLP, implying that the former contribution was about 1.25 times (0.486/0.388) higher than the latter. When examined separately, SP was still more effective than VLP in predicting OLP value. As in all seven linear regression equations, only one predictor of OLP was VLP (SS3), two predictors were SP (SS2 and SS7), and the remaining four predictors were VP together with SP (SS1, SS4, SS5, and SS6). Since three sample sites, SS2, SS3, and SS7, were all single predictors and could not be compared for predictive power of SP and VLP, these three sample sites were excluded. In the remaining four sample sites with VLP and SP as common predictors, the predictive power of SP for OLP was higher than that of VLP to varying degrees, especially the predictive power of SP for SS4 was about 3.1 times that of VLP (0.763/0.246). By comparing the audio/visual contribution ratio with sound level, and the audio/visual contribution ratio with the audio/visual preference ratio at the four remaining sample sites, it could be found that the trend of the audio/visual contribution ratio at each sample site was opposite to the direction of the average sound level (Figure 5) and was opposite to the direction of the audio/visual preference ratio (except for SS1) (Figure 6). In other words, the higher the average sound level or the audio/visual preference ratio, the smaller the audio/visual contribution ratio.

Table 4. Multiple linear stepwise regression analysis of OLP.

Sample Site (SS)	Regression Equation	Adjusted R^2
SS1	OLP = 2.245 + 0.462SP + 0.427VLP	0.616
SS2	OLP = 1.328 + 0.803SP	0.628
SS3	OLP = 0.356 + 0.918VLP	0.836
SS4	OLP = 1.367 + 0.763SP + 0.246VLP	0.853
SS5	OLP = −2.175 + 0.395SP + 0.394VLP	0.444
SS6	OLP = 1.020 + 0.615SP + 0.328VLP	0.665
SS7	OLP = 1.831 + 0.758SP	0.559
Total SS	OLP = 1.367 + 0.486SP + 0.388VLP	0.589

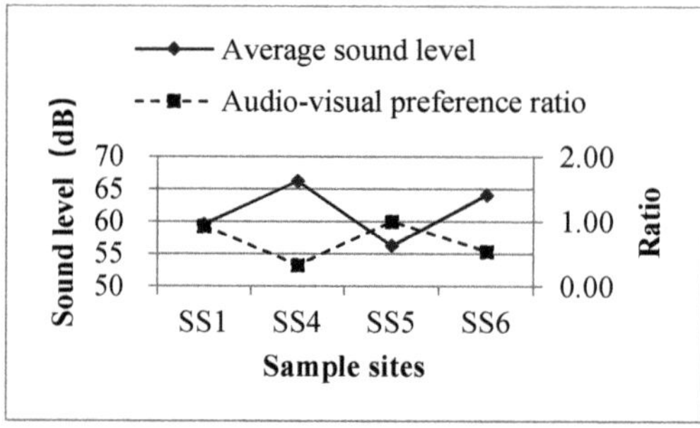

Figure 5. Ratio of audio/visual contribution to OLP and average sound level.

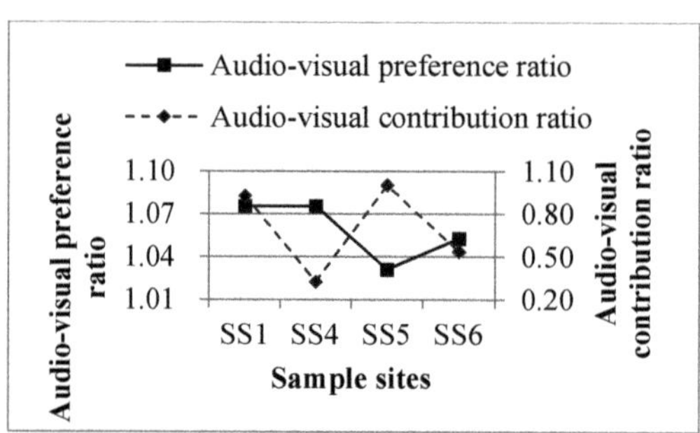

Figure 6. Ratio of audio/visual contribution to OLP and ratio of audio/visual preference.

3.4. Effect of Visitor's Individual Characteristics on OLP

To analyze the effects of visitor's individual characteristics on OLP, using the whole dataset (total sample sites), a backward multiple linear regression analysis was performed with VLP, SP, gender (G), age (A), educational background (E), occupation (O), visit frequency (F), and length of stay (L) as independent variables and OLP as the dependent variable. As showed in Table 5, when the adjusted R^2 reached its maximum value (0.593), only three individual characteristics, educational background (E), age (A), and gender (G), entered the multiple linear regression analysis, while occupation (O), visit frequency (F) and length of stay (L) were excluded. Educational background (E), age (A), and gender (G) showed negative correlation with OLP. Among them, only educational background (E) (sig = 0.034) could be used as an effective predictor of OLP (Impact coefficient = −0.103), while age (A) (sig = 0.059) and gender (G) (sig = 0.253) had no significant correlation with OLP.

Table 5. Multiple linear backward regression analysis of OLP.

Sample Site (SS)	Regression Equation	Adjusted R^2
Total SS	OLP = 2.487 + 0.489SP + 0.362VLP − 0.103E − 0.093A − 0.055G	0.593

Since the main task of this paper is to determine and compare the effects of VLP and SP on OLP, and the results of the regression analysis also show that the effects of population characteristics on OLP are small, the effects of population characteristics on VLP and SP, respectively, will not be analyzed in detail subsequently.

3.5. Influencing Factors of VLP and SP

To further analyze the main influencing factors of visitors' visual preference and auditory preference, Spearman correlation analysis was conducted between VLP and PVLA as well as between SP and PSA. As showed in Table 6, all 11 PVLA indicators had significant positive correlations with visual landscape preference at the 0.01 level, with visual harmony having the most important influence on VLP, color composition of landscape elements (color richness and color contrast), vegetation status (plant coverage and plant diversity), openness, relief, and naturalness occupying intermediate position, while the number of landscape buildings, plant level change, and neatness had the least impact on VLP. In this regard, Tveit et al. identified nine key visual concepts in affecting visual landscape quality: stewardship, coherence, disturbance, historicity, visual scale, imageability, complexity, naturalness, and ephemera [50]. Among these concepts, coherence (visual harmony), visual scale (openness), and naturalness were validated by this study. Another important concept validated was complexity, which in the present study can be referred to as color richness, color contrast, plant coverage, and plant diversity. A possible difference between the findings of Tveit et al. [50] and the present study concerns the indicator of stewardship (neatness), as neatness had the least impact on VLP among all visual perception attributes in the present study. For the indicator of neatness, participants may have integrated aesthetic values with the ecological values of the visual landscape [54], and different groups may have completely different opinions [55].

Table 6. Spearman correlation between visual landscape preference (VLP) and perceived visual landscape attributes (PVLA).

PVLA	Visual Harmony	Color Richness	Color Contrast	Plant Coverage	Plant Diversity	Openness	Relief	Naturalness	Number of Landscape Buildings	Plant Hierarchy Change	Neatness
VLP	0.529 **	0.491 **	0.488 **	0.465 **	0.450 **	0.449 **	0.427 **	0.417 **	0.411 **	0.410 **	0.410 **

** $p < 0.01$.

Spearman correlation analysis between SP and PSA showed that all perceived attributes except acoustic interference were significantly correlated with SP at the level of 0.01, and the correlation coefficients were ranked as follows: acoustic harmony, quietness, sound vitality, and acoustic richness (Table 7). Therefore, for the soundscape of urban parks, acoustic harmony and quietness are the most important concerns of visitors. Interestingly, no negative correlation was found between acoustic interference and SP, which is different from the results of Sevenant and Antrop [56]. A possible explanation is that traffic noise did not have a significant negative impact on park visitors because human sounds from leisure activities play an important role in constructing the relevant soundscape in urban recreation areas [57].

Table 7. Spearman correlation between soundscape preference (SP) and perceived soundscape attributes (PSA).

PSA	Acoustic Harmony	Quietness	Sound Vitality	Acoustic Richness	Acoustic Interference
SP	0.521 **	0.459 **	0.421 **	0.347 **	0.044

** $p < 0.01$.

4. Discussion

4.1. Comparison of the Contribution of Visual Landscape Preference and Soundscape Preference to Audiovisual Overall Landscape Preference

Most previous studies on landscape preference have focused on visual landscape preference or only explored the impact of soundscape perception on landscape preference [20,21,44,58], and few studies have juxtaposed visual landscape preference and soundscape preference to compare their contributions to audiovisual overall landscape preference. In terms of comparative research of audio-visual contribution, apart from the above-mentioned tests of landscape perception in laboratory settings [47], no relevant studies in field conditions have been reported to our knowledge. One of the main contents of this paper is to examine the findings of previous study under laboratory conditions that auditory preferences have a greater impact on landscape preference than visual preferences under certain conditions. The results of regression analysis showed the average contribution of the soundscape preference to audiovisual overall preference was about 1.25 times greater than that of visual landscape preference when differences between sample sites were not considered (data from seven sample sites were combined). Although this auditory/visual contribution ratio was not as high as that of the laboratory (4.5 times), it served as a valid support for the conclusion of the indoor study that "auditory preference has a greater impact on landscape preference than visual preference". Furthermore, this result reaffirmed the important influence of soundscape perception on landscape experience [21,39,59–61].

In addition, this study found that the audio/visual contribution ratio decreased as the average sound level or the audio/visual preference ratio increased. It can be inferred that the research conclusion of this paper that "soundscape preference has a higher impact on audiovisual overall landscape preference than visual landscape preference" applies mainly to park objects with high sound level or low soundscape quality. Whether this conclusion applies to park objects with higher soundscape quality needs to be explored in further studies.

It should be noted that although the researchers repeatedly reminded respondents to distinguish between visual, auditory and audio-visual integrated scenes before and during questionnaire implementation, respondents may not be able to completely separate visual scenes from auditory scenes under field environmental conditions. That is, the influence of auditory scenes cannot be completely shielded when evaluating visual scenes and vice versa. This may imply that there is some degree of crossover and mutual interference between visual landscape preferences, soundscape preferences, and audiovisual overall landscape preferences. To what extent such interference affects visitors' perceived landscape experience requires more in-depth research to evaluate.

4.2. Effect of Visitor's Individual Characteristics on OLP

The effect of individual characteristics on OLP was extremely limited compared to SP and VLP. Among all the population characteristics, only educational background was an important factor influencing OLP scores, and there was a significant negative correlation between educational background and OLP. This result was somewhat in line with Wang and Zhao's study [62], which found that education level and gender of respondents have a significant influence on preference assessment. In another study by Xu et al. [63], it was found that age and gender had an important impact on soundscape preference. However, this research found that age and gender had little important impact on OLP and were negatively correlated with OLP. It was revealed that the higher the education and the older the age, the lower the OLP value. Moreover, the value of OLP was lower for females than for males.

However, in any case, previous studies analyzing the influence of population characteristics on landscape preferences have focused on VLP and SP rather than OLP. Therefore, further research is needed to investigate the main population characteristics influencing OLP.

4.3. Main Influencing Factors of Visual Landscape Preference

4.3.1. Visual Harmony

The results of this study revealed that visual harmony had the strongest association with visual landscape preference among all the influencing factors of visual landscape preference. Visual harmony is mainly concerned with the harmony of the collocation (proportion) between different visual elements, and is an evaluation indicator that focuses more on the overall and comprehensive aesthetic perception [64], which may make visual harmony the closest to visual landscape preference to some extent and ultimately lead to the strongest correlation with visual landscape preference.

4.3.2. Neatness

Neatness is an indicator of the level of park management and maintenance. The results of Tveit et al. showed that maintenance management is one of the most important indicators of visual landscape evaluation [50]. A study by Zheng et al. showed that different groups had completely different opinions on neatness, with people from ecological backgrounds preferring wild natural landscapes and the general population preferring clean landscapes [55]. The respondents in this study mainly involved the general population, so neatness was second only to visual coordination in the ranking related to visual preference.

4.3.3. Color Composition

Color composition mainly refers to color richness and color contrast. This study found that, among these evaluation indicators affecting visual preference, the correlation between color composition and visual preference followed visual coordination and cleanliness as one of the most important indicators affecting visual landscape preference. Among them, color richness and color contrast ranked third and fifth, respectively (individual indicator ranking). The result is generally consistent with the findings of similar green spaces. In a study of Beijing Country Parks, Wang Ya-Juan found that the influence of color richness on landscape quality ranked second among seven landscape features [65]. In addition, the contribution of color contrast and color quantity ranked third and fourth, respectively, in the model for evaluating the aesthetic quality of rural green space landscape constructed by Shi Jiu-Xi et al. [66].

4.3.4. Naturalness

Most studies have shown that naturalness is one of the most important indicators of landscape preference [67], but some studies have also shown that the impact of naturalness on landscape preference is not always positive [68] and the relationship between the two is not always linearly positive [50]. In addition, the study of Sevenant and Antrop even stated that the degree of naturalness was not significantly indicative of overall preference for the landscape [56]. The present study has confirmed a significant influence on visual landscape preference, with perceived naturalness ranking fourth in affecting visual landscape preference.

4.3.5. Plant Diversity

Plant diversity is an important component of landscape diversity and is considered to be an important factor affecting landscape preference [69]. In this study, the impact of plant diversity ranked fifth in terms of its effect on visual landscape preference and was an important factor influencing visitors' visual landscape preference.

4.4. Main Influencing Factors of Soundscape Preference
4.4.1. Sound Comfort

To some extent, sound comfort is also a relatively holistic and integrated feeling, closer in meaning to soundscape satisfaction. Studies have shown that sound comfort is not only highly correlated with soundscape preference [70], but also has a significant positive correlation with landscape evaluation [71]. This study also proved that sound comfort played an important role in soundscape preference evaluation, and it was the primary factor affecting soundscape preference.

4.4.2. Quietness

Quietness and acoustic vitality are considered to be the two main dimensions that elicit emotional responses in soundscapes [72]. Quiet environment plays an important role in people's physical and mental recovery [73], so it is favored by people [74]. In addition, some studies have shown that quietness was closely related to landscape scenic beauty and recreational satisfaction [75], so quietness had an important impact on landscape perception. This study showed that the influence of quietness on soundscape preference was second only to sound comfort.

4.4.3. Acoustic Vitality

In this study, the effect of acoustic vitality ranked third after quietness as one of the two dimensions that evoke emotional responses to soundscapes. Acoustic vitality involves both sound organization (composition) and changes over time [72]. Several studies have shown a V-shaped distribution relationship between acoustic vitality and acoustic pleasantness, meaning that objects with high levels of acoustic vitality may be extremely pleasant or extremely unpleasant [76]. In other words, there is no simple linear relationship between acoustic vitality and acoustic pleasantness. However, this study showed a significant positive correlation between acoustic vitality and soundscape preference, which may be related to the fact that the park did not reach a high level of acoustic vitality.

4.4.4. Sound Richness

Sound richness and acoustic vitality are somewhat congruent in that both represent change, and sound richness will affect acoustic vitality to some extent. From the perspective of cognitive psychology, rich changes can provide possibilities for exploration, but richness also implies complexity, especially disorderly variation can increase cognitive difficulty, so moderate richness is preferred. However, it has also been shown that the higher the soundscape richness, the higher the soundscape pleasantness [77]. The present study showed a positive correlation between sound richness and soundscape preference, but the correlation between the two was relatively weak and ranked fourth.

4.4.5. Acoustic Interference

Interference was always seen as a negative indicator affecting the evaluation of landscape preference [43]. Interestingly, the acoustic interference in this study did not show a significant negative correlation with soundscape preference. A possible explanation is that the surrounding artificial noise, especially traffic noise, does not have a significant negative impact on visitors.

4.5. Inspiration for Park Landscape Planning and Design

Early park landscape planning and design generally focused on the design of visual elements and gave less consideration to visitors' soundscape experience. As a result, the overall satisfaction of the park is not very high. In view of how to improve visitors' satisfaction with park landscape, efforts can be made in the following aspects:

1. The quality of the acoustic environment should be regarded as an important design consideration and evaluation criterion. In the case of this study, soundscape preference had a greater impact on overall landscape preference than visual landscape preference in most sample sites. However, compared to the quality of the visual environment, the quality of soundscape environment was often in a relatively weak position, resulting in low visitor satisfaction with the overall landscape of the park. Therefore, the current soundscape environmental quality of parks needs to be optimized and improved.

2. The soundscape experience of visitors can be improved in four aspects: sound comfort, quietness, sound vitality, and sound richness. Priority should be given to enhancing the artistry and harmony of the combination of different soundscape elements to improve the acoustic comfort of the park (e.g., the promenade); optimizing the functional zoning of the park, making full use of topographic undulations and greenery isolation to improve the quietness of the overall park environment (e.g., the raining corridor and the elderly activity room); and, last but not least, introducing more waterscape sounds (fountains, water flow), birdsong and music that visitors love to enhance the sound vitality and soundscape richness of the park (e.g., the Longzhu Pavilion and the Zhongshan Square).

3. Further enhancing the visual experience of visitors. Specifically, it refers to increasing the coordination of visual elements (such as the isolation and coordination between the Seven Star Pool and the surrounding buildings of the park), improving the management and maintenance of the park (e.g., the elderly activity room), and optimizing the color design (improving color richness and color contrast, especially enriching the color of plants) (e.g., the Liberation Memorial Pavilion), enhancing the naturalness of the plant landscape, and increasing the diversity of plants (e.g., the promenade and the raining corridor).

5. Limitations and Future Work

This paper analyzed and quantitatively compared the influence of VLP and SP on OLP using seven sample sites of an urban forest park, and revealed the relationship pattern between audio/visual contribution ratio and sound level. However, this study only involved a single park with a relatively high degree of artificiality and a limited number of sample sites, so the subsequent study intends to include different types (different degrees of naturalization) of urban parks and more sample sites in order to investigate in depth the changes of the audio/visual contribution ratio and its relationship with different sound levels in different environments and scenes.

6. Conclusions

Landscape preference research has witnessed somewhat of a shift from a visual orientation to a more holistic and multisensory perspective. As part of the research effort towards multi-sensory landscape assessment, audio-visual interaction has received much research attention and has yielded fruitful results. However, the contribution of visitors' audiovisual perception to the overall landscape preference under on-site conditions has not been clarified. Therefore, with Zhongshan Park as an example, this paper investigated visitors' visual landscape preference, soundscape preference, and audiovisual overall landscape preference by using an on-site questionnaire survey. The main research findings were as follows: (1) Soundscape perception has an important impact on visitors' park landscape experience, and acoustic environment quality should be taken as an important design consideration and evaluation criterion. In the evaluation of most sample sites, the contribution of visitors' soundscape preference to the audiovisual overall preference

47. Gan, Y.; Luo, T.; Breitung, W.; Kang, J.; Zhang, T. Multi-sensory landscape assessment: The contribution of acoustic perception to landscape evaluation. *J. Acoust. Soc. Am.* **2014**, *136*, 3200–3210. [CrossRef]
48. Ren, X.; Kang, J. Interactions between landscape elements and tranquility evaluation based on eye tracking experiments. *J. Acoust. Soc. Am.* **2015**, *138*, 3019–3022. [CrossRef] [PubMed]
49. Gan, Y.; Luo, T.; Zhang, T.; Zhang, T.; Qiu, Q. Discussion on the 'objectivity' of subjective visual landscape evaluation: A case study of aesthetic landscape evaluation on the Houguang Lake of Wuhan. *Hum. Geogr.* **2013**, *28*, 58–63, 120. (In Chinese)
50. Tveit, M.; Ode, Å.; Fry, G. Key concepts in a framework for analysing visual landscape character. *Landsc. Res.* **2006**, *31*, 229–255. [CrossRef]
51. Ode, Å.; Tveit, M.S.; Fry, G. Capturing landscape visual character using indicators: Touching base with landscape aesthetic theory. *Landsc. Res.* **2008**, *33*, 89–117. [CrossRef]
52. Yao, Y.; Zhu, X.; Xu, Y.; Yang, H.; Sun, X. Assessing the visual quality of urban waterfront landscapes: The case of Hefei, China. *Acta Ecol. Sin.* **2012**, *32*, 5836–5845. (In Chinese)
53. Gong, P.; Liu, C.Q.; Gu, X.R. Research about Substitution Effects of Photograph and Animation Media in the Assessment of Urban Plant Visual Landscape. *Chin. Landsc. Arch.* **2017**, *8*, 19. (In Chinese)
54. Nassauer, J.I. Culture and changing landscape structure. *Landsc. Ecol.* **1995**, *10*, 229–237. [CrossRef]
55. Zheng, B.; Zhang, Y.; Chen, J. Preference to home landscape: Wildness or neatness? *Landsc. Urban Plan.* **2011**, *99*, 1–8. [CrossRef]
56. Sevenant, M.; Antrop, M. The use of latent classes to identify individual differences in the importance of landscape dimensions for aesthetic preference. *Land Use Policy* **2010**, *27*, 827–842. [CrossRef]
57. Hong, J.Y.; Jeon, J.Y. Influence of urban contexts on soundscape perceptions: A structural equation modeling approach. *Landsc. Urban Plan.* **2015**, *141*, 78–87. [CrossRef]
58. Hong, X.C.; Zhu, Z.P.; Liu, J. Perceived occurrences of soundscape influencing pleasantness in urban forests: A comparison of broad-leaved and coniferous forests. *Sustainability* **2019**, *11*, 4789. [CrossRef]
59. Ren, X.; Kang, J.; Liu, X. Soundscape perception of urban recreational green space. *Landsc. Arch. Front.* **2016**, *4*, 42–56.
60. Jiang, J.; Zhang, J.; Zhang, H.; Yan, B. Natural soundscapes and tourist loyalty to nature-based tourism destinations: The mediating effect of tourist satisfaction. *J. Travel Tour. Mark.* **2018**, *35*, 218–230. [CrossRef]
61. Jiang, B.; Xu, W.; Ji, W.; Kim, G.; Pryor, M.; Sullivan, W.C. Impacts of nature and built acoustic-visual environments on human's multidimensional mood states: A cross-continent experiment. *J. Environ. Psychol.* **2021**, *77*, 101659. [CrossRef]
62. Wang, R.H.; Zhao, J.W. Demographic groups' differences in visual preference for vegetated landscapes in urban green space. *Sustain. Cities Soc.* **2017**, *28*, 350–357. [CrossRef]
63. Xu, J.; Li, M.; Gu, Z.; Xie, Y.; Jia, N. Audio-Visual Preferences for the Exercise-Oriented Population in Urban Forest Parks in China. *Forests* **2022**, *13*, 948. [CrossRef]
64. Wu, J.S.; Yuan, T.; Wang, T. Preliminary theory of urban landscape esthetics based on three-dimensional landscape indicators. *Acta Ecol. Sin.* **2017**, *37*, 4519–4528. (In Chinese)
65. Wang, Y.J. *Study on the Visual Landscape Quality of Beijing Country Parks Based on SBE and SD Method*; Capital Normal University: Beijing, China, 2013. (In Chinese)
66. Shi, J.X.; Dong, J.W.; Zhang, Z.D.; Wang, X.M. Aesthetic quality evaluation of rural green space landscape. *J. Chin. Urban For.* **2013**, *11*, 5–8.
67. Svobodova, K.; Sklenicka, P.; Vojar, J. How does the representation rate of features in a landscape affect visual preferences? A case study from a post-mining landscape. *Int. J. Min. Reclam. Environ.* **2015**, *29*, 266–276. [CrossRef]
68. Lindhagen, A.; Hörnsten, L. Forest recreation in 1977 and 1997 in Sweden: Changes in public preferences and behaviour. *Forestry* **2000**, *73*, 143–153. [CrossRef]
69. Rogge, E.; Nevens, F.; Gulinck, H. Perception of rural landscapes in Flanders: Looking beyond aesthetics. *Landsc. Urban Plan.* **2007**, *82*, 159–174. [CrossRef]
70. Hong, J.Y.; Lee, P.J.; Jeon, J.Y. Evaluation of urban soundscape using soundwalking. In Proceedings of the 20th International Congress on Acoustics, Sydney, Australia, 23–27 August 2010.
71. Ji, X.R.; Lu, F.H.; Wang, Y.P. Acoustic comfort in university campuses' outdoor learning spaces—Based on 5 university libraries' outdoor space in Shanxi. *J. Appl. Acoust.* **2017**, *36*, 311–316. (In Chinese)
72. Davies, W.J.; Adams, M.D.; Bruce, N.S.; Cain, R.; Carlyle, A.; Cusack, P.; Hall, D.A.; Hume, K.I.; Irwin, A.; Jennings, P.; et al. Perception of soundscapes: An interdisciplinary approach. *Appl. Acoust.* **2013**, *74*, 224–231. [CrossRef]
73. Zhang, Y.; Kang, J.; Kang, J. Effects of soundscape on the environmental restoration in urban natural environments. *Noise Health* **2017**, *19*, 65. [PubMed]
74. Van den Berg, A.E.; Koole, S.L.; Van der Wulp, N.Y. Environmental preference and restoration:(How) are they related? *J. Environ. Psychol.* **2003**, *23*, 135–146. [CrossRef]
75. Liu, J.; Yu, S.S.; Wang, Y.J.; Zhang, J.Q. Research on the Interaction Effect between Landscape and Soundscape Experience in City Parks. *Chin. Landsc. Arch.* **2017**, *33*, 86–90. (In Chinese)
76. Hall, D.A.; Irwin, A.; Edmondson-Jones, M.; Phillips, S.; Poxon, J.E. An exploratory evaluation of perceptual, psychoacoustic and acoustical properties of urban soundscapes. *Appl. Acoust.* **2013**, *74*, 248–254. [CrossRef]
77. Liu, J.; Yang, L.; Zhang, X.W. Research on the Relationship between Soundscape Perception and Landscape Evaluation in Historical Block: A Case Study in the Three Lanes and Seven Alleys in Fuzhou. *Chin. Landsc. Arch.* **2019**, *35*, 35–39. (In Chinese)

Article

Improving Soundscape Comfort in Urban Green Spaces Based on Aural-Visual Interaction Attributes of Landscape Experience

Yuhan Shao [1], Yiying Hao [2], Yuting Yin [1,*], Yu Meng [1] and Zhenying Xue [1]

[1] College of Architecture and Urban Planning, Tongji University, Shanghai 200092, China
[2] Bureau Veritas, Glasgow G2 6BZ, UK
* Correspondence: yyin0326@tongji.edu.cn

Abstract: The importance of multi-sensory perception in constructing human landscape experiences has been increasingly emphasized in contemporary urban life. The aim of this study is to explore aural-visual interaction attributes that may influence people's perceived overall soundscape comfort in urban green spaces (UGSs). To achieve this, a total of 12 perceptive indicators were identified from the existing literature to evaluate people's perceived visual and acoustic attributes and types of sound sources, and their relations to the perceived soundscape comfort. 268 responses were obtained in a questionnaire-based survey conducted in five UGSs in Chengdu Outer Ring Ecological Zone. This was done whilst a typical objective acoustic indicator, sound level, was used as a mediator for potential changes on these relations within different sound level ranges. Results suggested that a low level of environmental sound does not correspond to higher ratings on the overall soundscape comfort. It was also found that the environmental sound level of 77 dBA was a turning point in the relation between people's soundscape comfort and its influential indicators in UGSs. A set of six models was then provided to describe the overall soundscape comfort and its contributing indicators in aural-visual interactions, respectively, in sound level ranges below and above 77dBA.

Keywords: urban green spaces; soundscape comfort; aural-visual attributes; perception; China

Citation: Shao, Y.; Hao, Y.; Yin, Y.; Meng, Y.; Xue, Z. Improving Soundscape Comfort in Urban Green Spaces Based on Aural-Visual Interaction Attributes of Landscape Experience. *Forests* **2022**, *13*, 1262. https://doi.org/10.3390/f13081262

Academic Editors: Xin-Chen Hong, Jiang Liu and Guang-Yu Wang

Received: 5 July 2022
Accepted: 6 August 2022
Published: 10 August 2022

Publisher's Note: MDPI stays neutral with regard to jurisdictional claims in published maps and institutional affiliations.

1. Introduction

Multi-sensory perception, spatial use pattern and the emotional responses of humans constitute their experiences in urban landscapes [1,2]. Urban green spaces (UGSs) have significant impacts on improving the physical [3], psychological [4] and social health [5,6] of urban residents. The design of UGSs has long been concerned with aesthetics, and there is enormous evidence of spatial form and environmental characteristics improving users' perceived visual quality [7,8]. Although more than 80 percent of our sensory input is visual [9], the importance of multi-sensory perception in constructing human landscape experiences has been increasingly emphasized [10,11], especially soundscape. Soundscape refers to the acoustic environment that people perceive or experience in a certain context [12]. It highlights the auditory properties of a landscape and sets the visual dominance in landscape perception research in a new context, which has now developed into a special focus on the cross effects of aural–visual interactions or soundscape–landscape relationships on human soundscape perception [13,14].

Overall soundscape comfort is comprised of people's perceptions of aural–visual environmental attributes [15,16]. Exploring the relevant indicators describing these perceived attributes and their mutual influences can enrich and improve the perceptive quality of UGSs and also provides cues for improving people's perceptions of soundscape comfort. However, to date, very few studies have evaluated soundscape and visual perception in combination in situ [13,17] nor constructed a full model [18] combining perceived aural–visual features and objective acoustic attributes in the application of improving people's soundscape comfort. This study intends to explore aural–visual indicators and their mutual interactions in influencing the overall soundscape perception of UGSs. Three groups of indicators measuring people's perceived visual attributes, aural attributes and perceived sound sources were identified to measure and improve people's perceived soundscape comfort in UGSs. This was done whilst a typical objective acoustic indicator [19], sound level, was used as a mediator. An on-site survey was conducted to explore whether and how these indicators contribute to people's overall perceived soundscape comfort, with the specific research objectives stated below:

1. To identify aural–visual interaction indicators related to the overall soundscape comfort in UGSs and investigate their influences.
2. To construct models indicating ways of improving overall soundscape comfort in UGSs based on the identified indicators.
3. To discuss the cross effects existing between people's perceived visual attributes, aural attributes and sound sources.

2. Influential Indicators of Soundscape Perception

2.1. Visual Perception Indicators in Landscape

Visual perception has long been taken as the major basis for evaluating the quality of urban green spaces in landscape studies [20]. Ode et al. (2008) cited biophilic theory, landscape aesthetic, information processing theory, spirit of place, prospect-refuge theory, topophilia, landscape heritage and restorative landscapes [21] as major theoretic bases for developing visual landscape concepts. Biophilic and prospect-refuge theory later constituted the foundation of visual preference research [22] which, as agreed by many scholars, can both be explained by an evolutionary perspective [23]. Prospect-refuge theory [24] proposes people feel most comfortable when they can observe what is happening around them while also being slightly protected. Kaplan and Kaplan then raised a preference matrix stating that coherence, complexity, legibility and mystery are important in creating environments since they can provide people with safety and comfort [25]. Another major and long-lasting theme responding to the question 'What is it that we like about landscape, and why do we like it?' [26] is the aesthetic approach to landscape. It is based on Maslow's hierarchy of needs [27] and composed of at least two interrelated components: cultural theory and biological theory [28].

Drawing from the theoretical literature, numerous studies have contributed to the search for environmental indicators linked with human visual landscapes [29]. A set of 40 commonly used affective descriptors in adaptation theory, such as complexity, naturalness, brightness, openness, coherence, were validated as indicators of visual landscape perceptions [30]. Similarly, complexity, coherence, disturbance, stewardship, imageability, visual scale, naturalness, historicity, and ephemera were identified as being capable of characterizing the visual landscape [21]. Grahn and Stigsdotter constructed perceived sensory dimensions (PSDs) based on existing evidence and a large-scale questionnaire survey determining eight of the most popular sensory dimensions of perception, including serenity, nature, space, species-richness, refuge, prospect, society and culture [31]. The PSDs were later widely applied in more than 60 studies [32] and proved to be an efficient tool in landscape assessment and design. However, evaluation frameworks do not normally consist of indicators solely measuring people's visual responses. When studies focus on the mutual influence of visual and aural aspects in landscape perceptions, these indicators can hardly provide adequate specific visual implications due to their inherent inter-relations with other perceptive aspects.

2.2. Aural Perception Indicators in Landscape

Soundscape has been used by a variety of disciplines to describe the relationship between a landscape and the composition of its sound [19]. In Southworth's work [16], it was first mentioned in urban planning literature addressing how the sounds of the built environment enhanced people's perception of space and their relationship to the activities occurring within cities [33]. There are three important directions in soundscape research aiming to investigate how people perceive the acoustic environment [34]. One is to objectively obtain soundscape information from analyzing sound recordings in terms of physical parameters [35,36] or in spectrograms [37] for regulating the environmental noise [38,39]. Moreover, traditional qualitative methods, such as questionnaires, interviews [40] and on-site observation [41,42], are used to acquire subjective soundscape information. Furthermore, the objective and subjective methods are also often combined in soundscape evaluations [43].

A large body of literature has also explored indicators that affect soundscape perceptions in different dimensions including comfort, tranquility, pleasantness, nature, gentleness, and appropriateness [44,45]. These were then integrated and put through principal component analysis for validation, resulting in a model later referred to as perceived affective quality (PAQ) [46], which raises eight important dimensions relevant to people's aural perceptions: pleasantness, vibrancy, eventfulness, chaos, annoying, monotony, uneventfulness and calm. The PAQs have been incorporated into the International Standards of Soundscape (ISO) [12] in order to unify soundscape perception descriptors and have been widely applied and validated in later soundscape studies [47]. Similar to visual landscape studies, most soundscape assessment frameworks contain indicators measuring human perceptions caused not only by the aural sensory dimension. As a result, the cross effect between audio-visual attributes has not been strictly clarified, and the results of relevant literature research are inconsistent.

2.3. Types of Sound Sources

Differing from perceptive indicators, sound source is an attribute affiliated with the definition of soundscape that investigates how the acoustic environment can be characterized by identifying what sound sources can be heard in the area and how dominant they are [12]. Scholars have characterized the sources of sounds into three major types: natural sounds that relate to non-organic elements of nature, such as waterfalls (geophony), organic but non-human sources, such as animals' copulatory vocalizations (known as biophony) and all environmental sounds generated by human sources (anthrophony), such as human voices or human activity-related sounds [15,48]. These were also incorporated into ISO [12] and formalized into three main types of sound sources that are recommended for inclusion in soundscape surveys: sounds of nature, sounds of technology and sounds of human beings. The latter two types are also referred to as mechanical and anthropogenic sounds in the relevant studies [49,50]. They are important in soundscape research since they provide potential cues implying necessary design interventions. For example, natural sounds are preferred while anthropogenic and mechanical sounds are less welcomed in urban parks [49]. In addition, the breeze significantly increases the likelihood of people giving a high comfort evaluation, while sounds from bikes or heavy vehicles significantly lower the likelihood [49].

2.4. Sound Level as Mediator

The design of soundscape developed from noise regulation. Earlier research and practices mostly emphasized the control and reduction of unwanted sound [38,39] in urban areas. However, later studies found that reducing sound level does not necessarily lead to an increase of acoustic comfort in urban areas [51,52]. Individuals' noise sensitivity [53], a stable personality trait that captures attitudes towards a wide range of environmental noises [54,55], and environmental attributes [56] both have impacts on the relation between sound level and people's perceived acoustic comfort. Evidence has shown that subjective evaluation of the acoustic environment relates well with the sound level when it is within a certain range [57,58], and the relation as well as the range varies with research contexts. For example, Yang and Kang found that the variation in acoustic comfort evaluations was almost negligible when the sound level was lower than around 73 dBA, whereas when the sound level was over 73 dBA for the mixed sound of fountains and demolition, the evaluation of acoustic comfort evidently decreased with the increase of the sound level [59].

In another study of soundscape perception in pocket parks, results also disclosed that influential indicators change significantly when the sound level passes 70 dBA [57].

2.5. This Study

To date, very few studies have constructed models [18] combining perceived aural–visual features and objective acoustic attributes in the application of improving people's soundscape comfort on the basis of understanding the influences of aural–visual interaction attributes on the overall soundscape comfort and the inter-relations between them.

This study aims to explore influential indicators of how people perceived overall soundscape comfort in UGSs. Through an on-site survey conducted in five UGSs in Chengdu Outer Ring Ecological Zone, it tries to construct models that can indicate ways of improving people's overall soundscape comfort according to these interaction indicators in perceived aural and visual aspects. Sound level is used as a mediator in the process to investigate whether and how these influential indicators and models change within different sound level ranges.

3. Methods

3.1. Study Area

This study chose typical UGSs within Chengdu Outer Ring Ecological Zone as study sites. Chengdu is a demonstration city for promoting the Park City agenda in China and has achieved evident progress since 2018. The Outer Ring Ecological Zone is one of the most important projects in this agenda and surrounded by over 30 UGSs of different scales and with different characteristics and functions. However, most UGSs now suffer from noise issues since the Outer Ring Ecological Zone is crossed by highways and is also affected by noise from railways and Tianfu airport. This has seriously affected people's landscape experience when using these green spaces and, hence, they have become appropriate research sites for this study.

Five UGSs (Figure 1, Table 1) with noise issues (sound level measured in the pilot study exceeded the limit of 70 dBA stipulated in the Standard of Sound Environment Quality [GB3096-2008] [60]), namely Bailuwan Wetland Park (UGS1), Guixi Ecological Park (UGS2), Jiangjia Art Garden (UGS3), Qinglong Lake Wetland Park (UGS4) and Tianfu Lotus Garden (UGS5), were selected as study sites to determine whether and how aural–visual soundscape stimuli can be modified to improve the overall soundscape comfort of green space users based on the following criteria: (1) first, study sites should be fully constructed and open to the public; (2) study sites should have different soundscape sources, visual and acoustic attributes; (3) study sites should contain at least one measurement spot with an open view and qualify for setting acoustic instruments. All selected study sites are dominated by natural characteristics and rich in botanic species.

Spots to conduct the questionnaire and the acoustic measurement were then selected within the five UGSs (Figure 1). Measurement spots should be able to present the visual and acoustic features of their represented UGSs. In addition, there should be a distance of an over three-minute walk from the entrance to the first measurement spot in each site to allow enough time for subjects to become mentally immersed in the setting [61,62]. The three-minute walking distance is also requested as the least interval within spots in the same site. In addition, there should be enough space for participants to wander around and experience the surrounding environment in each spot. With the consideration of the size of each site and the space's access to the public within each site, eight spots were chosen in UGS1, three in UGS2, eight in UGS3, one in UGS4 and two in UGS5.

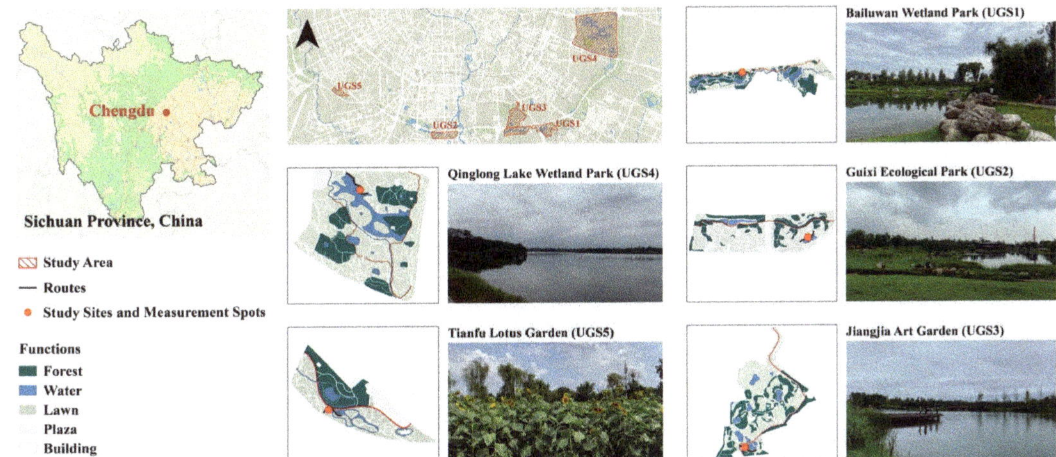

Figure 1. Site photos of measurement spots.

Table 1. General conditions of the selected study sites and measurement spots.

Study Sites	Area (ha)	Dominant Functions	Landscape Features
Bailuwan Wetland Park (UGS1)	275	Ecology and protection	The site is characterized by wetland, floating island and lotus pond. There is large scale of lawn in the site, decoration with stones, and seat provision. There are plenty of arbors grown at the rear of the site.
Guixi Ecological Park (UGS2)	159.3	Activity and entertainment	The site is characterized by grassland and lakes. The terrain of the site tilts towards the central. Arbors are at a distance from the centre of the site.
Jiangjia Art Garden (UGS3)	132	Activity and entertainment	The site is characterized by a large artificial lake. Herbals and shrubs are dominant within the site.
Qinglong Lake Wetland Park (UGS4)	1298	Ecology and protection	The site is characterized by large lakes, lawn and high and dense forests with diverse species of herbal, shrubs and arbors.
Tianfu Lotus Garden (UGS5)	48.8	Landscape and leisure	The site is characterized by flower fields and lawns.

3.2. Questionnaire

A total of 12 indicators were selected from existing research to measure people's perceived visual and acoustic attributes and types of sound sources in UGSs (Table 2). Four indicators, *aesthetic*, *openness*, *layering* and *order*, were selected from traditional visual landscape studies and PSDs [18,63] since other indicators, such as tranquility, society and culture, may not be triggered by only human visual perception. Naturalness was also excluded since green spaces are typical natural settings in urban areas. Only *vibrancy*, *harmony*, *eventfulness* and *monotony* were chosen as indicators of aural attributes because other indicators in PAQs [18] and international soundscape research standards [12], such as pleasantness and calm, may be influenced by environmental information other than acoustics. Furthermore, *harmony* was adopted from chaos to ensure semantic consistency [64] among indicators. Regarding sound sources, *natural*, *anthropogenic*, *mechanical* and *traffic sounds* were chosen as indicators based on pilot study results to investigate the dominant position of certain sounds as determined by their perceived occurrences and perceived loudness [65].

Table 2. Indicators of measuring visual and acoustic attributes and sound sources.

Indicators		Descriptions			Refs.
Visual attributes	Aesthetic (V1)	Visually appealing			[66]
	Openness (V2)	Visual and spatial openness			[67]
	Layering (V3)	The layer relationship between environmental elements			[68]
	Order (V4)	The organised state of environmental indicators			[69]
Acoustic attributes	Vibrancy (A1)	Sounds full of life			[70]
	Monotony (A2)	Not interesting, boring			[42]
	Harmony (A3)	Sounds arranged together in a pleasing way so that each part goes well with the others			[71]
	Eventful (A4)	Something important was happening in the sound			[72]
Sound sources	Traffic sound (S1)	Skateboards	Car horn	Rail	/
		Bicycle ring	Motor vehicles	Highway	
		Heavy vehicles	Airplane	Metro	
	Mechanical sound (S2)	Weeding	Construction	Construction	/
		Radio	Loudspeaker	Alarm	
	Anthropologic sound (S3)	Exercising	Cleaning	Whistling	/
		Walking	Parent–child activities	Smiling	
		Boating	Bells	Music	
		Talking	Pedaling	Singing	
	Natural sound (S4)	Insects	Tree murmur	Wind	/
		Croak	Twittering of birds	Ripple	
		Fish dive	Barking dogs	Rain	
The overall soundscape comfort (A)		/			/

12 indicators and the overall soundscape comfort were together designed into a questionnaire to investigate people's aural–visual perceptions of the five selected urban UGSs. The questionnaire contained two parts. The first part consisted of the descriptions (Table 2) of the 12 indicators measuring people's perceived visual and acoustic attributes and sound sources at the study sites. The second part had only one question evaluating people's overall perceived soundscape comfort of the environment. Both parts used a five-level Likert scale (1–5 representing strongly disagree to strongly agree).

3.3. Apparatus

Equivalent continuous A-weighted sound pressure level (LAeq) is one of the most common parameters in relevant acoustic research [73] for objectively describing the environmental sound level. In this study, it was selected as a mediator and measured by a multi-channel signal analyzer (AWA6290L+). The acoustic instrument is suitable for outdoor conditions with the advantages of having multi-channel receptors and high sensitivity; it has proved effective in measuring and recording outdoor environmental sound level and has been widely applied in academic research and practices [74,75]. In addition to the signal analyses, apparatus used in the field measurement also included two test microphones (AWA14423, frequency range 20–16,000 Hz, sensitivity about 40 mV/Pa), two preamplifiers (AWA14604, Integrated Circuits Piezoelectric, impedance conversion, frequency range 10–200 k Hz, Gain 1), two 80 mm diameter wind balls and two tripods (about 1.6 m high).

3.4. Data Collection

Previous studies suggested that people's life stage [76], educational background [77] and state of being alone or with company [78] have influenced their landscape perceptions. This study is especially concerned with the aural-visual perceptions of young people since they are more sensitive to surrounding environments than other age groups [79]. Therefore, a total of 20 university students, aged around 22 with normal or corrected-to-normal visual acuity, normal color vision and hearing, were recruited to evaluate each of the measurement spots in the five UGSs. The ratio of male participants to female was 1:1. The sample size was determined according to previous soundscape studies, which stated that the total number of valid responses should be over 100 [42]. All participants were trained before the formal survey.

The field survey was carried out on sunny days with an Air Quality Index less than 60 and wind speed below 5 m/s, in July 2021. Twenty recruited participants were divided into four groups, and each group was led by one researcher. Each participant group travelled through measurement spots in a different order to avoid having too many people appear at the same spot and time. To minimize disturbance caused by researchers, participants were asked to wander around within the scope of each measurement spot by themselves to experience their surrounding environments for five minutes [80] and then complete the questionnaire. Participants were asked to fill out the questionnaire survey in each of the 22 measurement spots, and a total of 380 responses were collected. The survey was ethically approved, and informed consent was obtained from each participant in advance. Noise data, such as an incomplete questionnaire or questionnaires with the same ratings for every indicator, were manually removed and the number of final valid responses was 268 (approximately 12 per spot).

While conducting the questionnaire survey, another group of researchers recorded the equivalent continuous A-weighted sound pressure level of each spot with a multi-channel signal analyzer to indicate environmental sound level. After the calibration using the sound level calibrator (AWA6021A, accuracy level 1), two amplifiers and microphones equipped with wind balls were placed at the center point of each spot, about 1.6 m above the ground, to record the environmental sound level. The sound level within each spot was recorded for three minutes.

3.5. Data Analysis

Potential relations between the 12 indicators describing the visual and acoustic attributes of urban UGSs and the overall soundscape comfort were explored based on the obtained questionnaire results. The analysis of sound level, retrieved from the recorded data in the acoustic instrument, was carried out using the SPL 1/3 octave bands analysis package in the AWA6229 6.0 software. Spectrum analysis was performed on audio files to calculate the equivalent continuous A-weighted sound pressure level (LAeq) of each spot for a consecutively recorded three minutes. The average LAeq of each measurement spot in a park was used as the LAeq of the site.

Responses of the 12 descriptive indicators and the overall soundscape comfort first went through reliability analysis to ensure the consistency of the obtained questionnaire results. Correlational analysis was then conducted between the overall soundscape comfort and the sound level to determine the turning point in this research context. Afterwards, the collected questionnaire responses were separated into two groups based on the disclosed turning point and analyzed by several different methods of multivariate analysis. Correlational analysis was also used in this step to identify indicators relevant to the overall soundscape comfort and linear regression models were formulated to establish stochastic relationships between the overall soundscape comfort and its influencing indicators. The interrelationships among people's perceived visual attributes, aural attributes and sound sources were then discussed accordingly.

4. Results

4.1. Manipulation Checks and Descriptive Analysis

Data were analyzed using SPSS V 25.0 and examined for internal consistency with Cronbach's alpha [81]. Calculation of internal consistency (Cronbach's α) was the preferred measure of inter-rater reliability when cases were rated in terms of an interval variable or interval-like variable, such as the Likert scale used in the questionnaire. The α values of the 12 indicators in the 22 measurement spots suggested sufficient internal consistency (Cronbach's $\alpha = 0.76 > 0.6$) and, hence, guaranteed the reliability of the obtained ratings.

The visual and acoustic evaluation results of a site were represented by calculating the average of the questionnaire responses at each spot within the site (Table 3), and the same method was applied for the LAeq results (Table 4). Results showed that UGS2 (3.86) and UGS3 (3.75) ranked the highest in the overall soundscape comfort, followed by UGS4 (3.67), UGS5 (3.55) and UGS1 (3.24). However, results of the overall soundscape comfort were not in line with the sound level measured by acoustic instrument. A ranking of sound level results from lowest to highest were UGS4 (69.3 dBA), UGS5 (74.5 dBA), UGS1 (75.9 dBA), UGS2 (78.8 dBA) and UGS3 (82.3 dBA). Figure 2.

Table 3. The arithmetic average of perceptive evaluation in the five study sites.

Indicators	UGS1	UGS2	UGS3	UGS4	UGS5
Aesthetic	3.29	4.16	3.86	4.00	3.72
Openness	3.77	4.33	3.99	4.20	3.31
Layering	3.19	3.81	3.94	3.60	3.45
Order	3.33	3.79	3.87	3.33	3.90
Vibrancy	3.10	3.40	3.13	3.40	3.10
Monotony	1.89	2.30	2.64	2.13	2.69
Harmony	3.30	3.98	3.95	3.67	3.59
Eventfulness	2.91	3.37	3.49	3.33	3.48
Traffic sound	4.43	3.00	3.33	3.13	3.86
Mechanical sound	1.26	1.88	2.01	1.73	2.24
Anthropologic sound	1.96	3.47	2.82	4.33	2.93
Natural sound	4.42	3.93	4.45	4.00	4.00
The overall soundscape comfort	3.24	3.86	3.75	3.67	3.55

Table 4. The arithmetic average of the sound pressure level (LAeq) in the different UGSs.

		UGS1	UGS2	UGS3	UGS4	UGS5
Total no. of sound measurement spots taken in the park area		8	3	8	1	2
LAeq (dBA)	Average	75.9	78.8	82.3	69.3	74.5
	Maximum	79.2	81.9	99.9	69.3	78.2
	Minimum	71.4	75.5	66.3	69.3	70.7

In respect to visual perception, though UGS2 and UGS3 still had relatively higher ratings, each of them had their own visual advantage. *Aesthetic* and *openness* were evident visual attributes of UGS2, while UGS3 was characterized by *layering* and *order*. The *order* of UGS4 scored low, but its other three visual indicators were rated above average. The results of UGS5 were just the opposite: *order* had higher ratings than *aesthetics*, *openness* and *layering*. UGS2 and UGS3 also had the highest ratings in the four aural indicators, with *harmony* being the obvious advantage for both UGS2 and UGS3. The *vibrancy* of soundscape was relatively good in UGS4, while UGS5 was generally characterized by the *eventfulness* of its environmental sounds. Among all the sites, UGS1 had the lowest ratings on almost all visual and aural indicators.

Five study sites showed significantly different attributes regarding sound sources, among which, *anthropologic sound* was dominant in UGS2 and UGS4. Also, UGS3 and UGS1 were both evidently characterized by *natural sound*, but they had different subdominant sound sources: *mechanical* and *traffic sounds*, respectively. *Mechanical* and *traffic sounds* were also prominent in UGS5, where a medium level of *natural* and *anthropologic sounds* could be perceived.

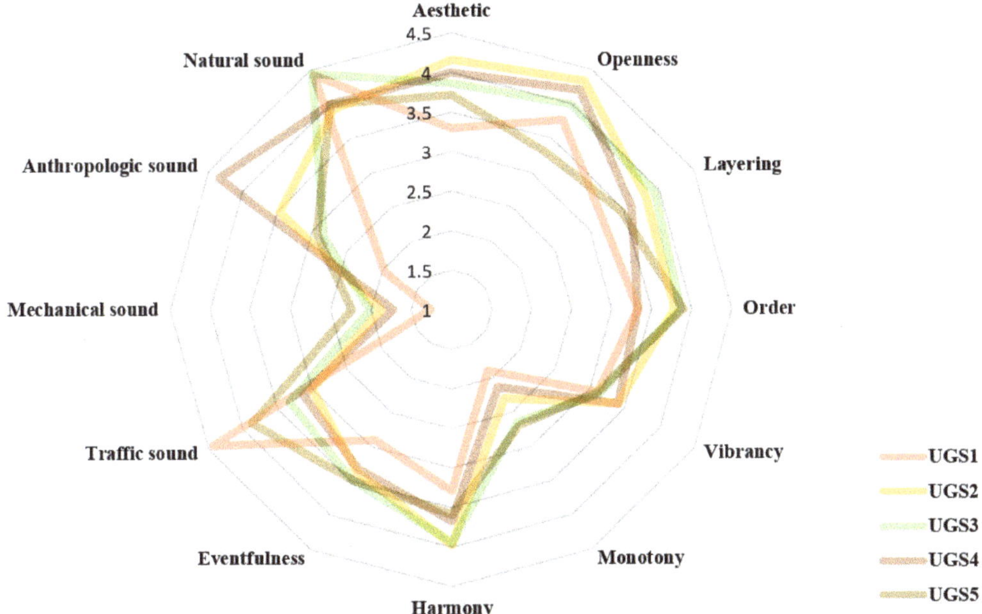

Figure 2. The evaluation results of the five study sites.

4.2. The Turning Point in the Relation between Sound Level and Soundscape Comfort

Inspired by previous studies [59] investigating the relation between soundscape perception and the sound level, this study proposed that the influential indicators of overall soundscape comfort may vary in different decibel ranges. Correlational analysis was carried out on all datasets, and results indicated there was an evident correlation between overall soundscape comfort and the sound level ($p < 0.01$). Further, scatter plot was used to present how the mean LAeq influenced the overall soundscape comfort. A turning point of 77 dBA was identified in the relation between sound level and soundscape comfort. The overall soundscape comfort slightly increased with sound level when it was above 77 dBA, while an obvious negative relation could be observed when sound level was lower than 77 dBA. Figure 3.

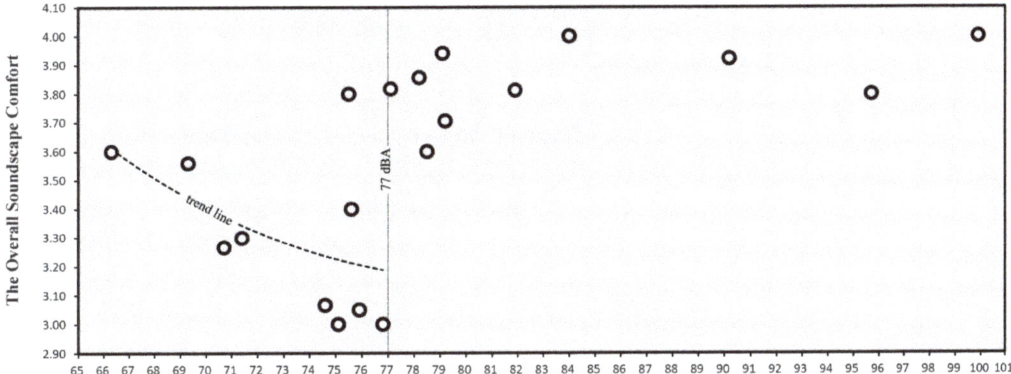

Figure 3. Scatter plot showing the relation between the overall soundscape comfort and sound pressure level.

Correlational analysis was conducted again to validate the groups of data above and below 77 dBA. Results showed that there was generally an evident negative correlation between the mean LAeq and the overall soundscape comfort when the sound level was under 77 dBA ($R = -0.335$, $p = 0.000 < 0.01$), but the correlation coefficient became rather low beyond this sound level ($R = 0.139$, $p = 0.117 > 0.05$). This indicates that when the sound level reached a certain value, which was 77 dBA in this study, subjects' evaluations varied significantly and became more unpredictable. Therefore, 77 dBA was found and validated as a turning point in this study, and the following analysis should be conducted within two groups, above and below 77 dBA, respectively.

4.3. Influences of Aural-Visual Perceptive Attributes in the Overall Soundscape Comfort

268 collected questionnaire responses and sound data measured at the 22 spots were used to formulate linear regression models to establish a stochastic model between the overall soundscape comfort perception and its influencing indicators: sound types, visual attributes and acoustic attributes. Pearson correlational analysis was also employed to reveal the mutual influences between the influencing indicators. Given that the identified turning point of sound pressure level was 77 dB, correlational analysis and regression analysis were taken in two groups of below and above 77 dB to explore potential relations.

4.3.1. Influential Indicators of the Overall Soundscape Comfort

Correlational analysis results showed that *mechanical* and *natural sounds* had no significant relation with the overall soundscape comfort in the below 77 dBA group, while *aesthetic* and *vibrancy* were not significantly correlated with the overall soundscape comfort in the other group. Thus, they were excluded in the regression analysis. Table 5.

Table 5. Correlational analysis between the overall soundscape comfort and other indicators.

	Groups	S1	S2	S3	S4	V1	V2	V3	V4	A1	A2	A3	A4
<77 dBA (n = 139)	Correlation (Pearson's r)	−0.560 **	0.049	0.313 **	0.147	0.661 **	0.498 **	0.345 **	0.326 **	0.218 **	0.285 **	0.656 **	0.396 **
	Significant (p)	0.000	0.570	0.000	0.084	0.000	0.000	0.000	0.000	0.010	0.001	0.000	0.000
>77 dBA (n = 129)	Correlation (Pearson's r)	−0.297 **	0.083	0.008	0.502 **	0.316 **	0.110	0.534 **	0.422 **	0.041	0.221 *	0.466 **	0.409 **
	Significant (p)	0.001	0.347	0.929	0.000	0.000	0.214	0.000	0.000	0.646	0.012	0.000	0.000

Indicator values with * and ** show significant differences with the overall soundscape comfort.

1. Sound types

In the under 77 dBA group, results showed that only the *traffic sound* ($p < 0.01$) was significantly related with the overall soundscape comfort, and a linear model with an R^2 of 0.314 appeared as:

$$A = 4.261 - 0.248 \times S1 \tag{1}$$

In the above 77 dBA group, both the *traffic* ($p < 0.01$) and *natural sounds* ($p < 0.01$) were confirmed to be relevant. A model with an R^2 of 0.350 was found as:

$$A = 3.185 - 0.109 \times S1 + 0.246 \times S4 \tag{2}$$

2. Visual attributes

The analysis of the below 77 dBA group showed that two indicators describing visual environment attributes, *aesthetic* ($p < 0.01$) and *openness* ($p < 0.01$), were evidently related to the overall soundscape comfort. The model generated, with an R^2 of 0.491, was as follows:

$$A = 1.870 + 0.289 \times V1 + 0.118 \times V2 \tag{3}$$

In the above 77 dBA group, the other two indicators of visual aspects, *layering* and *order*, appeared to be closely related to the overall soundscape comfort. A model was constructed, and its R^2 was 0.332:

$$A = 2.758 + 0.182 \times V3 + 0.097 \times V4 \tag{4}$$

3. Acoustic attributes

Correlational analysis results suggested that people's acoustic perceptions evaluating the *monotony* ($p < 0.01$), *harmony* ($p < 0.01$) and *eventfulness* ($p < 0.01$) levels of the surrounding environment were significantly influential indicators of the overall soundscape perception. A model describing their relations was calculated with an R^2 of 0.483 in the dataset under 77 dBA:

$$A = 1.701 + 0.072 \times A2 + 0.354 \times A3 + 0.076 \times A4 \tag{5}$$

In another group above 77 dBA, relevant indicators were the same, but the final model ($R^2 = 0.342$) appeared to be different:

$$A = 2.427 + 0.081 \times A2 + 0.209 \times A3 + 0.112 \times A4 \tag{6}$$

Referring to similar soundscape studies, an R^2 over 0.3 [82,83] provides sufficient reliability for the linear model. The constructed models above describing the relation between the overall soundscape perception and its influencing indicators were, thus, effective. The estimated coefficient values for all indicators and model fitting information are listed in Table 6.

Table 6. Model fit and estimation of B coefficients of models.

Groups		Model Fit (R^2)	Attribute	Estimate B	Standard Error	t-Value	p-Value
Sound types	<77 dBA	0.314	Constant	4.261	0.126	33.771	0.000
			S1	−0.248	0.031	−7.919	0.000
	>77 dBA	0.350	Constant	3.185	0.170	18.706	0.000
			S1	−0.109	0.025	−4.360	0.000
			S4	0.246	0.035	7.132	0.000
Visual attributes	<77 dBA	0.491	Constant	1.870	0.129	14.452	0.000
			V1	0.289	0.036	8.056	0.000
			V2	0.118	0.031	3.796	0.000
	>77 dBA	0.332	Constant	2.758	0.142	19.452	0.000
			V3	0.182	0.034	5.390	0.000
			V4	0.097	0.033	2.966	0.004
Acoustic attributes	<77 dBA	0.483	Constant	1.701	0.145	11.761	0.000
			A2	0.072	0.027	2.652	0.009
			A3	0.354	0.043	8.229	0.000
			A4	0.076	0.029	2.625	0.010
	>77 dBA	0.342	Constant	2.427	0.183	13.246	0.000
			A2	0.081	0.023	3.605	0.000
			A3	0.209	0.046	4.581	0.000
			A4	0.112	0.032	3.474	0.001

4.3.2. Cross Effects of Visual and Acoustic Attributes and Sound Sources

Correlational analysis was also conducted to explore the cross effects of visual and acoustic attributes and sound sources. It was found that the mutual influences between three perceptive aspects appeared to be different in the below and above 77 dBA groups. In the group with sound level below 77 dBA, 42 pairs of indicators were found to be significantly related to each other. In general, there were evident cross effects between visual and aural indicators, most influences being positive. The effects of sound sources on aural and visual indicators, however, varied. *Anthropologic* and *natural sounds* were found to be positively correlated with visual and aural indicators, while negative relations were observed between the other two types of sound sources and visual and aural indicators. Indicator pairs (29) with significant mutual effects were far fewer when sound level was over 77 dBA. Obvious positive relations were also revealed within visual indicators, which could hardly be found among aural indicators. Also, compared to the below 77 dBA group, indicator pairs with negative relations evidently decreased. For both groups, no mutual influences were found between the four types of sound sources. Figure 4.

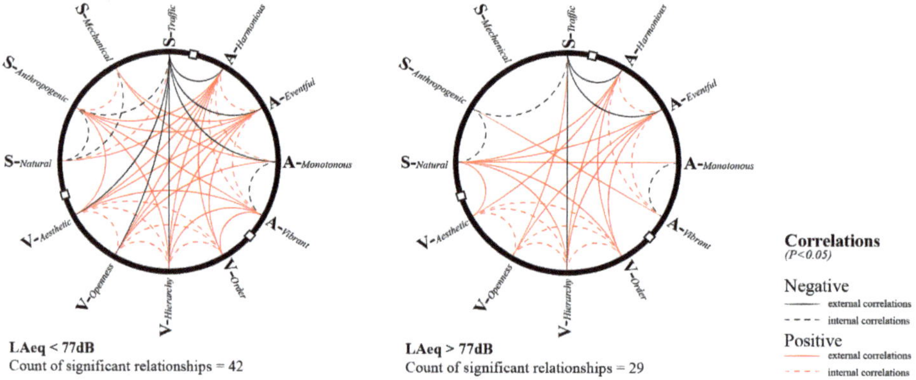

Figure 4. Cross effects of visual and acoustic attributes and sound sources.

5. Discussion and Conclusions

5.1. The Effects of Sound Level on the Overall Soundscape Comfort

This research found that a low level of environmental sound does not corresponding to higher ratings on the overall soundscape comfort. This, again, implies that importance of visual landscape and specific sound attributes in affecting people's perceived soundscape comfort [15]. A typical example is UGS3, which has high level of environmental sound as well as high ratings on the aural-visual attributes and the overall soundscape comfort rating. This means that loud sounds do not necessarily have a negative impact on people's auditory and visual perceptions [84]. In addition, UGS4 has the lowest sound level but its comfort rating only ranks at the third among the five sites. This may be due to the dominance of *mechanical* and *human sounds* on the site, as when the two exceed certain ranges, they are often associated with negative human perceptions [85].

Another important finding is that the environmental sound level of 77 dBA was a turning point in the relation between people's soundscape comfort and its influential indicators in UGSs. The turning point of 77 dBA was different from any similar evidence in previous studies. For example, Yang and Kang found that the subjective evaluation of the sound level generally related well with the mean LAeq, especially when the sound level was below 73 dBA [59]. Below and above the point of 70 dBA was also found to be influential to indicators contributing to human soundscape perception in pocket parks [57]. However, the specific decibel value of turning points reasonably varied in the evaluated environments and participant groups, as it linked with human aural–visual perceptions.

5.2. Different Influential Mechanism below and above 77dBA

In the analysis conducted on the two groups of below and above 77 dBA, it was found that there were obvious differences in terms of specific relevant indicators and their weights in contributing to overall soundscape comfort. *Traffic sound* was found to have negative effects on the overall soundscape comfort in both ranges below and above 77 dBA. The overall soundscape comfort can also be positively adjusted by *natural sound* when the sound level was above 77 dBA. This suggested that the benefit of sound sources to overall soundscape comfort only worked when the sound level passed 77 dBA. Aural perceptive indicators with influences on the overall soundscape comfort were the same in the ranges below and above 77 dBA, which meant that the overall soundscape comfort could be stably promoted by the increase of *openness, order* and *layering*. However, visual attributes contributing to the overall soundscape comfort were completely different in the ranges below and above 77 dBA. Research findings indicated that *aesthetic* and *openness* were important to people's perceived soundscape comfort when the environmental sound level was below 77 dBA, while when the sound level was over 77 dBA, *order* and *layering* appeared to be more important. People's ability to perceive visual aesthetics may have been disrupted by higher levels of ambient sound, when the sense of order and layering of the visual landscape could help them combat interference brought on by the loudness of sound.

5.3. Mutual Influences between Aural and Visual Aspects in UGSs

Audio-visual interactions are proved to be essential in human perceived soundscape comfort in this study. Results disclosed not only whether and how visual attributes, aural attributes and sound sources interact with each other (external correlation), but also how indicators in each perceptive aspect internally influence with each other (internal correlation). It was found that aural attributes, *harmonious* and *eventful* are positively related with all of the four visual indicators, *openness, aesthetic, hierarchy* and *order*, while *vibrance* is only influenced by the last three. Congruent positive effects were also found within the four visual indicators, which is in line with findings in most visual preference studies [86]. Previous studies exploring audio-visual interactions mostly concerned with the different types of sounds and their effects on people perceived visual settings with different landscape characteristics, or vice versa. Very few of them adopted indicators describing people's perceived aural–visual qualities as this study does, possibly due to the difficulty of using subjective indicators to provide direct cues for design improvements. For example, it was proved that the more urban the visual setting, the more negative the sound ratings [87]. The importance of audio-visual interactions is also confirmed in adjusting soundscape perception in China's protected areas [88], and in constructing tranquil spaces [89]. Among those studies discussing the mutual influences between environmental visual and aural attributes, the main focus was on specific sound sources (i.e., children shouting, lawn mowing and birds) [41] and visual contents, such as plants, flora, water and typological features [89]. The universal assessment tool for UGSs soundscape comfort which this study seeks to deliver, therefore, has the advantage of flexible application but also has its limitations in developing specific design improvements.

5.4. Limitations

Though research outcomes appeared to be reasonable and were mostly in line with previous findings, limitations did exist in the study in terms of the experimental design and data analysis. Given that the sound pressure level needed to be measured in a fixed point, the questionnaire survey was also conducted in the designated measurement spots. The different pattern of participants wandering to experience the spots within the site may have led to differences in the evaluation results. In addition, the field survey was carried out in July; the relatively high temperature may have affected people's perceptions [90] as well. The accuracy of the perceptive results may have also been constrained by the number and the specific group (university students) of participants. The open areas in UGS2, UGS3 and UGS5 were much smaller than the other two sites; fewer measurement spots were selected for them. The different numbers of measurement spots and different sizes of open areas in each park are also believed to be influential. As for the data analysis, the overall soundscape comfort was rated with a five-point verbal scale (rating between 1–5), while the sound pressure level was measured by professional apparatus (results between 60–88 dBA). Considering that the range of the sound level was not fixed, no standardization was used to process the data, which may have slightly influenced the identification of the turning points. Moreover, research findings provide no direct relationship between human perceptions and design interventions, thus the identified indicators and the adjusted model can only be used as auxiliary tools in landscape practices. Relevant studies should further explore effective ways of translating people's aural–visual landscape experience into design languages.

5.5. Conclusions

This study explored influential indicators of human soundscape comfort and the influences of aural-visual interaction attributes on people's overall perceived soundscape comfort in urban green spaces. Research findings not only indicated environmental perceptive cues related to people's perceptions of overall soundscape comfort, but a set of six models was also constructed to show how the overall soundscape comfort could be improved through these influential indicators. The outcome of this study not only implies possible design interventions for delivering high-quality urban green spaces, but also extends the applications of audio–visual perceptions in the design of urban public spaces. These can further contribute to enriching human spatial experiences from the perspective of multi-sensory design interventions.

Author Contributions: Conceptualization, Y.Y., Y.H. and Y.S.; methodology, Y.Y. and Y.S.; software, Y.M. and Z.X.; validation, Y.Y. and Y.H.; formal analysis, Y.Y., Y.M. and Z.X.; investigation, Y.M. and Z.X.; resources, Y.S. and Y.H.; data curation, Y.Y. and Y.H.; writing—original draft preparation, Y.Y.; writing—review and editing, Y.Y. and Y.H.; visualization, Y.M. and Z.X.; supervision, Y.S.; project administration, Y.S.; funding acquisition, Y.S. All authors have read and agreed to the published version of the manuscript.

Funding: This research was funded by Nature Science Foundation of China: 51808393; Shanghai Post-doctoral Excellence Program: 2021357; Restorative Urbanism Research Center (RURC), Joint Laboratory for International Cooperation on Eco-Urban Design, Tongji University (CAUP-UD-06); Foreign Expert Program, Ministry of Science and Technology, China (G2022133023L).

Institutional Review Board Statement: The study was conducted in accordance with the Declaration of Helsinki and approved by the Ethics Committee of Tongji University (protocol code 2020tjdx075 and date of approval 09/11/2020).

Informed Consent Statement: Informed consent was obtained from all subjects involved in the study.

Conflicts of Interest: The authors declare no conflict of interest.

References

1. Shao, Y.; Xue, Z.; Yin, Y. Exploration of Public Participation Method from the Humanistic Perspective: Perception Stimulation. *Landsc. Archit.* **2020**, *27*, 116–122.
2. Wei, F. "Sensewalk Mapping" as a Method for Teaching Multi-sensorial Landscape Perception. *Landsc. Archit.* **2021**, *28*, 96–106.
3. Lee, A.C.; Maheswaran, R. The health benefits of urban green spaces: A review of the evidence. *J. Public Health* **2011**, *33*, 212–222. [CrossRef] [PubMed]
4. Lee, A.C.K.; Jordan, H.C.; Horsley, J. Value of urban green spaces in promoting healthy living and wellbeing: Prospects for planning. *Risk Manag. Healthc. Policy* **2015**, *8*, 131. [CrossRef] [PubMed]
5. Hong, A.; Sallis, J.F.; King, A.C.; Conway, T.L.; Saelens, B.; Cain, K.L.; Fox, E.H.; Frank, L.D. Linking green space to neighborhood social capital in older adults: The role of perceived safety. *Soc. Sci. Med.* **2018**, *207*, 38–45. [CrossRef] [PubMed]
6. Maas, J.; Van Dillen, S.M.; Verheij, R.A.; Groenewegen, P.P. Social contacts as a possible mechanism behind the relation between green space and health. *Health Place* **2009**, *15*, 586–595. [CrossRef] [PubMed]
7. Tang, Z.; Liu, B. Progress in Visual Landscape Evaluation. *Landsc. Archit.* **2015**, *09*, 113–120.
8. Shu, X.; Shen, X.; Zhou, X.; Hao, P. Research on Strategies and Determinants of Nature Education Environment Design Based on Landscape Perception Theory. *Landsc. Archit.* **2019**, *26*, 48–53.

9.   Rock, I.; Harris, C.S. Vision and touch. *Sci. Am.* **1967**, *216*, 96–107. [CrossRef]
10. Hong, X. Progress of Contemporary Urban Soundscape Researches. *Landsc. Archit.* **2021**, *28*, 65–70. [CrossRef]
11. Hong, X.; Huang, Z.; Wang, G.; Liu, J. Modeling Long-Term Perceived Soundscape in Urban Parks: A Case Study of Three Urban Parks in Vancouver, Canada. *Landsc. Archit.* **2022**, *29*, 86–91. [CrossRef]
12. The International Organization for Standardization. *PD ISO/TS 12913-2:2018 Acoustics—Soundscape—Part 2: Data Collection and Reporting Requirements*; ISO Copyright Office: Geneva, Switzerland, 2018.
13. Jo, H.I.; Jeon, J.Y. Overall environmental assessment in urban parks: Modelling audio-visual interaction with a structural equation model based on soundscape and landscape indices. *Build. Environ.* **2021**, *204*, 108166. [CrossRef]
14. Nitidara, N.P.A.; Sarwono, J.; Suprijanto, S.; Soelami, F.N. The multisensory interaction between auditory, visual, and thermal to the overall comfort in public open space: A study in a tropical climate. *Sustain. Cities Soc.* **2022**, *78*, 103622. [CrossRef]
15. Liu, J.; Kang, J.; Behm, H.; Luo, T. Effects of landscape on soundscape perception: Soundwalks in city parks. *Landsc. Urban Plan.* **2014**, *123*, 30–40. [CrossRef]
16. Southworth, M.F. The Sonic Environment of Cities. Doctoral Dissertation, Massachusetts Institute of Technology, Boston, MA, USA, 1967.
17. Jo, H.I.; Jeon, J.Y. The influence of human behavioral characteristics on soundscape perception in urban parks: Subjective and observational approaches. *Landsc. Urban Plan.* **2020**, *203*, 103890. [CrossRef]
18. Xiang, Y.; Hedblom, M.; Wang, S.; Qiu, L.; Gao, T. Indicator selection combining audio and visual perception of urban green spaces. *Ecol. Indic.* **2022**, *137*, 108772. [CrossRef]
19. The International Organization for Standardization. *PD ISO/TS Acoustics–Soundscape–Part 1: Definition and Conceptual Framework*; ISO Copyright Office: Geneva, Switzerland, 2014.
20. Biljecki, F.; Ito, K. Street view imagery in urban analytics and GIS: A review. *Landsc. Urban Plan.* **2021**, *215*, 104217. [CrossRef]
21. Ode, Å.; Tveit, M.S.; Fry, G. Capturing landscape visual character using indicators: Touching base with landscape aesthetic theory. *Landsc. Res.* **2008**, *33*, 89–117. [CrossRef]
22. Palmer, S.E.; Schloss, K.B.; Sammartino, J. Visual aesthetics and human preference. *Annu. Rev. Psychol.* **2013**, *64*, 77–107. [CrossRef]
23. Kaplan, R.; Kaplan, S. *The Experience of Nature: A Psychological Perspective*; Cambridge University Press: New York, NY, USA, 1989.
24. Ruddell, E.J.; Hammitt, W.E. Prospect refuge theory: A psychological orientation for edge effect in recreation environments. *J. Leis. Res.* **1987**, *19*, 249–260. [CrossRef]
25. Kaplan, S. Aesthetics, affect, and cognition: Environmental preference from an evolutionary perspective. *Environ. Behav.* **1987**, *19*, 3–32. [CrossRef]
26. Appleton, J. *The Experience of Landscape*; Wiley Chichester: New York, NY, USA, 1996.
27. Maslow, A.; Lewis, K. Maslow's hierarchy of needs. *Salenger Inc.* **1987**, *14*, 987–990.
28. Bourassa, S.C. Toward a theory of landscape aesthetics. *Landsc. Urban Plan.* **1988**, *15*, 241–252. [CrossRef]
29. Schirpke, U.; Tappeiner, G.; Tasser, E.; Tappeiner, U. Using conjoint analysis to gain deeper insights into aesthetic landscape preferences. *Ecol. Indic.* **2019**, *96*, 202–212. [CrossRef]
30. Russell, J.A.; Lanius, U.F. Adaptation level and the affective appraisal of environments. *J. Environ. Psychol.* **1984**, *4*, 119–135. [CrossRef]
31. Grahn, P.; Stigsdotter, U.K. The relation between perceived sensory dimensions of urban green space and stress restoration. *Landsc. Urban Plan.* **2010**, *94*, 264–275. [CrossRef]
32. Stoltz, J.; Grahn, P. Perceived sensory dimensions: An evidence-based approach to greenspace aesthetics. *Urban For. Urban Green.* **2021**, *59*, 126989. [CrossRef]
33. Meng, Q.; Kang, J. Effect of sound-related activities on human behaviours and acoustic comfort in urban open spaces. *Sci. Total Environ.* **2016**, *573*, 481–493. [CrossRef]
34. Brown, A.L. Advancing the concepts of soundscapes and soundscape planning. In Proceedings of the Conference of the Australian Acoustical Society (Acoustics 2011), Gold Coast, Australia, 2–4 November 2011.
35. Barber, J.R.; Burdett, C.L.; Reed, S.E.; Warner, K.A.; Formichella, C.; Crooks, K.R.; Theobald, D.M.; Fristrup, K.M. Anthropogenic noise exposure in protected natural areas: Estimating the scale of ecological consequences. *Landsc. Ecol.* **2011**, *26*, 1281–1295. [CrossRef]
36. Farina, A.; Pieretti, N.; Piccioli, L. The soundscape methodology for long-term bird monitoring: A Mediterranean Europe case-study. *Ecol. Inform.* **2011**, *6*, 354–363. [CrossRef]
37. Pijanowski, B.C.; Farina, A.; Gage, S.H.; Dumyahn, S.L.; Krause, B.L. What is soundscape ecology? An introduction and overview of an emerging new science. *Landsc. Ecol.* **2011**, *26*, 1213–1232. [CrossRef]
38. Harris, C.M. *Handbook of Acoustical Measurements and Noise Control*; McGraw-Hill New York: New York, NY, USA, 1991.
39. Kliuchko, M.; Heinonen-Guzejev, M.; Vuust, P.; Tervaniemi, M.; Brattico, E. A window into the brain mechanisms associated with noise sensitivity. *Sci. Rep.* **2016**, *6*, 39236. [CrossRef] [PubMed]
40. Liu, J.; Kang, J.; Luo, T.; Behm, H.; Coppack, T. Spatiotemporal variability of soundscapes in a multiple functional urban area. *Landsc. Urban Plan.* **2013**, *115*, 1–9. [CrossRef]
41. Liu, J.; Kang, J.; Luo, T.; Behm, H. Landscape effects on soundscape experience in city parks. *Sci. Total Environ.* **2013**, *454*, 474–481. [CrossRef] [PubMed]
42. Kang, J.; Zhang, M. Semantic differential analysis of the soundscape in urban open public spaces. *Build. Environ.* **2010**, *45*, 150–157. [CrossRef]

43. Jeon, J.Y.; Lee, P.J.; You, J.; Kang, J. Perceptual assessment of quality of urban soundscapes with combined noise sources and water sounds. *J. Acoust. Soc. Am.* **2010**, *127*, 1357–1366. [CrossRef]
44. Axelsson, Ö.; Nilsson, M.E.; Hellström, B.; Lundén, P. A field experiment on the impact of sounds from a jet-and-basin fountain on soundscape quality in an urban park. *Landsc. Urban Plan.* **2014**, *123*, 49–60. [CrossRef]
45. Liu, J.; Wang, Y.; Zimmer, C.; Kang, J.; Yu, T. Factors associated with soundscape experiences in urban green spaces: A case study in Rostock, Germany. *Urban For. Urban Green.* **2019**, *37*, 135–146. [CrossRef]
46. Axelsson, Ö.; Nilsson, M.E.; Berglund, B. A principal components model of soundscape perception. *J. Acoust. Soc. Am.* **2010**, *128*, 2836–2846. [CrossRef]
47. Navickas, E. Urban Soundscape: The Relationship between Sound Source Dominance and Perceptual Attributes along with Sound Pressure Levels. Bachelor Thesis, University of Groningen, Groningen, The Netherlands, 2020.
48. Kang, J.; Schulte-Fortkamp, B.; Fiebig, A.; Botteldooren, D. Mapping of soundscape. *Soundscape Built Environ.* **2016**, *161*, 36.
49. Tse, M.S.; Chau, C.K.; Choy, Y.S.; Tsui, W.K.; Chan, C.N.; Tang, S.K. Perception of urban park soundscape. *J. Acoust. Soc. Am.* **2012**, *131*, 2762–2771. [CrossRef]
50. Zhang, X.; Ba, M.; Kang, J.; Meng, Q. Effect of soundscape dimensions on acoustic comfort in urban open public spaces. *Appl. Acoust.* **2018**, *133*, 73–81. [CrossRef]
51. Kang, J.; Yang, W. Soundscape in urban open public spaces. *World Archit.* **2002**, *144*, 76–79.
52. Schulte-Fortkamp, B. The quality of acoustic environments and the meaning of soundscapes. In Proceedings of the 17th International Conference on Acoustics, Rome, Italy, 2–7 September 2001.
53. Moreira, N.M.; Bryan, M. Noise annoyance susceptibility. *J. Sound Vib.* **1972**, *21*, 449–462. [CrossRef]
54. Ellermeier, W.; Eigenstetter, M.; Zimmer, K. Psychoacoustic correlates of individual noise sensitivity. *J. Acoust. Soc. Am.* **2001**, *109*, 1464–1473. [CrossRef] [PubMed]
55. Zimmer, K.; Ellermeier, W. Psychometric properties of four measures of noise sensitivity: A comparison. *J. Environ. Psychol.* **1999**, *19*, 295–302. [CrossRef]
56. Schreckenberg, D.; Griefahn, B.; Meis, M. The associations between noise sensitivity, reported physical and mental health, perceived environmental quality, and noise annoyance. *Noise Health* **2010**, *12*, 7. [CrossRef]
57. Wang, X. *The Optimization Study of the Soundscape for the Urban Park Green Space*; Tianjin University: Tianjin, China, 2014.
58. Maristany, A.; López, M.R.; Rivera, C.A. Soundscape quality analysis by fuzzy logic: A field study in Cordoba, Argentina. *Appl. Acoust.* **2016**, *111*, 106–115. [CrossRef]
59. Yang, W.; Kang, J. Acoustic comfort evaluation in urban open public spaces. *Appl. Acoust.* **2005**, *66*, 211–229. [CrossRef]
60. Ministry of Ecology and Environment of the People's Republic of China. *Environmental Quality Standard for Noise*; China Environmental Science Press: Beijing, China, 2008.
61. Jo, H.I.; Jeon, J.Y. Effect of the appropriateness of sound environment on urban soundscape assessment. *Build. Environ.* **2020**, *179*, 106975. [CrossRef]
62. Aumond, P.; Can, A.; De Coensel, B.; Ribeiro, C.; Botteldooren, D.; Lavandier, C. Global and continuous pleasantness estimation of the soundscape perceived during walking trips through urban environments. *Appl. Sci.* **2017**, *7*, 144. [CrossRef]
63. Motoyama, Y.; Hanyu, K. Does public art enrich landscapes? The effect of public art on visual properties and affective appraisals of landscapes. *J. Environ. Psychol.* **2014**, *40*, 14–25. [CrossRef]
64. Jo, H.I.; Jeon, J.Y. Perception of urban soundscape and landscape using different visual environment reproduction methods in virtual reality. *Appl. Acoust.* **2022**, *186*, 108498. [CrossRef]
65. Liu, J.; Xiong, Y.; Wang, Y.; Luo, T. Soundscape effects on visiting experience in city park: A case study in Fuzhou, China. *Urban For. Urban Green.* **2018**, *31*, 38–47. [CrossRef]
66. Lee, L.-H. Perspectives on landscape aesthetics for the ecological conservation of wetlands. *Wetlands* **2017**, *37*, 381–389. [CrossRef]
67. Plit, J.; Myga-Piątek, U. The degree of landscape openness as a manifestation of cultural metamorphose. *Quaest. Geogr.* **2014**, *33*, 145–154. [CrossRef]
68. Liu, Y.; Li, L. Mountainous city featured landscape planning based on GIS-AHP analytical method. *ISPRS Int. J. Geo-Inf.* **2020**, *9*, 211. [CrossRef]
69. Zhang, G.; Yang, J.; Wu, G.; Hu, X. Exploring the interactive influence on landscape preference from multiple visual attributes: Openness, richness, order, and depth. *Urban For. Urban Green.* **2021**, *65*, 127363. [CrossRef]
70. Kawai, K.; Kojima, T.; Hirate, K.; Yasuoka, M. Personal evaluation structure of environmental sounds: Experiments of subjective evaluation using subjects' own terms. *J. Sound Vib.* **2004**, *277*, 523–533. [CrossRef]
71. Brown, A.; Kang, J.; Gjestland, T. Towards standardization in soundscape preference assessment. *Appl. Acoust.* **2011**, *72*, 387–392. [CrossRef]
72. Liu, J.; Yang, L.; Xiong, Y.; Yang, Y. Effects of soundscape perception on visiting experience in a renovated historical block. *Build. Environ.* **2019**, *165*, 106375. [CrossRef]
73. Nilsson, M.; Botteldooren, D.; De Coensel, B. Acoustic indicators of soundscape quality and noise annoyance in outdoor urban areas. In Proceedings of the 19th International Congress on Acoustics, Madrid, Spain, 2–7 September 2007.
74. Song, X.; Lv, X.; Yu, D.; Wu, Q. Spatial-temporal change analysis of plant soundscapes and their design methods. *Urban For. Urban Green.* **2018**, *29*, 96–105. [CrossRef]
75. Cao, X.; Hsu, Y. The Effects of Soundscapes in Relieving Stress in an Urban Park. *Land* **2021**, *10*, 1323. [CrossRef]

76. Lyons, E. Demographic correlates of landscape preference. *Environ. Behav.* **1983**, *15*, 487–511. [CrossRef]
77. Hartig, T.; Staats, H. Linking preference for environments with their restorative. *Landsc. Res. Landsc. Plan. Asp. Integr. Educ. Appl.* **2005**, *12*, 279.
78. Staats, H.; Jahncke, H.; Herzog, T.R.; Hartig, T. Urban options for psychological restoration: Common strategies in everyday situations. *PLoS ONE* **2016**, *11*, e0146213. [CrossRef] [PubMed]
79. Kurakata, K.; Mizunami, T.; Sato, H.; Inukai, Y. Effect of ageing on hearing thresholds in the low frequency region. *J. Low Freq. Noise Vib. Act. Control* **2008**, *27*, 175–184. [CrossRef]
80. Bahalı, S.; Tamer-Bayazıt, N. Soundscape research on the Gezi Park–tunel square route. *Appl. Acoust.* **2017**, *116*, 260–270. [CrossRef]
81. Hinton, P.R. *Statistics Explained*; Routledge: London, UK, 2014.
82. Yang, M. *A Review of Regression Analysis Methods: Establishing the Quantitative Relationships between Subjective Soundscape Assessment and Multiple Factors*; Universitätsbibliothek der RWTH Aachen: Aachen, Germany, 2019.
83. Lionello, M.; Aletta, F.; Kang, J. A systematic review of prediction models for the experience of urban soundscapes. *Appl. Acoust.* **2020**, *170*, 107479. [CrossRef]
84. Hong, J.Y.; Lam, B.; Ong, Z.-T.; Ooi, K.; Gan, W.-S.; Kang, J.; Yeong, S.; Lee, I.; Tan, S.-T. The effects of spatial separations between water sound and traffic noise sources on soundscape assessment. *Build. Environ.* **2020**, *167*, 106423. [CrossRef]
85. Green, M.; Murphy, D. Environmental sound monitoring using machine learning on mobile devices. *Appl. Acoust.* **2020**, *159*, 107041. [CrossRef]
86. Hong, J.Y.; Jeon, J.Y. Designing sound and visual components for enhancement of urban soundscapes. *J. Acoust. Soc. Am.* **2013**, *134*, 2026–2036. [CrossRef]
87. Viollon, S.; Lavandier, C.; Drake, C. Influence of visual setting on sound ratings in an urban environment. *Appl. Acoust.* **2002**, *63*, 493–511. [CrossRef]
88. Xu, X.; Wu, H. Audio-visual interactions enhance soundscape perception in China's protected areas. *Urban For. Urban Green.* **2021**, *61*, 127090. [CrossRef]
89. Pheasant, R.J.; Fisher, M.N.; Watts, G.R.; Whitaker, D.J.; Horoshenkov, K.V. The importance of auditory-visual interaction in the construction of 'tranquil space'. *J. Environ. Psychol.* **2010**, *30*, 501–509. [CrossRef]
90. Guan, H.; Hu, S.; Lu, M.; He, M.; Mao, Z.; Liu, G. People's subjective and physiological responses to the combined thermal-acoustic environments. *Build. Environ.* **2020**, *172*, 106709. [CrossRef]

Article

Effects of Soundscapes on Human Physiology and Psychology in Qianjiangyuan National Park System Pilot Area in China

Peng Wang [1], Youjun He [1], Wenjuan Yang [1], Nan Li [2,3] and Jiaojiao Chen [2,*]

1 Research Institute of Forestry Policy and Information, Chinese Academy of Forestry, Beijing 100091, China
2 Economic Research Institute, International Centre for Bamboo and Rattan, Beijing 100102, China
3 China Eco-Culture Association, Beijing 100102, China
* Correspondence: cjj@icbr.ac.cn

Abstract: The development of China's national parks is still in the initial stage, and few scholars have studied the effects of soundscapes on human physiology and psychology from the perspective of the auditory senses in national parks. In this study, the Qianjiangyuan National Park System Pilot Area was taken as the research subject, physiological indicators of subjects were collected through a biopAC-MP150 multi-channel physiological instrument data platform, and the subjective psychological response of soundscapes was measured using a Likert scale. The results showed that the sound of water had the most significant effect on the heart rate and respiratory rate of the subjects. Agricultural sound had the greatest impact on the skin conduction levels, while conversation had the least overall impact on human physiology. There were significant differences in comfort, excitement, and significance among the different soundscapes ($p < 0.001$). The sounds of insects are more likely to elicit feelings of comfort and excitement, while the sounds of birds are more likely to arouse curiosity. No significant correlation was observed between the physiological indices and psychological indices. The study on the effects of different soundscapes on human physiology and psychology in China's national parks will provide a basis for the decision makers of national parks to formulate more effective planning, design, and management policies regarding soundscapes.

Keywords: soundscape; national parks; human physiology; human psychology

Citation: Wang, P.; He, Y.; Yang, W.; Li, N.; Chen, J. Effects of Soundscapes on Human Physiology and Psychology in Qianjiangyuan National Park System Pilot Area in China. *Forests* **2022**, *13*, 1461. https://doi.org/10.3390/f13091461

Academic Editors: Xin-Chen Hong, Jiang Liu and Guang-Yu Wang

Received: 30 August 2022
Accepted: 6 September 2022
Published: 11 September 2022

Publisher's Note: MDPI stays neutral with regard to jurisdictional claims in published maps and institutional affiliations.

1. Introduction

The concept of soundscapes was first proposed by Raymond Murray Schafer, a Canadian musician and environmentalist [1], known as "the pioneer of acoustic culture research". The soundscape emphasizes the perception and understanding of the sound environment by individuals or society [2,3]. Although the acoustic environment and landscape have the same acoustic source components, there are some differences between them. Acoustic environment, which is mainly defined by the physical properties of sounds, refers to the sound produced by various creatures in the environment or venue for communication purposes. In contrast, a soundscape is mainly defined by human perception, human experience, and sound reconstruction. The essence is the interaction between human and the acoustic environment [4]. Under the influence of physical attributes, the perceptual features of various types of soundscapes are obviously different. Chen et al. believed that pleasure is a comprehensive index to measure the perceptual features of a soundscape, and pointed out that the factors determining pleasure mainly included physical attributes, visual scenes, humanistic value, and individual characteristics [5]. Based on existing research results [6–9], it can be seen that a soundscape with more natural elements such as birdsong and water sound provides a higher overall pleasure, while that with more noise elements such as traffic and machinery offer a lower overall pleasure. The prevailing theory of soundscape stems from two sources. First, the cultural needs of humans require an auditory input to generate different aesthetic feelings. People's perception of the landscape does not depend solely on vision, but also on the overall sensory responses such as taste, sight, hearing, touch, and smell. An overemphasis on vision comes at the expense of other sensory organs. Second, agricultural sounds and traffic noises of the industrial age have triggered the desire to focus on the acoustic environment, and social development has created a closer relationship between humans and nature. Contemporary soundscape research has mainly focused on related theories and methods of the International Organization for Standardization (ISO). In 2014, 2018, and 2019, the ISO issued the International Standards for soundscapes, which specify the definition, data acquisition, and evaluation standards of soundscapes [3,10,11]. According to the definition of the ISO, a soundscape is an acoustic environment as perceived or experienced and/or understood by a person or people, in context. As a whole, the relationship between soundscape and the human body is very close [12].

Studies on the soundscape of national parks began in the late 1970s in the United States. The National Park Service (NPS, USA) established the Natural Sounds and Night Skies Division (NSNSD), which is committed to protecting, maintaining, and restoring the soundscapes of national parks. The protection and management of soundscapes emphasize the silence of nature or wilderness. The management and control of soundscapes in national parks have focused on the noise caused by human interference [13]. Park et al. constructed a data model of a soundscape and pedestrian space in Rocky Mountain National Park, and found that 49.6% of the tourist group enjoyed natural sounds for 15 min if the sound was below the 35 dB noise threshold [14]. Marin et al. adopted the dose–response method to analyze the relationship between the tourists' motivation to visit national parks and the degree of acceptance of different sounds, revealing that the acceptance of tourists grew with the quietness of the environment [15]. Beal et al. investigated the perception of noise by campers in Australia's Queensland National Park and reported that agricultural sound was the most unacceptable sound type to humans [16].

In recent decades, the number of people with suboptimal health has rapidly increased; therefore, there is an increasing desire for access to nature and more outdoor space, especially during the COVID-19 lockdown period. A growing number of researchers have turned their attention to the relationship between natural soundscapes and people's health. Hume and Ahtamad studied the influence of different soundscapes on people's moods and emotional stimulation. They reported that sounds with a positive perception affected the respiratory rates, while sounds with a negative perception impacted the electrical activity of the skin. This direct relationship between pleasantness and respiratory rate response is greater in males than in females. The electromyography readings increased in both males and females with unpleasant sound clips [17]. Medvedev et al. investigated the effects of soundscapes on the physiological indices after stress stimulation and a rest period, confirming that the sound environment had a positive effect on people's health and well-being [18]. Erfanian et al. summarized the psychophysiological implications of soundscapes. They found that although some studies found a link between the physiological effect of soundscapes and perceptual attributes, most of the research merely focused on one or two physiological reactions. Unfortunately, this link has not been fully verified [19]. Wang et al. studied the impact of soundscape perception behavior on vision and confirmed that sound frequency, sound preference, and auditory satisfaction had an impact on visual perception [20]. Li et al. used heart rate (HR), heart rate variability (HRV), and other indices to explore the tendency of the physiological indices and subjective recoverability in environmental audio–visual interaction [21]. Such studies have paved the way for future research on soundscapes and human physiology and psychology.

China did not establish national parks until relatively recently, so national parks are still an original concept in this country. There are few studies on the health benefits of soundscapes, and most related soundscape studies are conducted in urban areas such as parks, campuses, and urban forests [22]. For example, Zhang et al. selected three types of urban forests and analyzed the ecological health functions of urban forests through the continuous monitoring of six environmental indices: temperature, relative air humidity, wind speed, CO_2 concentration, air negative oxygen ion concentration, and noise [23]. Weng et al. invited 30 college students to participate in audiovisual evaluations in the campus landscape of the Fujian Agriculture and Forestry University in order to explore the impact of the green soundscape on mood and attention [24]. Yang et al. examined the differences in the psychology of tourists with different voice preferences, who visited the ancient town of Lijiang [25]. Chen et al. studied the health care benefit of sound in forest therapy. By analyzing human psychology and emotional state in forest sound, they believed that birdsong, water sounds, and sounds of wind blowing leaves could exert a high positive impact on human psychology [26]. Song et al. selected two environments, urban and forest, and confirmed that forest sound therapy had a moderating effect on the negative emotions of tired working ladies, and forest sound could alleviate fatigue symptoms, but the effect of short-term forest recuperation was limited [27]. However, few scholars have studied the effects of soundscapes on human physiology and psychology from the perspective of auditory senses in national parks.

Therefore, in order to better reveal the effects of different types of soundscapes in the national park on the human body, by taking the Qianjiangyuan National Park System Pilot Area (hereinafter referred to as the system pilot area) as the research subject, two research questions were proposed: (1) Are there any differences in the physiological and psychological effects of different soundscapes on the public in system pilot areas? (2) What are the effects of soundscapes on public physiology and psychology in the system pilot areas? By analyzing the changes in the physiological indices and the data of the psychological indices, this study explored the association of soundscapes with human physiology and psychology in the system pilot area. The results will further reveal the differences in the physiological and psychological effects of various soundscapes and provide new insights for the evaluation and restorative soundscape planning and design of national parks. In addition, this study could help managers master the benefits of soundscapes in various regions, and provide theoretical support and data basis for the management of ecosystems in the future by soundscape analysis and ecological acoustic methods.

2. Materials and Methods

2.1. Target Area

As one of the 10 pilot areas of the national park system in China, the system pilot area covered three regions: the Gutian Mountain National Nature Reserve, the Qianjiangyuan Provincial Scenic Area, and the Qianjiangyuan National Forest Park. Located in Western Zhejiang Province (East longitude 118°01′–118°37′, North latitude 28°54′–29°30′), the target area lies at the junction of Zhejiang, Anhui, and Jiangxi. With a land area of approximately 252.38 km^2, it can be divided into four functional areas: core protection area, ecological conservation area, recreational exhibition area, and traditional use area (Figure 1). Among them, the acreage of the core protection area, ecological conservation area, recreational exhibition area, and traditional use area is 72.33 km^2, 135.80 km^2, 8.12 km^2, and 36.13 km^2, respectively.

The system pilot area is a typical vegetation transition zone between South China and North China, which has complex terrains, dense forests, and crossed river valleys. It is home to some ancient and extensive low-elevation, mid-subtropical, evergreen, broad-leaved forests, which are rare throughout the world. This area covers four towns (Suzhuang, Changhong, Hetian, and Qixi) and 72 natural villages (19 administrative villages), with a total population of 9744 [28].

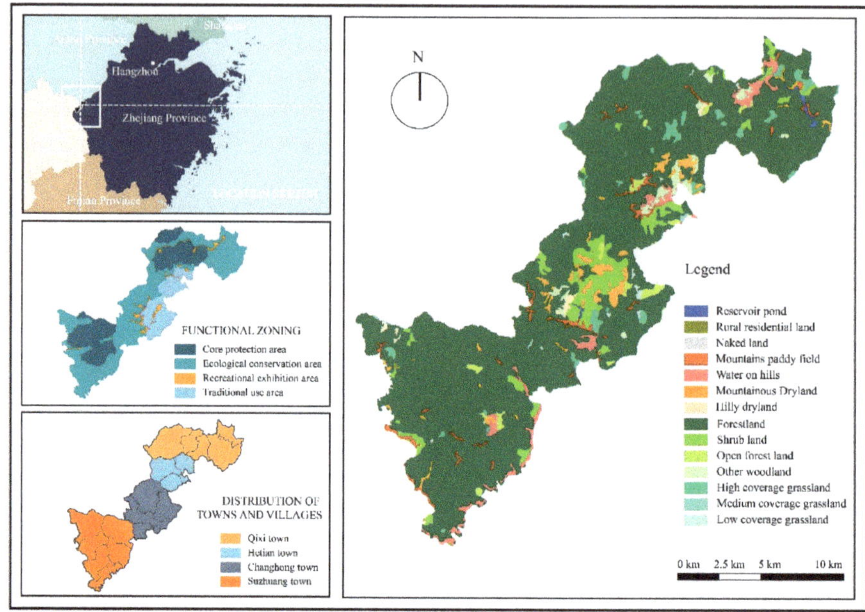

Figure 1. Status of the system pilot area. Note: Land use data from the Resources and Environmental Science and Data Center, Chinese Academy of Sciences (https://www.resdc.cn/, accessed on 3 September 2022). Functional zoning data were obtained from the Qianjiangyuan National Park Administration.

2.2. Method and Process

2.2.1. Sound Collection and Material Selection

The existing physiological test equipment is not suitable for outdoor use due to changeable terrains, large altitude difference, a forest coverage rate of 81.7%, and a complex natural environment. Therefore, this study used audios for indoor physiological and psychological tests to exclude natural conditions (basic conditions of the transformation and formation of natural factors rather than human factors such as terrains, climates, etc.) and other factors (human interference factors for outdoor experiments such as heavy equipment). The audios used in this study were obtained on sunny and breezy days from September to November 2020, and the audio data were acquired at 8:00–10:30 and 15:00–17:00. A Sony PCM m 10 recording pen was placed on a tripod 1.5 m above the ground. The microphone was kept vertical to the main wind direction of the recording. The typical sampling method was used in the system pilot area to ensure the representativeness of the soundscape in light of factors such as forests, rivers, other ecosystems, bird biodiversity, community distribution patterns, and recreational spaces. The data were collected along the public recreational route, taking into account the typical ecological areas and important public living places. Fifty-three typical areas were selected as sample plots for sound collection, and the samples were marked as T1, T2, T3...T53. The sound collection was repeated three times in each plot for a duration of 20–30 s. Among the 53 samples, T6, T11 and T32 had the largest sound levels, which were 70.71 dB(A), 66.9 dB(A), and 62.64 dB(A), respectively. T12, T16, T25, and T52 had the smallest sound levels, which were all 30 dB(A). According to the geographical coordinates of the samples, the Kriging interpolation method was used to generate a spatial distribution map of the sound pressure levels in ArcGIS (Figure 2). Considering the various sound interference factors, the selected sound material was clipped to about 10 s.

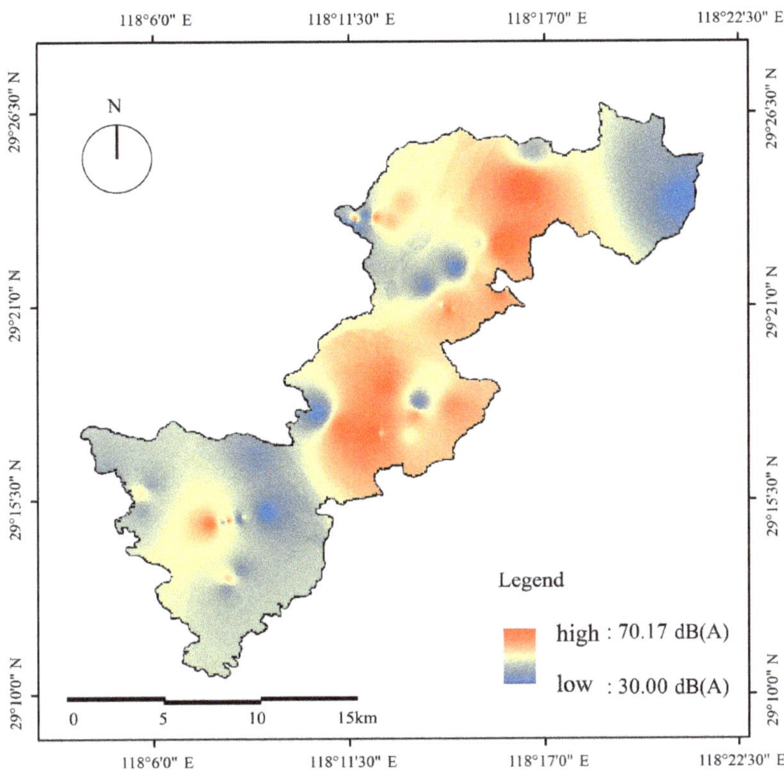

Figure 2. The spatial distribution of sound levels in the soundscape of the system pilot area.

2.2.2. Selection of Study Participants

Previous studies have shown that college students are ideal subjects for participating in landscape evaluations [29,30]. In this study, 96 undergraduates, M.A students, and PhD students from the School of Economics, School of Humanities and Arts of Renmin University of China and the Chinese Academy of Forestry were randomly chosen as volunteers for the experiment. Due to sensor failures and the participants' last-minute withdrawals, 64 volunteers (25 males and 39 females) participated in the experiment. Specifically, 26 undergraduate students, 34 master students, and four doctoral students who studied ecology, forestry, art, design, economics, finance, and literature, etc. were involved. The age bracket was mainly between 21 and 25 (Table 1). To ensure the validity of the physiological monitoring and psychological response data, the study participants were asked about their physical conditions, and their answers were recorded before the start of the experiment.

Table 1. The demographic characteristics of the study participants.

Project	Group	Number of Subjects	Percentage (%)
Gender	Male	25	39%
	Female	39	61%
Age	18–20 years old	23	36%
	21–25 years old	38	59%
	26–30 years old	3	5%
Education Level	Undergraduates	26	41%
	Master's Degree Candidate	34	53%
	Doctoral Candidate	4	6%
Place of Origin	Eastern Region	29	45%
	Central Regions	11	17%
	Western Regions	24	38%
Major	Art Major	19	30%
	Forestry Major	26	40%
	Other Majors	19	30%
Universities and Colleges	Chinese Academy of Forestry	38	59%
	Renmin University of China	26	41%

2.2.3. Classification of Forest Soundscapes

This study investigated the types of soundscapes spanning the sounds of birds, insects, running water, and wind-blown leaves to mixed natural sounds as well as unnatural sounds such as footsteps, conversations, frolicking, agricultural sounds, and traffic sounds. To classify soundscapes, the authors consulted nine experts in ecology, acoustics, landscape architecture, humanities, etc. The classification was repeated three times to ensure the accuracy of the identification of the sound sources. When multiple sounds were superimposed, the loudness, frequency, and duration of the sound determined its final classification.

After consultation, the sound landscape of the system pilot area was finally divided into three types: natural sounds, artificial sounds, and mixed natural and artificial sounds (Table 2). Among them, natural sounds mainly include insect sounds, bird sounds, water sounds, and mixed natural sounds.

Table 2. The classification of soundscapes.

Project	Primary Classification	Secondary Classification
Soundscape of the national park	Natural sounds	Insect sounds Birdsong Water sounds Mixed natural sounds
	Artificial sounds	Agricultural sounds Conversation
	Mixing natural and artificial sounds	—

2.2.4. Selection of Physiological and Psychological Indices

Selection of the physiological indices. The physiological indices include heart rate, respiratory rate, and skin conductance level. They were measured by a BIOPAC physiological multi-conductance meter (model MP150) manufactured by the BIOPAC company in the United States. The modules for collecting electrical signals included electrocardio (ECG), electroencephalogram (EEG), electrogastrogram (EGG), electromyography (EMG), electro-oculogram (EOG), and electroretinogram (ERS), etc. The heart rate is the number of pulse beats per minute. A change in heart rate reflects the activity level of the sympathetic nervous system and the parasympathetic nervous system. When an individual is in a state of rest or relaxation, the function of the parasympathetic nervous system increases, and the heart rate slows down. When an individual is in a state of excitement or stress, the excitability of the sympathetic nervous system increases, that of the parasympathetic nervous system decreases, and the heart rate increases. In a natural environment, an increase in the heart rate of the subjects indicates that the subject is excited and happy [31]. The respiratory rate is the number of breaths per minute, and studies show that this index is affected by the environment. When a person is in a pleasant or unpleasant mental state, their respiratory rate is significantly different from the norm [31]. Electrical skin activity refers to changes in the function of the sweat glands of the skin, reflecting the activity of the sympathetic nerves. This activity is used as an index of emotional and cognitive loading [32]. The increased secretion of the sweat glands increases the electrical conductivity and the skin conductance levels. The skin conductance levels decrease when a person is mentally relaxed [33]. In general, the heart rate and skin conductivity reflect the human body's autonomous activities, and both are affected by the environment. A person's respiratory rate is significantly different when the human body is in either a happy and unhappy state. As the above physiological indices provide insight into the potential physiological and psychological processes of subjects, they are important indices of emotional stimulation.

Selection of the psychological indices. ISO 12913 (2014, 2018, 2019) provides eight adjective attributes for the evaluation of a soundscape-based pleasantness and eventfulness model. Referring to the existing research results [31,32,34,35], combined with the Profile of Mood State (POMS), the scale was compiled by Grow J.R. and revised by Professor Zhu Beili of East China Normal University in 1994 [36]. Finally, the comfort, excitement, and distinctiveness of a certain sound were selected as the evaluation indices. The Likert scale (5, 4, 3, 2, 1) was used for the evaluation. Before scoring the subjects' psychological state, the staff explained the process to the subjects. As the subjects were well-educated college students, they had a good understanding of the scoring method of the Likert scale. To prevent the recent psychological state of the subjects from affecting the objectivity of the research results, all subjects were measured using the SCL-90 scale, which is a popular tool for assessing the college students' mental health [37]. The results show that the scores of each factor of the test group were at the normal level, indicating that the recent mental health status of the test group was good. Hence, the subjects in the test group were suitable for this sound test.

2.3. Process

This research experiment was carried out in the Standard Laboratory of the Chinese Academy of Forestry. One staff member and one participant entered the laboratory during each experiment. The laboratory experiment was conducted on sunny days from December 2020 to January 2021 at 8:00–10:00 and 15:00–17:00 to minimize the influence of external noise. The laboratory's sound insulation effect ensured that the whole process of the experiment was quiet and the indoor temperature was kept between 18 °C and 22 °C. The experiment adopted a ThinkPad X1 laptop (screen size: 14 inches; manufacturer: Lenovo; location: Shenzhen, Guangdong, China) and a JBL-FLIP6 speaker (power was 20–40 W, and the vocal channel was 2.0; manufacturer: Harman; location: Dongwan, Guangdong, China). As the experiment was in the peak of the COVID-19 epidemic, this study did not use conventional earphones, but used speakers to play soundscape samples given the disinfection requirements of epidemic prevention and control, and the large number of subjects (96 subjects). The volume of the loudspeaker was consistent with the sound level dB(A) monitored by field recording. The laboratory temperature was between 20 °C and 25 °C, and the relative humidity was 51%–56% RH.

The process of the research experiment consisted of four stages (Figure 3): preparation and introduction, pre-investigation, data collection, and subjective evaluation. During the first stage, participants provided their name, gender, age, college, major, hometown, and other basic information before the test. The participants were asked about their physical condition, especially their mental health, medication, and hearing. Then, they were briefly informed of the content and procedures of the research experiment, which helped to reduce any stress and increase their acceptance of the proceedings.

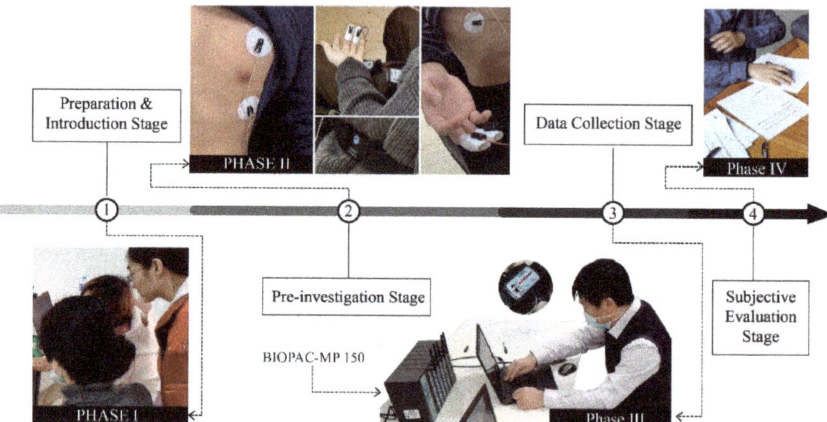

Figure 3. A schematic diagram of the experiment flow.

In the second stage, participants had physiological sensors with disposable electrode pads attached to their skin after it was sanitized with alcohol and saline. During the data collection period, they were asked to remain relaxed and still and refrain from talking.

In the third stage, the sound materials were supplemented with a photo of the landscape. In order to highlight the typicality and representativeness of the sound sample, this study seamlessly connected and scrolled the samples of 10 s, with a total duration of 60 s. The playback mode was seamless rolling playback, and the experimental time was controlled by the subjects. There was a silent sample between every two sound samples. The baseline data were the physiological data of the silent samples, with the time controlled by the subjects.

In the fourth stage, participants evaluated the psychological indices (comfort, excitement, and distinctiveness) of the sound based on their own subjective judgement.

2.4. Data Processing and Analysis

All data collected in this study were analyzed by Microsoft Excel 2010 and IBM SPSS Statistics 21. Physiological data, which were defined as the physiological data of the audio samples minus the physiological data of previously silent samples, were analyzed and exported using the BIOPAC-MP150 platform. The main statistical methods were one-way ANOVA and multiple comparisons. Psychological data came from the questionnaire survey. In order to ensure the quality of the data, the authors conducted a reliability and validity analysis.

The results showed that Cronbach's alpha was 0.756, falling between 0.6 and 0.8, which suggested the internal consistency of the data. The results of the Kaiser–Meyer–Olkin (KMO) and Bartlett's test of sphericity were obtained by factor analysis. The KMO was 0.666, close to 0.7. The chi-square value of Bartlett's test of sphericity was 2600.654, and the degrees of freedom were 3, which were significant at the 95% or even 99% confidence level, respectively, and indicated the high effectiveness of the data. In addition, a Pearson correlation analysis was used to study the correlation between the variables.

3. Results

3.1. Effects of Different Soundscapes on Human Physiology

3.1.1. Effects on Heart Rate Changes

Differences in the heart rate of the subjects were observed between different types of soundscapes in the system pilot area (Table 3). Through a single-factor analysis of variance, significant differences were observed between the subjects' heart rate data for the seven soundscapes ($p < 0.05$). Compared with the silent environment, the heart rates of the participants increased by 1.28 bpm, 1.62 bpm, 2.45 bpm, 2.43 bpm, 0.59 bpm, 2.15 bpm, and

1.65 bpm, respectively, when they listened to sounds of insects, birds, water, mixed natural sounds, conversations, agricultural sounds, and mixed natural and unnatural sounds. Overall, the subjects' heart rates showed the largest increase when they heard water sounds and mixed natural sounds and the smallest increase when they heard conversations.

Table 3. A comparison of the physiological indices of the subjects for different soundscapes.

Soundscape Category	Heart Rate Changes (bpm)	Skin Conductance Level Changes (µs)	Respiratory Rate Changes (resp)
Overall difference	0.035 *	0.135	0.182
	Differences between soundscape types		
Insect sounds	1.28 ± 1.05 [ab]	−3.51 ± 1.11 [a]	0.87 ± 0.35 [a]
Birdsong	1.62 ± 1.11 [bc]	−3.60 ± 0.99 [a]	0.97 ± 0.72 [a]
Water sounds	2.45 ± 0.42 [a]	−3.68 ± 1.22 [a]	1.37 ± 0.35 [a]
Agricultural sounds	2.15 ± 0.14 [abc]	−4.27 ± 0.20 [a]	1.25 ± 0.29 [a]
Conversation	0.59 ± 1.44 [c]	−2.98 ± 1.17 [a]	0.84 ± 0.74 [a]
Mixed natural sounds	2.43 ± 0.55 [c]	−4.14 ± 0.18 [a]	1.36 ± 0.39 [a]
Mixing natural and artificial sounds	1.65 ± 0.84 [abc]	−3.81 ± 0.56 [a]	1.03 ± 0.43 [a]

Note: * represents the significant difference at the level of 0.05. Data = mean ± SD (standard deviation). Data with different letters in the same column are significantly different.

3.1.2. Effects on Human Skin Conductance Level Changes

The skin conductance of the subjects responded differently to different types of forest soundscapes in the pilot area (Table 3). Compared with the silent environment, their skin conductance level decreased by 3.51 µs, 3.60 µs, 3.68 µs, 4.14 µs, 2.98 µs, 4.27 µs, and 3.81 µs when they listened to sounds of insects, birds, water, mixed natural sounds, conversations, agricultural sounds, and mixed natural and unnatural sounds, respectively. Overall, the skin conductance level showed the largest decrease when the participants heard agricultural sounds and the smallest decrease when they heard conversations.

3.1.3. Effects on Respiratory Rate Changes

Differences were observed in the subjects' respiratory rates for different types of soundscapes in the system pilot area (Table 3). Compared with the silent environment, their respiratory rates increased by 0.87 resp, 0.97 resp, 1.37 resp, 1.36 resp, 0.84 resp, 1.25 resp, and 1.03 resp when they heard sounds of insects, birds, water, mixed natural sounds, conversations, agricultural sounds, mixed natural, and unnatural sounds, respectively. Overall, the respiratory rate showed the most significant increases when the participants listened to water sounds and the smallest increase when they listened to conversations.

3.2. Effects of Different Soundscapes on Human Psychology

A single-factor analysis of variance was performed on the subjects' psychological response data for the seven different types of soundscapes (Table 4). The results showed significant differences in comfort, excitement, and distinctiveness for the seven soundscapes ($p < 0.001$). Comfort (3.92), excitement (3.32), and distinctiveness in response to insect sounds (2.91) had the highest scores (psychological cognitive score), and comfort (1.92), excitement (2.14), and distinctiveness (1.88) in response to conversation sounds had the lowest scores. The overall scores in terms of the comfort of natural sounds such as insects, birds, mixed natural sounds, and water sounds were higher than those of agricultural sounds, conversation sounds, natural, and unnatural sounds. Similar trends were observed for the index of excitement. However, there were differences in the distinctiveness of the seven soundscapes. Bird sounds were more likely to cause subjects to experience the distinctiveness of this sound. Overall, natural sound received higher scores than unnatural and mixed sounds in the matrix of comfort, excitement, and distinctiveness.

Table 4. The effects of different soundscapes on human psychology.

Soundscape Category	Comfort Index	Excitement Index	Distinctiveness Index
Overall difference	0.000 **	0.000 **	0.000 **
	Differences between soundscape types		
Insect sounds	3.92 ± 0.34 [c]	3.32 ± 0.32 [e]	2.91 ± 0.19 [d]
Birdsong	3.68 ± 0.58 [a]	3.29 ± 0.45 [ab]	2.92 ± 0.28 [b]
Water sounds	3.33 ± 0.45 [ab]	3.08 ± 0.33 [a]	2.73 ± 0.25 [a]
Agricultural sounds	2.48 ± 0.58 [c]	2.28 ± 0.31 [e]	2.26 ± 0.17 [d]
Conversation	1.92 ± 0.37 [c]	2.14 ± 0.20 [de]	1.88 ± 0.21 [cd]
Mixed natural sounds	3.43 ± 0.41 [c]	2.80 ± 0.08 [cd]	2.55 ± 0.23 [bc]
Mixing natural and artificial sounds	2.61 ± 0.59 [b]	2.63 ± 0.35 [bc]	2.39 ± 0.278 [b]

Note: ** represents the significant difference at the level of 0.01. Data = mean ± SD (standard deviation). Data with different letters in the same column are significantly different.

3.3. Correlation between Physiological Variation and Psychological Response

The test results showed a significant correlation between heart rate and skin conductance level (p = 0.018), and between heart rate and respiratory rate (p = 0.001). The correlation coefficient between heart rate and skin conductance was −0.839, and the significant level was above 0.05. The correlation coefficient between heart rate and respiratory rate was 0.948, and the significant level was above 0.01. In addition, there was a significant correlation between comfort and excitement (p = 0.001), and between comfort and distinctiveness (p < 0.001). Their correlation coefficients were 0.956 and 0.968, respectively. Both passed the significance test at 1%. Moreover, excitement was significantly correlated with distinctiveness (p < 0.001), and the correlation coefficient was 0.977, which passed the significance test at 1%. No significant correlation was observed between the physiological indices and psychological indices. Hence, heart rate has a significant negative correlation with the skin conductance level. The correlations between all of the other indices were strongly positive (Table 5).

Table 5. The correlation between the physiological changes and psychological indices.

	Skin Conductance Level Changes	Respiratory Rate Changes	Comfort Index	Excitement Index	Distinctiveness Index
Heart rate changes	−0.839 *	0.948 **	0.369	0.241	0.383
Skin conductance level changes		−0.749	−0.194	−0.002	−0.196
Respiratory rate changes			0.135	0.000	0.122
Comfort index				0.956 **	0.968 **
Excitement index					0.977 **

Note: ** indicates the significance level of 1%, * indicates the significance level of 5%.

4. Discussion

4.1. Effects of Soundscapes in System Pilot Area on Physiology

This study found significant changes in the heart rate (p < 0.05) of subjects when they listened to different types of soundscapes in the system pilot areas of national parks in China. However, the changes in the respiratory rates and skin conductance levels were not significantly different. The subjects had increased heart rates and respiratory rates as well as decreased skin conductance levels when they were exposed to various sounds.

The increase in the heart rate indicates that the sympathetic nerve activity increased, resulting in the excitation or happiness of the subjects. Hume and Ahtamad recorded the heart rate of 80 participants who listened to different sounds and found that after listening to pleasant sounds, the participants' heart rates increased [17]. Natural soundscapes play a positive role in changes in heart rates. Water sounds had the most profound influence on the heart rate, and conversation sounds the least influence. The system pilot area is an important area under water source protection and contains the source of the Qiantang River, Zhejiang Province's "Mother River" and "water tower", which has nurtured the budding Chinese civilization in the Qiantang River Basin. The region is rich in water resources, and known for the numerous rivers, lakes, and waterfalls. This directly affects the relationship between water sounds and the human heart rate. Buxton et al. surveyed the natural soundscapes in the 221 locations of 68 U.S. national parks. They arrived at the conclusion that water sounds were the most effective at increasing positive emotions and maintaining health [38]. Their conclusions were consistent with that of Barton and Pretty: the natural environment, especially the water environment, provides important health benefits for the public [39]. This means that not only national parks but also soundscapes have ecological functions in humans. The results in this study confirm the conclusions of Buxton, Barton, and Pret-ty et al. on the natural soundscape, but they are also incomplete to some extent. The main reason is that although the study area is a national park, there are a certain number of inhabitants in it, and the places they live in are relatively concentrated. Therefore, artificial soundscapes such as conversation sound and agricultural sound has also been the focus of research. This study confirms that the impact of conversation sounds on heart rate variation is minimal, but this has not been fully confirmed as it is affected by the positioning of national parks in different countries (there are no or few inhabitants in most national parks).

Skin conductance activity is an electrical phenomenon that occurs due to functional changes in the skin sweat glands. It reflects the activity of the sympathetic nerves and is an index of emotional and cognitive load [40]. The changes in the skin conductance level of the subjects for different types of soundscapes indicate that agricultural sound had the most significant impact and conversation sound had the least impact on the skin conductance response. The pilot area has a population of 9744 and covers many reservoirs such as Bijiahe Reservoir and Qixi Reservoir. Most local villagers make a living through agricultural and forestry production. Agricultural sound is related to water conservancy projects and vehicles as well as other agricultural and livestock activities. These sounds are loud, disruptive, and last for extended periods of time, affecting the skin conductance response of the subjects. Li et al. conducted a tourist survey on the soundscape of Meiling National Forest Park. They found that loud noise irritated people and a very quiet soundscape made people uneasy and nervous [41]. Zhu et al. reported significant differences in the skin conductance levels of subjects in different forest environments, and unnatural sounds (e.g., footsteps) had a significant impact on the skin conductance level [33]. Although the research results of Li and Zhu et al. on artificial soundscapes could not fully support the conclusions of this study, they were sufficient to confirm that soundscapes with various artificial elements exerted a more typical impact on human skin conductance activity. Therefore, policy makers should ensure relaxing and pleasant experiences for visitors through soundscape improvement, in order to plan forest trails and tourist routes of national parks.

The analysis of the effect of different types of soundscapes indicated that a mixture of water and natural sounds had the largest influence on the subjects' respiratory rates and conversation sounds had the least influence. Cui et al. established a soundscape comfort model for the Zhongshan Scenic Area. They reported that a soundscape with a moderate sound level was comfortable for people, and an overly noisy soundscape made people irritable [42]. Bradley et al. found happy people and unhappy people had stark differences in their respiratory rate [43]. The research results of Gomez and Danuser showed that the respiratory rate increased when people listened to music after listening to noises [44]. They also found that compared with a silent environment, natural sounds evoke more pleasant feelings and increase the respiratory rates. This study concluded that it is not comprehensive to explore the effects of a national park soundscape on respiratory rate changes from the perspective of various types of soundscapes. According to the research results, although the change in respiratory frequency caused by water sound was the largest while that caused by conversation sound was the smallest, it does not mean that natural sounds are more likely to cause changes in respiratory frequency than artificial sounds. This is mainly because the impact of natural soundscapes such as insect songs and bird chirps on the change in respiratory frequency is also small as a whole. Therefore, referring to the research conclusions of Bradley, Gomez, and Danus-er et al., this study believes that the impact of soundscapes on respiratory frequency may be related to the volume of soundscape and emotional elements, which needs to be further demonstrated in future studies.

4.2. Effects of Soundscapes in System Pilot Area on Psychology

Significant differences in the three psychological indices (comfort, excitement, and distinctiveness) ($p < 0.001$) were observed among the different types of soundscapes. In general, the sounds of insects, birds, water, mixed natural sounds, and other natural sounds of the forest soundscapes had a positive impact on human psychology. The sound of insects evoked feelings of comfort and excitement, while the sound of birdsong was more likely to intrigue subjects. Our research results are consistent with those of previous studies [45]. Forest sounds that include insect sounds as the main component have a relatively high sound frequency, and those with bird sounds as the main component have a high variability and melodiousness. The pilot area is known as a "Mysterious Primitive Forest and Paradise of Wild Bird". According to local biodiversity statistics (data provided by the National Park Service), it is home to 238 species of wild birds in 17 orders and 63 families in the pilot area. For instance, Syrmaticus ellioti, a national I-level key protected species, is abundant here. In addition, bird species such as Lophura nycthemera, Streptopelia orientalis, and Carrulax canorus are common. Wang et al. analyzed the psychological indices of college students in urban areas, woodlands, flower fields, and near water bodies and found that the values of the psychological indices were greatly varied in woodlands and water landscapes and less variable in urban areas [46]. This finding indicates that the blue–green natural space has a positive impact on human psychology. The sound of water has therapeutic effects [47]; it masks noises while increasing the sense of tranquility and calms emotions in people [48].

This study preliminarily proved that the soundscapes of national parks dominated by natural sounds can lift people's mood and deliver a more pleasant experience, and invoke a more pleasant feeling. This study attempts to reveal the different psychological responses to sounds when external interference is excluded. For future research, it is necessary to use controllable simulations and more advanced technologies to achieve a more comprehensive evaluation.

4.3. Effects of Different Types of Soundscapes in National Parks on Human Body

Physiological and psychological data show that the soundscapes of national parks may be related to the people's physical state. The soundscapes of national parks can change people's physiological and psychological states and have an impact on people's physical health to a certain extent. Compared with a silent environment, soundscapes increase heart rates, reduce the skin conductance level, and increase respiratory rates. In addition, we found significant differences in psychological indices such as excitement and distinctiveness between the different types of soundscapes. This result indicates that the participants' sympathetic nervous system excitement level increased, and the excitability of the parasympathetic nervous system decreased, resulting in a relatively pleasant and excited mental state. This conclusion is consistent with Hao et al.'s findings that the heart rates increased and the skin conductance levels decreased when people heard pleasant sounds [31]. Annerstedt et al. explored the physiological recovery status of subjects in different soundscape environments by collecting their cardiovascular data and saliva cortisol. The results showed that the subjects' parasympathetic nerve excitability significantly changed in a natural acoustic environment, indicating that this environment is conducive to people's physiological recovery and a potential connection exists between the natural soundscape and stress recovery [45]. This study confirms, from a physiological perspective, that sound stimulation related to the forest environment has better health benefits than a silent environment or visual stimulation [49].

On the other hand, there were differences in the health benefits of different types of soundscapes. The sound of water and mixed natural sounds had the largest impact on the health benefits, which may be related to the landscape features of the pilot area. The pilot area is rich in water resources, and most of the roads are surrounded by rivers and waterfalls; thus, the sound of the water is heard in most areas. In recent years, water landscapes have played an increasingly important role in natural therapy. Tedja et al. reported that listening to the sound of running water helped people to meditate and facilitated the relaxation of the body and mind [50]. This shows that waterscapes (water sound) require attention in the future planning and management of the soundscape of national parks. In addition, the system pilot area sets to leverage its water resources and sounds to develop a forest-centered therapy center. Mixed natural sounds generally contain more than two types of sounds such as the rhythmic sounds of birds, insects, water, and wind. Therefore, a soundscape characterized by loudness, a strong rhythm, and high variability inspires excitement. Buxton et al. argue that the sound of running water helps to regulate emotions and maintain health, and bird sounds reduce stress and annoyance [38], which is consistent with the results of this study. From the aspect of psychological response, insects and bird sounds are more abundant than other sounds; therefore, they have a larger influence on the health benefits than quiet soundscapes and monotonous, noisy, and unnatural sounds. Ratcliffe et al. conducted semi-structured interviews with 20 subjects and found that birdsong was the most suitable natural sound type for releasing stress and restoring attention, which was of significant potential for stress recovery [51]. Our study also verified the decompression theory of Ulrich et al. [52], who observed that a positive environment substantially relaxed people, reduced the sense of pressure, and provided a positive physiological response.

4.4. Research Prospects and Deficiencies

Few systematic studies have been conducted on the improvement in human physiology and psychology by soundscapes in national parks. In this study, the health effects of various forest soundscapes were clarified by investigating the psychological and physiological changes caused by different forest soundscapes in the system pilot area. This experiment only used a soundscape evaluation, and not a visual evaluation, to reveal the different reactions of the subjects after they heard forest sounds without external interference. However, this approach has some limitations. For instance, Ratcliffe et al. found that the soundscape of green spaces had higher health benefits, which might be related to the high vegetation coverage [51]. Watts et al. found that the higher health benefits of blue spaces were related to the restorative benefits of water sounds as well as visual landscape features [53]. In addition, due to the limitations of the existing physiological and psychological indicators, the existing research results cannot fully cover the impact of soundscapes on the human body. Subsequent research will incorporate more typical physiological indicators such as the heart rate variability to assess the effects of soundscape more comprehensively.

Limitations exist in this study as it was restricted by the existing objective experiments. Although ecological validity of the subjects' response to sample stimulation was ensured in the research on soundscape perception in a simulated indoor environment, the complex acoustic environment was inevitably driven by other factors, which exerted a certain impact on the results. Therefore, although our conclusions enjoyed statistical significance, whether it has universal significance needs further verification. In later research, more advanced technical means will be used to create a controllable simulated natural environment to produce a more comprehensive evaluation. The researchers of this study are aware of the necessity for further research and advanced technology to create a more realistic environment for simulation. Some researchers believe that the results of indoor experiments differ from those obtained in a natural environment. For example, Ishiyama analyzed the adverse effects of traffic noise on subjects, and observed that the irritability of the subjects caused by noise was related to the number of sound samples. This study selected college students as subjects, which also have a demographic bias. In the future, the authors will invite a wider scope of stakeholders such as residents of national park communities or foreign tourists to participate in this soundscape research. In addition, high-frequency sounds were more likely to cause the subjects to feel irritable [54]. In the future, our group plans to explore the use of virtual reality to investigate audiovisual forest landscapes in national parks. In addition, the authors will collaborate with ornithologists and entomologists as well as scientists of other fields to analyze the effects of different bird and insect sounds on physiology and psychology.

5. Conclusions

This study took the Qianjiangyuan National Park System Pilot Area as the object, analyzed the impact of soundscape stimulation on human physiological changes and psychological responses, and discussed the relationship between physiology and psychology based on soundscape stimulation. The following conclusions were drawn. (1) Significant changes in the heart rate ($p = 0.003$) of subjects exposed to different types of soundscapes. However, the changes in the respiratory rate and skin conductance level were not significantly different. The soundscape influenced the physiological responses of the subjects to varying degrees. The heart rate and the respiratory rate increased, and the skin conductance level decreased. The sound of water had the most significant influence on the heart rate and respiratory rate. Agricultural sound substantially affected the skin conductance level, and the sound of conversation had the least overall impact on the subjects' physiology. (2) The soundscape caused changes in the subjects' psychology to different degrees. Different types of soundscapes caused significant differences in comfort, excitement, distinctiveness, and other psychological indices. The sound of insects made people feel comfortable and excited, and the sounds of birds made them feel curious. Unnatural sounds resulted in the lowest scores. (3) No significant correlation was observed between the physiological indices and psychological indices. Except for a significant negative correlation between the change in heart rate and the change in the skin conductance level, all other indicators presented a positive correlation.

This study clarified the influence of different types of soundscapes in national parks on human physiological and psychological changes, and revealed the interrelationship between physiological changes and psychological responses, thus promoting the inclusion of the effects of soundscapes on the human body in national park management. Based on the research results, the following two suggestions are put forward: (1) Make full use of the existing terrain, optimize the hearing experience of national parks through the occlusion function of the terrain, and create an attractive soundscape; and (2) birdsong and underwater sounds have certain therapeutic functions. We suggest planning tree species in key areas that can attract birds and bees, optimizing plant configuration, and creating a natural soundscape through biological means. By exploring the impact of different types of soundscapes on the psychological and physiological changes, this study proves that the soundscapes of national parks have an impact on people's health. In future research, we will continue to explore the impact of mixed sounds (e.g., the mixed sounds of birds, those of insects, and those of other creatures) on people's health and investigate the audio–visual interaction in national parks, paying attention to both audio signals and visual behavior.

Author Contributions: Conceptualization, P.W. and J.C.; Methodology, P.W.; Software, W.Y.; Validation, N.L.; Formal analysis, Y.H.; Investigation, W.Y.; Resources, N.L.; Data curation, W.Y.; Writing—original draft preparation, P.W. and J.C.; Writing—review and editing, P.W. and J.C.; Visualization, N.L.; Supervision, J.C.; Project administration, N.L.; Funding acquisition, P.W. All authors have read and agreed to the published version of the manuscript.

Funding: This research was funded by the National Natural Science Foundation of China (52008389) and the Fundamental Research Funds of ICBR (Grant No. 1632021030).

Data Availability Statement: Not applicable.

Acknowledgments: All authors are very grateful to the Qianjiangyuan National Park Administration in China who provided the data. We also thank three classmates (Le Li, Haodong Liu, and Zhiqiang Gao) for their help on the paper.

Conflicts of Interest: The authors declare no conflict of interest.

References

1. Raymond, M.S. *The Tuning of the World*; Knopf: New York, NY, USA, 1977; pp. 274–275.
2. Truax, B. *The World Soundscape Project's Handbook for Acoustic Ecology*; A.R.C. Publications: Vancouver, BC, Canada, 1978; pp. 6–16.
3. *ISO 12913-1: 2014*; AcousticsSoundscape-Part 1: Definition and Conceptual Framework. ISO: Geneva, Switzerland, 2014. Available online: https://www.iso.org/obp/ui/#iso:std:iso:12913:-1:ed-1:v1:en (accessed on 3 September 2022).
4. Xu, X.Q.; Yang, R.; Newman, P.; Taft, D. A review of national park sound-scape studies. *Chin. Landsc. Archit.* **2016**, *32*, 25–30.
5. Chen, K.A.; Lu, J.; Yang, X.L.; Berglund, B. An experimental study on the dimension number of park Soundscape perception attributes. *Noise Vib. Control.* **2009**, *29*, 132–137.
6. Hong, X.C.; Wang, G.Y.; Liu, J.; Lan, S.R. Cognitive persistence of soundscape in urban parks. *Sustain. Cities Soc.* **2019**, *51*, 17–26. [CrossRef]
7. Hong, X.C.; Liu, J.; Wang, G.Y.; Jiang, Y.; Wu, S.T.; Lan, S.R. Factors influencing the harmonious degree of soundscapes in urban forests: A comparison of broad-leaved and coniferous forests. *Urban For. Urban Green.* **2019**, *39*, 18–25. [CrossRef]
8. Bian, J.Y.; Leng, J.H.; Zhao, Z.; Wang, Y.B. Research on urban residents' perception and evaluation of greenway soundscape. *For. Resour. Manag.* **2021**, *1*, 111–120.
9. He, M.; Pang, H. Research and progress of acoustic landscape. *J. Landsc. Archit.* **2016**, *5*, 88–97.
10. *ISO/TS 12913-2:2018*; AcousticsSoundscape-Part 2: Data Collection and Reporting Requirements. ISO: Geneva, Switzerland, 2018. Available online: https://www.iso.org/obp/ui/#iso:std:iso:ts:12913:-2:ed-1:v1:en (accessed on 3 September 2022).
11. *ISO/TS 12913-3:2019*; AcousticsSoundscape-Part 3: Data analysis. ISO: Geneva, Switzerland, 2019. Available online: https://www.iso.org/obp/ui/#iso:std:iso:ts:12913:-3:ed-1:v1:en (accessed on 3 September 2022).
12. Wang, Y.Z. How to move towards auditory culture: The spatial and temporal breakthrough of auditory sense and the discussion of aesthetic subjectivity. *Cult. Stud.* **2018**, *1*, 17–28.
13. Miller, N.P. US National Parks and management of park soundscapes: A review. *Appl. Acoust.* **2008**, *69*, 77–92. [CrossRef]
14. Park, L.; Lawson, S.; Kaliski, K.; Newman, P.; Gibson, A. Modeling and Mapping Hikers' Exposure to Transportation Noise in Rocky Mountain National Park. *Park Sci.* **2010**, *26*, 59–64.
15. Marin, L.D.; Newman, P.; Manning, R.; Vaske, J.J.; Stack, D. Motivation and Acceptability Norms of Human-Caused Sound in Muir Woods National Monument. *Leis. Sci.* **2011**, *33*, 147–161. [CrossRef]
16. Beal, D.J. Campers' attitudes to noise and regulation in Queensland national parks. *Aust. Parks Recreat.* **1994**, *30*, 38–40.
17. Hume, K.; Ahtamad, M. Physiological responses to and subjective estimates of soundscape elements. *Appl. Acoust.* **2013**, *74*, 275–281. [CrossRef]
18. Medvedev, O.N.; Shepherd, D.; Hautus, M.J. The restorative potential of soundscapes: A physiological investigation. *Appl. Acoust.* **2015**, *96*, 20–26. [CrossRef]
19. Erfanian, M.; Mitchell, A.J.; Kang, J.; Aletta, F. The Psychophysiological Implications of Soundscape: A Systematic Review of Empirical Literature and a Research Agenda. *Int. J. Environ. Res. Public Health* **2019**, *16*, 3533. [CrossRef] [PubMed]
20. Wang, P.; Zhang, C.; Xie, H.; Yang, W.; He, Y. Perception of National Park Soundscape and Its Effects on Visual Aesthetics. *Int. J. Environ. Res. Public Health* **2022**, *19*, 5721. [CrossRef] [PubMed]
21. Li, Z.Z.; Ba, M.H.; Kang, J. Effect of soundscape on physiological indices of audio-visual interactions in urban environments. *Sustain. Cities Soc.* **2021**, *75*, 103360. [CrossRef]
22. Song, J.W.; Ma, H.; Yin, F. Review of Soundscape. *Noise Vib. Control.* **2012**, *32*, 16–20.
23. Zhang, Z.Y.; Ye, B.; Yang, J.; Chen, Q.J.; He, Q.J.; Dong, J.H. Monitoring of Dynamic Changes of Ecological Health Functions of Urban Forests in Hangzhou. *J. Northwest For. Univ.* **2014**, *29*, 31–36+255.
24. Weng, Y.X.; Zhu, Y.J.; Dong, J.Y.; Wang, M.H.; Dong, J.W. Effects of Soundscape on Emotion and Attention on Campus Green Space—A Case Study of Fujian Agriculture and Forestry University. *Chin. Landsc. Archit.* **2021**, *37*, 88–93.
25. Yang, L.L.; Zhang, J.; Xu, Y.F.; Chen, X. Soundscape subjective evaluation based on tourists' preference: A case study of Dayan ancient town. *J. Appl. Acoust.* **2020**, *39*, 625–631.
26. Chen, F.P.; Wang, Y.Q.; Li, H. Study on mental health function of forest natural soundscape. *For. Econ.* **2016**, *38*, 95–99.
27. Song, C.; Li, Y.; Zhang, Y.J.; Wu, J.P. Effect of forest therapy on mental health of fatigued working women. *Environ. Occup. Med.* **2022**, *39*, 168–173.

28. Development and Reform Commission of Zhejiang Province. Master Plan for the Pilot Area of Qianjiangyuan National Park System (2016–2025). People's Government of Zhejiang Province 2017. Available online: https://www.zjzwfw.gov.cn. (accessed on 16 July 2022).
29. Stamps, A. Demographic Effects in Environmental Aesthetics: A Meta-Analysis. *J. Plan. Lit.* **1999**, *14*, 155–175. [CrossRef]
30. Wang, Y.; Chen, X.F. Application of psychophysical method in evaluation of foreign forest landscapes. *Sci. Silvae Sin.* **1999**, *5*, 110–117.
31. Hao, Z.Z.; Wang, C.; Xu, X.H.; Wang, Y.Y.; Zhang, C.; Duan, W.J.; Wang, Z.Y.; Jin, Y.B. Effect of Shenzhen Urban Forest Soundscape on Psychological and Physiological Changes of Human Beings. *J. Northwest For. Univ.* **2019**, *34*, 231–239.
32. Li, Z.Z.; Kang, J. Sensitivity analysis of changes in human physiological indices observed in soundscapes. *Landsc. Urban Plan.* **2019**, *190*, 103593. [CrossRef]
33. Zhu, Y.J.; Weng, Y.X.; Fu, W.C.; Dong, J.W.; Wang, M.H. Effects of soundscape perception on health benefits of forest parks: A case study of Fuzhou National Forest Park. *Sci. Silvae Sin.* **2021**, *57*, 9–17.
34. Romero, V.R.; Maffei, L.; Brambilla, G.; Ciaburro, G. Modelling the soundscape quality of urban waterfronts by artificial neural networks. *Appl. Acoust.* **2016**, *111*, 121–128. [CrossRef]
35. Romero, V.R.; Ciaburro, G.; Brambilla, G.; Garzón, C. Representation of the soundscape quality in urban areas through colours. *Noise Mapp.* **2019**, *6*, 8–21. [CrossRef]
36. Zhang, W.Y. Applicability of POMS on Female college students during physical entertainment. *J. Xi'an Phys. Educ. Univ.* **2009**, *26*, 508–512.
37. Zhan, L.Y.; Lian, Q.; Wang, F. A study on the changes of Freshmen 'mental health in an agricultural and Forestry University from 2010 to 2014. *Chin. J. Sch. Health* **2016**, *37*, 708–710+715.
38. Buxton, R.T.; Pearson, A.L.; Allou, C.; Fristrup, K.; Wittemyer, G. A synthesis of health benefits of natural sounds and their distribution in national parks. *Proc. Natl. Acad. Sci. USA* **2021**, *118*, 2013097118. [CrossRef] [PubMed]
39. Barton, J.; Pretty, J.N. What is the best dose of nature and green exercise for improving mental health? A multi-study analysis. *Environ. Sci. Technol.* **2010**, *44*, 3947. [CrossRef] [PubMed]
40. Li, Z.Z.; Kang, J.; Ba, M. Influence of distance from traffic sounds on physiological indices and subjective evaluation. *Transp. Res. Part D Transp. Environ.* **2020**, *87*, 102538. [CrossRef]
41. Li, H.; Wang, Y.Q.; Chen, F.P. Evaluation of Tourist Survey of Soundscape in Meiling National Forest Park. *Sci. Silvae Sin.* **2018**, *54*, 9–15.
42. Cui, Z.H.; Yang, X.Y. Evaluation of soundscape in the core area of Dr. Sun Yat-sen's Mausoleum in Nanjing. *J. Nanjing For. Univ.* **2019**, *43*, 121–127.
43. Bradley, M.M.; Lang, P.J. Affective reactions to acoustic stimuli. *Psychophysiology* **2010**, *37*, 204–215. [CrossRef]
44. Gomez, P.; Danuser, B. Affective and physiological responses to environmental noises and music. *Int. J. Psychophysiol.* **2004**, *53*, 91–103. [CrossRef]
45. Annerstedt, M.; Jönsson, P.; Wallergård, M.; Johansson, G.; Karlson, B.; Grahn, P.; Hansen, A.M.; Währborg, P. Inducing physiological stress recovery with sounds of nature in a virtual reality forest-results from a pilot study. *Physiol. Behav.* **2013**, *118*, 240–250. [CrossRef]
46. Wang, Q.; Zhang, Y.L.; Zhao, R.L.; Niu, L.X. Study on the Effects of Four Campus Green Landscapes on College Students' Physiological and Psychological Indices. *Chin. Landsc. Archit.* **2020**, *36*, 92–97.
47. White, M.P.; Pahl, S.; Ashbullby, K.; Herbert, S.; Depledge, M. Feelings of restoration from recent nature visits. *J. Environ. Psychol.* **2013**, *35*, 40–51. [CrossRef]
48. Watts, G.R.; Pheasant, R.J. Tranquillity in the Scottish Highlands and Dartmoor National Park—The importance of soundscapes and emotional factors. *Appl. Acoust.* **2015**, *89*, 297–305. [CrossRef]
49. Deng, L.; Luo, H.; Ma, J.; Huang, Z.; Sun, L.X.; Jiang, M.Y.; Zhu, C.Y.; Li, X. Effects of integration between visual stimuli and auditory stimuli on restorative potential and aesthetic preference in urban green spaces. *Urban For. Urban Green.* **2020**, *53*, 126702. [CrossRef]
50. Tedja, Y.W.; Tsaih, L. Water soundscape and listening impression. *J. Acoust. Soc. Am.* **2015**, *138*, 1750. [CrossRef]
51. Ratcliffe, E.; Gatersleben, B.; Sowden, P.T. Bird sounds and their contributions to perceived attention restoration and stress recovery. *J. Environ. Psychol.* **2013**, *36*, 221–228. [CrossRef]
52. Ulrich, R.S.; Simons, R.F.; Losito, B.D.; Fiorito, E.; Miles, M.; Zelson, M. Stress recovery during exposure to natural and urban environments. *J. Environ. Psychol.* **1991**, *11*, 201–230. [CrossRef]
53. Watts, G.R.; Pheasant, R.J.; Horoshenkov, K.V.; Ragonesi, L. Measurement and subjective assessment of water generated sounds. *Acta Acust. United Acust.* **2009**, *95*, 1032–1039. [CrossRef]
54. Ishiyama, T.; Hashimoto, T. The impact of sound quality on annoyance caused by road traffic noise: An influence of frequency spectra on annoyance. *JSAE Rev.* **2000**, *21*, 225–230. [CrossRef]

Article

Temporal and Spatial Characteristics of Soundscape Ecology in Urban Forest Areas and Its Landscape Spatial Influencing Factors

Yujie Zhao [1], Shaowei Xu [1], Ziluo Huang [1], Wenqiang Fang [1], Shanjun Huang [1], Peilin Huang [1], Dulai Zheng [1], Jiaying Dong [2], Ziru Chen [3], Chen Yan [1], Yukun Zhong [1] and Weicong Fu [1,4,*]

1   College of Landscape Architecture and Art, Fujian Agriculture and Forestry University, 15 Shangxiadian Rd., Fuzhou 350000, China
2   School of Architecture, Clemson University, Clemson, SC 29634, USA
3   College of Architecture and Urban Planning, Fujian University of Technology, 33 Xuefunan Rd., Fuzhou 350118, China
4   Collaborative for Advanced Landscape Planning, Faculty of Forestry, The University of British Columbia, Vancouver, BC V6T 1Z4, Canada
*   Correspondence: weicongfufj@163.com

Abstract: We explored the spatial and temporal characteristics of the urban forest area soundscape by setting up monitoring points (70 × 70 m grid) covering the study area, recorded a total of 52 sound sources, and the results showed that: (1) The soundscape composition of the park is dominated by natural sounds and recreational sounds. (2) The diurnal variation of sound sources is opposite to that of temperature, 6:00–9:00 is the best time for the public to perceive birdsong, and after 18:00, the park is dominated by insect chirps. (3) The PSD (power spectral density) and the SDI (soundscape diversity index) of the park are greatly affected by public recreation behaviors, and some recreation behaviors may affect the vocal behavior of organisms such as birds. (4) Spaces with high canopy density can attract more birdsong and recreational sounds in summer, and the combination of "tree + lake" can attract more birdsong. Vegetation has a significant dampening effect on traffic sound. (5) Landscape spatial elements, such as the proportion of hard ground, sky, trees, and shrubs, have a significant impact on changes in the PSD, the SDI and different kinds of sound sources. The research results provide effective data support for improving the soundscape of urban forests.

Keywords: sound source composition; ecosystem acoustics; temporal and spatial change; landscape space

Citation: Zhao, Y.; Xu, S.; Huang, Z.; Fang, W.; Huang, S.; Huang, P.; Zheng, D.; Dong, J.; Chen, Z.; Yan, C.; et al. Temporal and Spatial Characteristics of Soundscape Ecology in Urban Forest Areas and Its Landscape Spatial Influencing Factors. *Forests* **2022**, *13*, 1751. https://doi.org/10.3390/f13111751

Academic Editor: Paloma Cariñanos

Received: 30 September 2022
Accepted: 19 October 2022
Published: 24 October 2022

Publisher's Note: MDPI stays neutral with regard to jurisdictional claims in published maps and institutional affiliations.

1. Introduction

Urban forests are important places for the public to get a close experience with nature, and the ways to improve the quality of urban forest recreation have received increasing attention and become the focus of many scholars [1,2]. As the construction of urban forests has entered the stage of quality improvement, managers have shifted from the creation of visual landscapes to a comprehensive improvement of the sensory experience [3–6], and soundscape is an important part of the senses. Ecosystem sounds create a soundscape comprised of acoustic periodicities and frequencies emitted from the ecosystem's biophysical entities. These sounds are acoustic signals that reflect the dynamics of biological, social, and physical systems of a landscape. Soundscape ecology is "the study of systematic relationships between humans, organisms, and their sonic environment". As an important resource ranked only second to the visual landscape, the soundscape has gradually received attention in the optimization design of urban forests [7,8]. Understanding and making proper use of the sounds in the environment around us are directly related to the livability of living environments and the quality of life of residents.

The soundscape varies in different landscape spaces. Scholars have conducted various studies on urban forests and other public open spaces, which mainly focused on soundscape changes [9–13] and quality evaluation [14–20]. Ali Jahani et al. assessed the role of birdsong components in the psychological recovery of urban visitors and developed a decision support system as a practical tool [21]. Hong Xinchen et al. explored the impact of soundscape-driving factors on the perception of birdsong and established a perceived birdsongs model (PBM) that effectively simulated the process from soundscape information to perceived birdsongs in urban forests. The methodology developed could be useful for soundscape assessment and conservation in urban forests [22]. Liu Jiang et al. showed that landscape features greatly affect the public's perception of soundscapes and human-made sounds dominate urban soundscapes in the time-space dimension, but birdsongs and certain geophysical sounds also play an important role in soundscape [23]. Hao Zezhou et al. studied the variation characteristics and influencing factors of birdsongs and insect chirps in urban forests, as well as analyzed the activity patterns of urban biomes based on soundscape records [24]. Scholars have achieved fruitful achievements in the protection of green space biodiversity, improvement of public recreational benefits, and the reduction in noise; however, the existing research mainly uses the soundwalk method or sets up monitoring points in typical recreational areas to collect and analyze the acoustic data, and therefore, the results of studies are of weak guiding significance to the whole park.

We record sounds autonomously at multiple places in the park throughout the day and providing a window on the ecological spatiotemporal continuum, thereby provide a tool to capture the data needed to assess ecological integrity over time. This survey was able to visualize temporal patterns of landscape acoustic signals throughout the day to examine changes in the soundscape over time. Our study uses Hot Spring Park, located in the core area of Fuzhou city, during the summer as an example. To provide references for the management and optimization of the public space soundscape in urban core areas, monitoring sites in Hot Spring Park were selected by dividing grids and we focused on solving the following problems: (1) identify the structural features of urban forest sound sources, (2) analyze the spatial and temporal variation characteristics of the park soundscape, and (3) explore the differences in soundscape characteristics and influencing factors of various landscape spaces.

2. Materials and Methods

2.1. Study Location

Fuzhou was awarded the title of the "Chinese National Forest City" in 2017 and the "National Forest Tourism Model City" in 2018. Hot Spring Park is located in the city center of the intersection of "Riding Greenway" and "Inner River Greenway" in Fuzhou, which belongs to the core of urban forest [25]. The Hot Spring Park covers an area of about 10 hm^2(119°18′34″ E~119°18′55″ E, 26°5′43″ N~26°5′59″ N). The park contains high species richness, with over 2000 big trees and 200 kinds of plants being planted here, and contains dozens of plant community structures, such as palm forest, evergreen coniferous forest, evergreen broad-leaved forest and deciduous broad-leaved forest [26]. With the increase in development in the city, Hot Spring Park has become a place for exercise, leisure, and entertainment for the public. In recent years, Hot Spring Park has undergone continuous upgrades and transformations, which provides the possibility for future soundscape integration.

2.2. Soundscape Monitoring and Data Acquisition

2.2.1. Soundscape Monitoring

According to previous studies, the feasible distance between recording equipment is 100 m [27–29]. We divided Hot Spring Park into grids that were equally distanced at 70 × 70 m [27] and selected 31 monitoring points (Figure 1). After our pre-experiment in July 2021, we randomly selected 3 sunny and breezy days for acoustic data collection (contains the weekend) so that our experiment would be more consistent with the actual situation of the park soundscape. Monitoring points were collected by placing a Sony PCM-D100 recording device in the center of a grid; if the monitoring point was inaccessible, an appropriate offset was made. Sampling occurred between 6:00 and 22:00 (park opening hours), and the sampling frequency was set to 44,100 Hz. The data resolution is set to 16 bit and the audio format was WAV. After sampling, we randomly intercepted 2 min of sound files within one hour [11], and had an average of 992 min of sound clips per day.

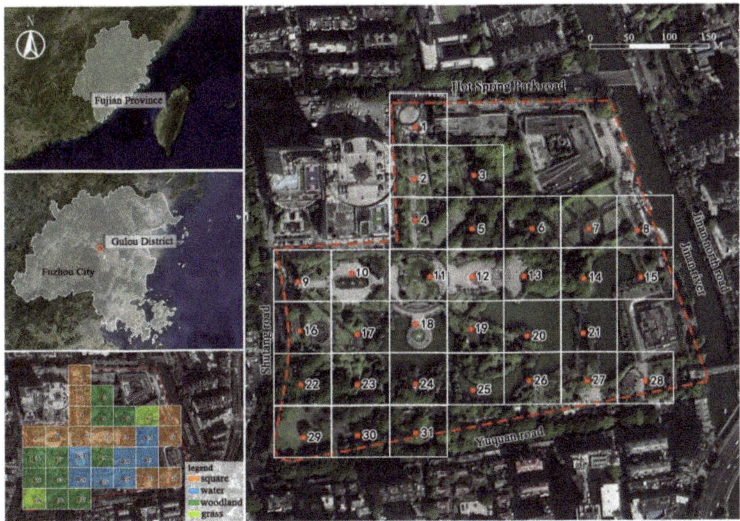

Figure 1. Grid and location of monitoring points (the numbers represent our monitoring points).

2.2.2. Panoramic Photos Collection

To analyze the landscape spatial characteristics of the monitoring site, a panoramic camera (Insta360 One X, panoramic image example shown in Figure 2) was placed during sound collection.

Figure 2. Panoramic overview of 31 monitoring point.

2.3. Data Processing and Analysis

2.3.1. Sound Source Analysis

To identify and classify the types of sound sources in urban parks, Adobe Audition was used in combination with spectrograms. Sound events can be visualized through the spectrogram, where brighter and more intense colors indicate a sound with greater power [30]. In our experiment, each acoustic signal appeared at the end of vocalization as one vocalization event [31] (Figure 3). In the statistics of audio, different kinds of sound sources in the same audio can be counted separately in frequency and time length, and sound types can overlap with each other.

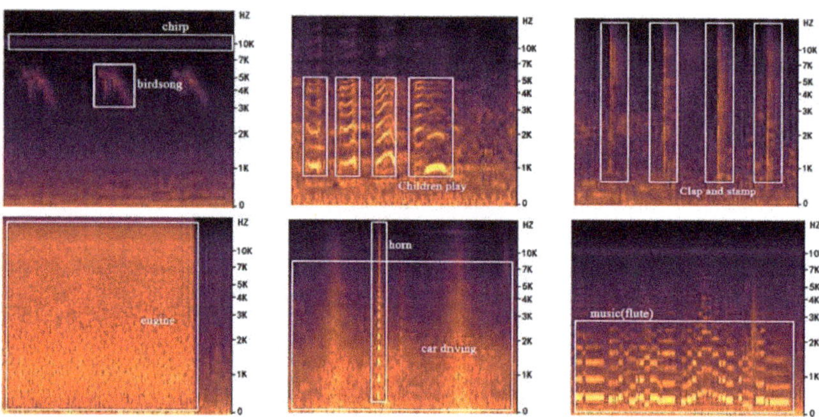

Figure 3. Schematic diagram of sound spectrum recognition in each frequency band.

2.3.2. Power Spectral Density

Different entities in the soundscape produce sounds at different frequencies [11]. Sound signals depict frequency on the vertical axis of the spectrum and time on the horizontal axis. The shading indicates the intensity (the power spectral density, PSD) of the sound signal at a particular frequency and time. Calculating the power spectral density can reflect the intensity of all the collected sound signals more objectively, and can also calculate the intensity of sound signals in different frequency intervals. The PSD, measured in w/kHz, was created by Welch in 1967 and represents the physical quantity of the power of an acoustic signal in relation to its frequency [32]. It is usually used to study acoustic signals of random vibration. The power of the sound can reflect the change of sound amplitude; that is, the greater the power, the greater the amplitude, the greater the amplitude, the greater the loudness, the greater the loudness, the stronger the sound is perceived by people [33]. Matlab was used to calculate the PSD in the range of 1–11 kHz to explore its temporal and spatial variations. In this study, the frequency of the sound sources observed was concentrated above 1 kHz, and data in the lowest interval, 0–1 kHz, was excluded as this can be largely attributed to human operation.

2.3.3. The Soundscape Diversity Index

The soundscape diversity index (SDI), used to evaluate the diversity of sound elements in the environment, was developed as an improved Simpson Diversity Index by scholars [34,35]. The main expression is as follows:

$$SDI = 1 - \sum_{i=1}^{S} \left(\frac{n}{N}\right)^2 \tag{1}$$

where n and N are the total number of perceived occurrences of a particular sound, I, and all sounds, S, in the soundscape sample, respectively. The *SDI* ranges between 0 and 1, and the greater the value, the more diverse the soundscape.

The SDI could "objectively" reflect soundscape characteristics and illustrates the overall soundscape perception characteristics [34], we calculated the SDI of each monitoring point to explore the spatial and temporal variation characteristics of park soundscape diversity.

2.3.4. Image Semantic Segmentation

To begin image semantic segmentation, a total of 770 pictures of Fuzhou city park were collected by camera (Contains both regular and panoramic photos) and randomly divided into training and test set according to the ratio of 10:1. Based on the classification of ADE20K dataset, images (including ordinary images and panoramic images) were manually annotated and input into DeeplabV3+ for training. The segmentation accuracy of the model reached 66.022% (MIoU = 66.022%), among which the segmentation accuracy of the water surface reached 84.53%. DeeplabV3+ is a mainstream network model in the field of image semantic segmentation, with multi-scale convolution layer and encoding-decoding dual modules [36]. It can achieve fine segmentation of the structure contained in the images [37]. As an encoding module, the DeeplabV3+ network can achieve hierarchical nested extraction of target features and multi-scale context information extraction, whereas the decoding module can integrate the low-level features and high-level abstract features generated in the DeeplabV3+ backbone network, and its structure is shown in Figure 4 [38]. Based on the DeepLabV3+ network structure, the urban park landscape labels are divided into 9 categories: water, trees, shrubs, ground cover, people, garden sketches, buildings, hard ground, and sky.

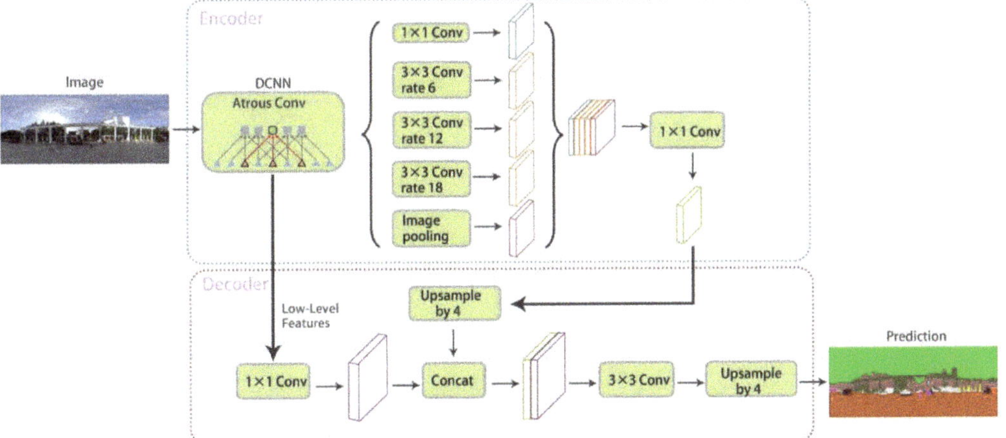

Figure 4. DeepLab V3+ model structure diagram [36].

3. Results

3.1. Composition Characteristics of the Urban Forest Soundscape in Summer

Through the discrimination of audio files by manually listening combined with Adobe Audition software. A total of 52 types of sound sources were identified and then, based on the methods of previous studies, divided into four primary classifications: natural sound, traffic sound, management sound, and recreation sound (Table 1.) [2,39–41]. The data of sound source types (including frequency and duration of different types of sound sources) in the monitoring points were averaged to represent the composition of sound sources in the park. A total of 60,527.6 s of sound source data were counted, with a total of 11,202 times. In terms of duration, natural sounds accounted for 28.51% of the overall, with birdsong and insect chirps being the majority (87.94%). Traffic sounds accounted for 9.63%, and consisted mainly of car driving sounds (78.49%). Management sounds accounted for 19.99%, with engine, fountain machinery and park radio sounds being the majority (88.90%). Recreational sounds accounted for 41.87%, and consisted mainly of music and conversation (70.13%). In terms of frequency, natural sounds accounted for 43.29% of the overall, with birdsong being the majority (96.74%); traffic sounds accounted for 5.27%; management sounds accounted for 10.40%; and recreational sounds accounted for 41.04%.

Table 1. Classification of park sound source types.

Primary Classification	Expound	Secondary Classification	Sound Element
natural sound	produced by natural elements, and no interference from human activities.	plant sound	leaves
		animal sounds	birdsong, insects, pets, birds flapping wings, frogs
		natural phenomenon sound	wind
traffic sound	traffic sound around the park	traffic sound	car driving, car engine, tram driving, horn, airplane roaring, motorcycle, bus arrival, bus announcement, brake, siren, ambulance, bicycle bell
management sound	The sound of park management	maintenance sound	engine, steel plate, digging, electric drill, sweeping, door opening, trailer, beat, shovel, weed, fountain machinery, sprinkler
		device sound	park radio
recreation sound	sound from park activities and social interactions	activity sound	play badminton, clap and stomp, skateboard, jump rope, key shaking, slap the ball, portable radio, music, run and jump, walk,
		social sound	conversation, children playing, sing, reunion, cry, laugh, sneeze, cough, whistle, shout

3.2. Diurnal Variation Characteristics of the Soundscape

3.2.1. Diurnal Variation Characteristics of Sound Sources

Overall, the soundscape presented a "U"-shaped change feature (Figure 5). Natural sounds occurred more frequently between 6:00 and 9:00 and lasted for a long time. Birdsong dominated the park soundscape, with Heron calls being the main bird soundscape present at noon. Natural sounds lasted longer at night and consisted mainly of insect chirps. Traffic sound dominated the park soundscape between 12:00 and 14:00 in duration. From 15:00 to 16:00, management sounds occupied a dominant position in the park soundscape, with the sounds of engines, park radio, sweeping, and mechanical fountains accounting for a large proportion. The sound of recreation appeared more frequently and lasted longer before 12:00 and after 16:00. The sounds of music and conversation in the morning made a greater contribution to the duration, and the sounds of people clapping, stomping, playing badminton, and walking made a greater contribution to the frequency. The sound of children playing, running and jumping accounted for a considerable proportion after 16:00.

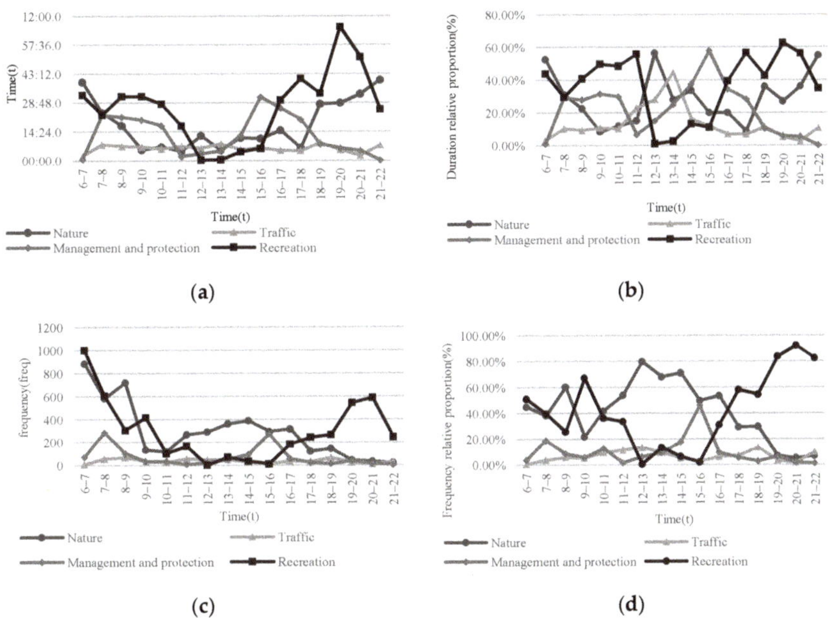

Figure 5. Diurnal variation characteristics of duration, frequency and relative proportion of various sound sources; (**a**) diurnal variation of sound duration; (**b**) diurnal variation characteristics of relative proportion of sound duration; (**c**) diurnal variation of sound frequency; (**d**) diurnal variation characteristics of relative proportion of sound frequency.

3.2.2. Diurnal Variation Characteristics of the PSD

The PSD of the soundscape during the summer in Hot Spring Park were distributed between 0 and 1.2 w/kHz (Figure 6). Except for the sounds from birds, insects, horns, and brakes, most of the sound sources were included in the range of 1–3 kHz, mainly management and recreation sound. We found that power of the sound in the range of 1–3 kHz fluctuated and rose after sunrise. The peak in power between 16:00 and 17:00 and was greatly affected by the sounds of children playing, running, jumping, engine and wind. Between 20:00 and 22:00, the PSD decreased and gradually stabilized. The sound of birds, insects, and frogs were mainly in the 3–9 kHz frequency band. Some artificial sounds, such as horns, music, people playing badminton, clapping and stomping, would also reach frequencies above 3 kHz. Sound power in the 3–9 kHz band exhibited a very similar pattern to 1–3 kHz.

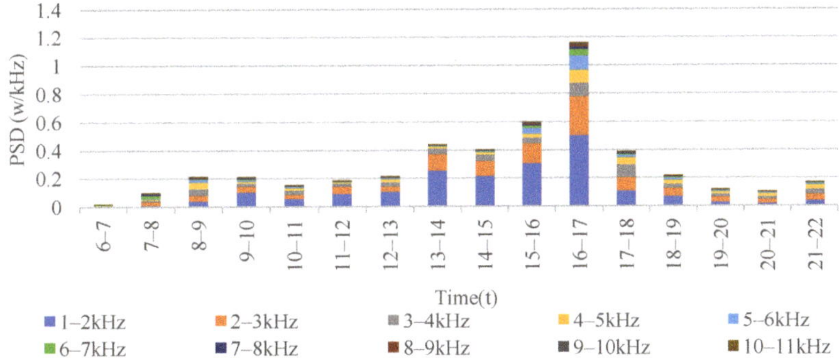

Figure 6. Diurnal variation of acoustic power spectral density in 10 frequency ranges.

The relative proportion and diurnal variation characteristics of each frequency sound are shown in Figure 7. Sound power decreases with the increase in the frequency. The sound power of 1–3 kHz, mainly from recreation and management sound, accounted for more than 50% of the park's power. The relative proportion of sound power fluctuated and rose after sunrise, and gradually decreased after reaching the maximum between 13:00 and 14:00 noon. However, the proportion of sound power above 3 kHz showed the opposite trend.

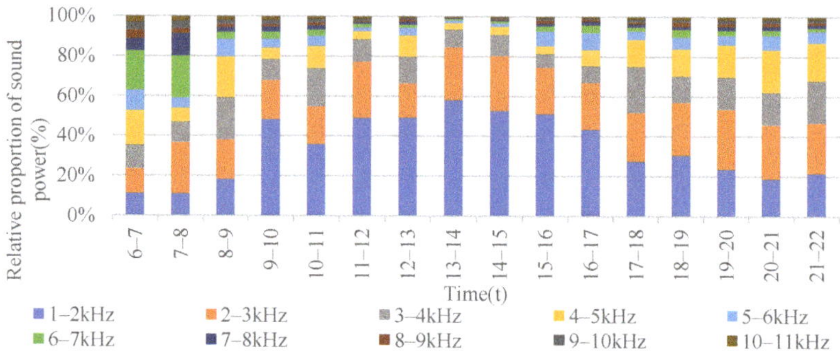

Figure 7. Daily variation of the relative proportion of sound power in each frequency band.

3.2.3. Diurnal Variation Characteristics of the SDI

The SDI was relatively high during the morning and night, and reached a low level between 12:00 and 15:00. After 15:00, the SDI gradually increased, and reached a peak of 0.827 between 17:00 and 18:00, and then gradually decreased once again (Figure 8).

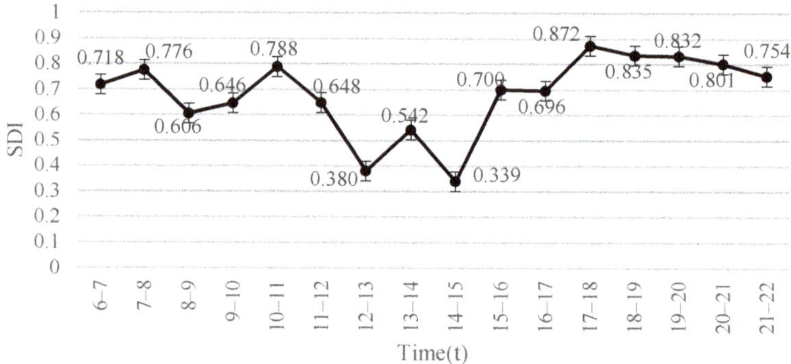

Figure 8. Characteristics of daily variation of the SDI.

3.3. Spatial Variation Characteristics of the Soundscape

3.3.1. Spatial Variation Characteristics of Soundscape Composition

The spatial variation characteristics of soundscape composition are shown in Figure 9. Natural sounds were evenly distributed in the park between 6:00 and 8:00. The frequency and duration of natural sounds were relatively low at monitoring points 12, 13, 27, 28 (square), and 29 (lawn along the street), these monitoring points experienced more recreational sounds during this period. Between 8:00 and 10:00, natural sounds were concentrated in the southeast area of the park, with monitoring point 14 near the lake experiencing the most sounds. Between 10:00 and 16:00, the frequency of natural sounds were highest in the children's activity area, *Ficus microcarpa* by the lake, and the monitoring points with high canopy closure around them, mainly birdsong. The natural sound at the relatively open lakeside promenade in the southern part of the park and the monitoring points in the northern bamboo garden had a longer duration and consisted of mainly insect chirps.

Between 16:00 and 18:00, the natural sounds at monitoring point 13 (*Ficus microcarpa* by the lake) and the southern lakeside promenade had high frequency and long duration. After 18:00, the frequency of natural sounds in most areas of the park decreased, and the vocal behavior of birds decreased., mainly insect chirps.

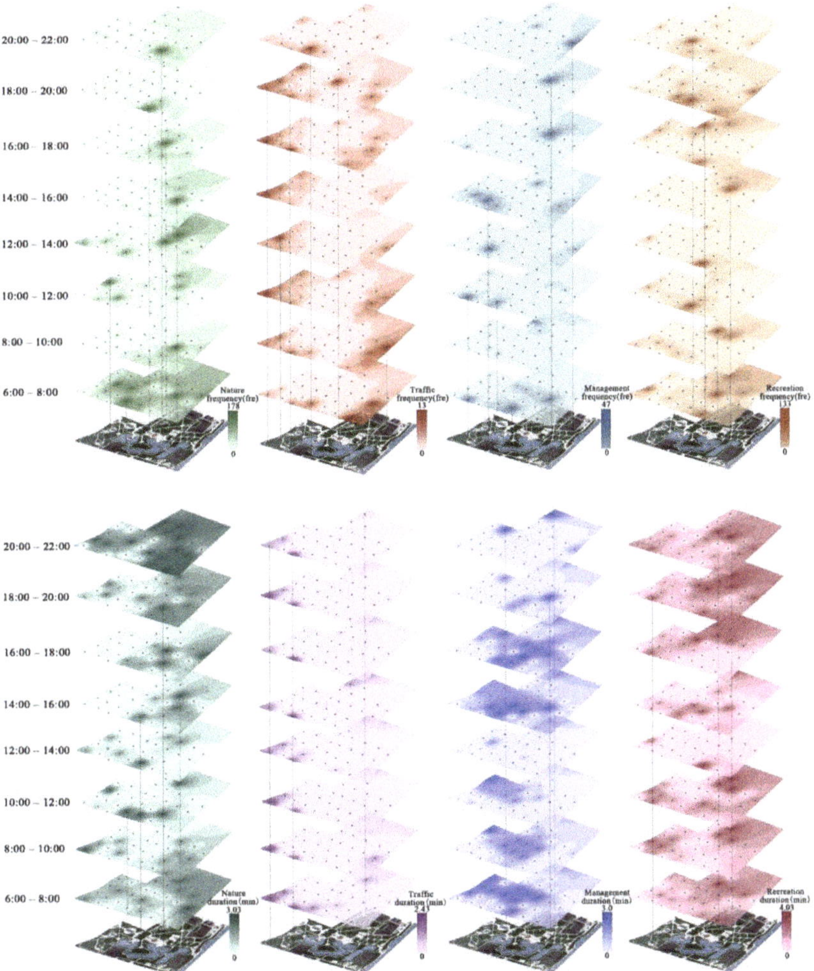

Figure 9. Spatial distribution of natural, traffic, management, and recreational sounds.

The frequency of traffic sound was higher at the park boundary. These monitoring points were mostly open spaces with single vegetation structure, were close to the main road and were greatly affected by traffic sound. The Aquarius Square had a high frequency of traffic sounds due to the entry of an external tram. Traffic sound affected an area up to 140 m from the park boundary, and the sound of horns could reach up to 210 m in the park. Monitoring points 16, 30, 26, and 27 were separated from surrounding roads by buildings or complex vegetation structure, even though the road was closer, the monitoring points were less affected by traffic sound.

Management sound was relatively random in the park. Between 6:00 and 12:00, the monitoring sites of the arched fountains have a longer sound duration, the sound of fountain machinery accounted for a large proportion. Between 14:00 and 18:00, the sound duration in the central square and the surrounding lake area was longer and consisted mainly of the engine. After 18:00, the main sound at the entrance of the park was the sound of the park radio.

Recreation sound was relatively evenly distributed between 6:00 and 8:00, and was either concentrated in the central square or a large hard area of the node. Areas that had a higher frequency of sound from badminton being played or the sound of clapping and stomping were Ficus microcarpa Garden, the south entrance square, the central Vase Square, and the north Bamboo Garden. Sound from the boulevard and children's activity area lasted for a long time and mainly included the sound of music, children's playing, and conversation. Between 8:00 and 16:00, the sound of music and conversation accounted for a large proportion, and their distribution was concentrated in a certain hard ground and tree covered environment. After 18:00, the sound of recreation covered the entire park again. The duration of the recreational sound increased significantly. The water features in the park did not play a significant role in attracting recreational sound.

3.3.2. Distribution Characteristics of the PSD

Figure 10 shows the distribution of the PSD in Hot Spring Park during the summer. The maximum sound power of the park (PSD = 1.09 w/kHz) was located at the riverside promenade on the east side of the park, which was affected by the beating sound of the external residential area, the sound of electric drills and the sound of traffic. Meanwhile, due to the open space, the monitoring point was located at the tuyere, which was greatly affected by the wind. The entrance on the north side of the park was close to the main road and was affected by traffic sounds such as cars driving and horns, the sound of park radio in entrance, and the sound of children playing during school closing time at dusk. The sounds of children playing also contributed a large amount of power, with a PSD of 0.8 w/kHz.

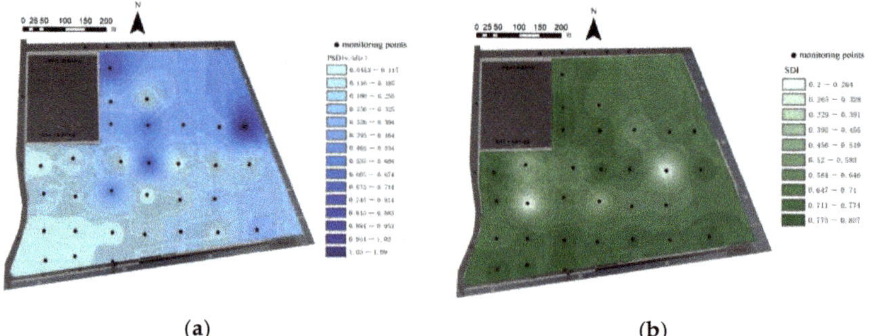

(a)   (b)

Figure 10. (**a**) Spatial distribution characteristics of the PSD; (**b**) spatial distribution characteristics of the SDI.

The PSD of monitoring point 12—Aquarius square, monitoring point 18—circular fountain and monitoring point 5—shady path were relatively high. In Aquarius square and circular fountain, the landscape space had the following characteristics: low basal plants, high basal constructions, without covering, both open space, plant enclosure was relatively low, and greatly affected by the sounds of engine and various recreational activities. Shady path in the basement of higher plants, the basal construction surface of low, high degree of vegetation cover, it is a semi-enclosed space, where the sound of children playing is of high power. In the southern part of the park, the sound energy was less than 0.2 W/kHz, and was mainly from the sounds of footsteps and conversation.

3.3.3. Distribution Characteristics of the SDI

As shown in Figure 10, monitoring points 12 and 13 were in open spaces with a large proportion of the basal construction and no coverage, which created a diverse sound environment for recreation. Monitoring points 2 and 4 were close to the north entrance of the park and were in the main passage for the public to enter the park. The sound of recreation was abundant at monitoring points 2 and 4. In addition, management sounds such as floor sweeping, were incorporated due to the daily management of the park. Monitoring point 24 was close to the lake, with complex vegetation and a superior natural acoustic environment. Monitoring point 31 was an open entrance plaza on the south side of the park, which was rich in recreational sound. Monitoring points 28 and 29 were close to the road and the traffic sound was rich. Meanwhile, monitoring point 28 was located next to the running track in the park, so the recreational sound was relatively rich and the SDI was high.

Monitoring point 14 was at the lakeside rest node based on the construction and water. The surrounding vegetation structure was simple, contained less area for movement, and the attraction of natural sound and recreational sound was insufficient. There was an artificial bird's nest with prominent bird singing at monitoring point 17, and most of the public gathered to bird watch. The lack of other recreational behaviors led to a low SDI.

3.4. Influencing Factors of Landscape Space of the Soundscape

SPSS 26.0 software was used for correlation analysis and regression analysis to explore the relationship between landscape elements and the SDI, the PSD, and the frequency and duration of various sounds.

As shown in Table 2., the proportion of hard ground had a significant positive correlation with the SDI and the PSD, while the proportion of shrubs had a significant negative correlation with the PSD. Through linear regression analysis, the listed landscape factors can explain 55.5% of the change in the PSD ($R^2 = 0.555$), and there was no collinearity between variables (VIF < 10), and the samples are independent of each other (DW = 2.336) (Table 3). Our results showed that the proportion of hard ground, sky, trees, and shrubs had a significant impact on the change in the PSD ($p < 0.05$) (Table 4). The greater the proportion of hard ground, the lower the proportion of sky, trees, and shrubs, the higher the PSD.

$$PSD = 0.776 + 0.01 \times X_1 - 0.031 \times X_6 \tag{2}$$

Table 2. Correlation between the PSD, the SDI and the proportion of landscape spatial elements.

Indicators	Hard Ground Ratio	Building Ratio	Sky Ratio	Arbor Ratio	Tourists Ratio	Shrub Ratio	Ground Cover Ratio	garden Ornaments Ratio	Water Ratio
SDI	0.364 *	0.008	−0.069	0.147	−0.160	0.136	−0.002	0.233	−0.227
PSD	0.361 *	0.133	0.084	−0.272	0.032	−0.362 *	−0.062	0.122	0.008

* At the 0.05 level (two tailed), the correlation is significant.

Table 3. Summary of multiple linear regression models of the PSD and proportion of landscape elements.

Model	R	R^2	Adjusted R^2	Estimated Standard Error	Durbin-Watson
1	0.745 [a]	0.555	0.355	0.16873	2.336

R^2 reflects the variance explained by the regression equation as a percentage of the variance of the dependent variable. Durbin-Watson reflects autocorrelation of independent variables.

Listed landscape factors can explain 43.2% of the change in the SDI (*R*2 = 0.432), and the samples were independent of each other and there was no collinearity (VIF < 10, DW = 2.152) (Table 5). It was found that with the higher proportion of hard ground, the higher the diversity of the spatial soundscape and richer the soundscape (Table 6).

$$SDI = -0.053 + 0.013 \times X_1 \tag{3}$$

Table 4. Analysis of landscape elements affecting the PSD.

Model		Unstandardized Coefficients		Standardized Coefficients	t	Significance	Collinearity Statistics	
		B	Standard Error	Beta			Tolerance	VIF
1	(constant)	0.766	0.355		2.157	0.043		
	X1 Hard ground ratio	0.010	0.005	0.520	2.123	0.046	0.371	2.698
	X2 Building ratio	0.001	0.012	0.012	0.068	0.947	0.680	1.471
	X3 Sky ratio	−0.017	0.006	−1.161	−2.826	0.010	0.132	7.585
	X4 Arbor ratio	−0.015	0.005	−1.153	−3.066	0.006	0.157	6.359
	X5 Tourists ratio	−0.043	0.074	−0.115	−0.585	0.565	0.579	1.727
	X6 Shrub ratio	−0.031	0.013	−0.405	−2.420	0.025	0.793	1.261
	X7 Ground cover ratio	0.009	0.007	0.309	1.237	0.230	0.357	2.804
	X8 Garden ornaments ratio	0.234	0.231	0.187	1.015	0.322	0.657	1.521
	X9 Water ratio	0.012	0.007	0.317	1.670	0.110	0.617	1.620

Table 5. Summary of multiple linear regression models of the SDI and proportion of landscape elements.

Model	R	R^2	Adjusted R^2	Estimated Standard Erro	Durbin-Watson
1	0.657 [a]	0.432	0.176	0.152	2.152

R^2 reflects the variance explained by the regression equation as a percentage of the variance of the dependent variable. Durbin-Watson reflects autocorrelation of independent variables.

Table 6. Analysis of landscape elements affecting the SDI.

Indicators	Unstandardized Coefficients		Standardized Coefficients	t	Significance	Collinearity Statistics	
	B	Standard Error	Beta			Tolerance	VIF
(constant)	−0.053	0.320		−0.167	0.869		
X1 Hard ground ratio	0.013	0.004	0.804	2.905	0.009	0.371	2.698
X2 Building ratio	0.008	0.010	0.167	0.818	0.423	0.680	1.471
X3 Sky ratio	0.005	0.005	0.426	0.917	0.370	0.132	7.585
X4 Arbor ratio	0.008	0.004	0.711	1.672	0.110	0.157	6.359
X5 Tourists ratio	−0.122	0.066	−0.409	−1.847	0.080	0.579	1.727
X6 Shrub ratio	0.013	0.012	0.210	1.110	0.280	0.793	1.261
X7 Ground cover ratio	0.006	0.007	0.280	0.992	0.333	0.357	2.804
X8 Garden ornaments ratio	−0.119	0.208	−0.119	−0.574	0.572	0.657	1.521
X9 Water ratio	0.000	0.006	0.017	0.079	0.938	0.617	1.620

Table 7 further explores the relationship between the proportion of each landscape space element and various sound sources. Our results showed that, in terms of sound frequency, the proportion of hard ground had an extremely significant negative correlation with the frequency of natural sound. There was a significant negative correlation between the proportion of buildings and the frequency of recreational sound, while the proportion of trees has a significant positive correlation with the frequency of recreational sound.

There was a significant positive correlation between the proportion of trees, shrubs, garden sketches and the frequency of management sound.

Table 7. Correlation between various sound sources and the proportion of landscape space elements.

Indicators	Hard Ground Ratio	Building Ratio	Sky Ratio	Arbor Ratio	Tourists Ratio	Shrub Ratio	Ground Cover Ratio	Garden Ornaments Ratio	Water Ratio
Natural frequency	−0.537 **	−0.043	−0.109	0.157	−0.191	−0.079	0.318	0.018	0.287
Traffic frequency	−0.007	0.330	−0.301	0.217	0.154	0.023	0.249	−0.058	−0.233
Management frequency	−0.099	−0.213	−0.295	0.366 *	−0.030	0.521 **	0.173	0.413*	−0.060
Recreational frequency	−0.023	−0.427 *	−0.208	0.381 *	−0.299	−0.146	0.236	0.161	−0.032
Natural duration	−0.574 **	−0.432 *	−0.156	0.252	−0.443 *	0.002	0.572 **	−0.086	0.233
Traffic duration	−0.047	0.637 **	−0.370 *	0.314	0.218	−0.248	0.227	−0.243	−0.155
Management duration	0.285	0.193	0.424 *	−0.619 **	0.276	−0.190	−0.456 *	−0.203	−0.026
Recreational duration	−0.021	−0.250	−0.540 **	0.503 **	0.066	0.050	0.402 *	0.195	−0.305

* At the 0.05 level (two tailed), the correlation is significant; ** at the 0.01 level (two tailed), the correlation is significant.

In terms of duration of vocalization, the proportion of hard ground, buildings, and people had a significant negative correlation with the duration of natural sound. Likewise, the proportion of ground cover had a significant positive correlation with the duration of natural sound. There was a significant positive correlation between the proportion of buildings and the duration of traffic sound, and a significant negative correlation between the proportion of sky and the duration of traffic sound. There was a significant positive correlation between the proportion of sky and the duration of the sound of management and protection, while the proportion of trees and ground cover had a significant negative correlation with the duration of the sound of management and protection. There was a significant negative correlation between the proportion of sky and the duration of recreational sound, but the proportion of trees and ground cover had a significant positive correlation with the duration of recreational sound.

4. Discussion

4.1. Diurnal Characteristics of the Soundscape

We found that natural sounds appeared most frequently between 6:00 and 9:00 and mainly consisted of birdsong. This verified the dawn chorus behavior of birds after sunrise [11,42]. However, the dusk chorus was not found in our monitoring, which may have been related to the different urban environments where the monitoring site was located. The increase in sound of recreation at dusk may also impact on the vocalization of birds. The long duration of nocturnal insect chirping was consistent with previous studies [43]. Traffic sounds dominated the park soundscape at noon as the high temperatures limited the frequency of natural and recreational sounds. Recreational sounds dominated the park soundscape before 12:00 and after 16:00, which was consistent with previous studies [44], and showed that the summer temperatures had a great influence on the recreational behavior of the public.

In the summer park, sounds in the 1–3 kHz frequency band accounted for more than 50% of the overall sound power of the park. The overall daily change was in the shape of "U", with the valley appearing between 14:00 and 15:00 at noon. The diurnal variation characteristics of sound power were in good agreement with the temperature variation characteristics. In this study, playing, running, and jumping in the park during the late afternoon, when children were walking through or staying in the park after school, resulted in high levels of sound power in the park, and increased levels of sound power at dusk when the park engine and wind were affected. We found that the nighttime sound power of urban parks was lower, which was different from previous research results [31]. This could possibly be due to the prohibition of activities such as square dancing during the epidemic period, as well as children's activities at night were distributed in fewer areas (mostly concentrated in children's activity areas), the frequency and duration of children playing sound are low, and the power presented is small, which lead to lower sound power at night. Similarly, previous studies found that biological sounds were mainly distributed above 2 kHz [45,46], with vocal peaks appearing in the early morning [13]. We found that high-frequency and low-frequency sounds exhibited a similar diurnal variation pattern, we also found that many recreational sounds, such as the sound of people playing badminton, clapping, stomping, the sound of children playing, occupied the high-frequency band of biological sound and showed previously unseen higher power on the spectrogram, these recreational behaviors may have an impact on biological vocalizations. It was speculated that previous studies were mostly in the suburban forest and the sound of recreation had less influence on it. In terms of park soundscape diversity, the SDI showed a trend of higher in the morning and evening and lower in the noon, which was the same as the time distribution trend of recreational sound. The recreational behavior of the public had a great impact on the diversity of the park soundscape. Therefore, the SDI can be used as one of the indicators to measure the diversity of recreation behaviors in urban parks, and can be used as the basis for the evaluation of park recreation activities in subsequent studies.

4.2. Spatial Variation Characteristics of the Soundscape

The frequency and duration of natural sounds were lower in squares and parks along the street, but higher in forests with high canopy closure and lakeside forest areas. Our results confirmed that trees are an important factor in determining urban biodiversity [47–49], and that the physical attributes of tree, such as larger canopies and side branches, make big trees more attractive to birds [50]. Therefore, areas with high vegetation canopy closure and the combination of "big trees + lake water" can contribute to a suitable habitats for birds in summer, also enables the public to hear birdsong at noon in summer (such as egret and ardeola bacchus).

Traffic sounds had a higher frequency and longer duration at the park boundary. Most areas within 140 m from the park boundary were affected by traffic sounds, and vehicle horns were able to reach as far as 210 m in the park. The relatively open area close to the traffic road was influenced by the traffic sounds, and the open terrain was conducive to sound transmission, so the sound power is higher. By comparing monitoring points close to the road, we found that multi-layered plant configuration (especially shrub layer construction and utilization of plants with low branching points) and increased density of vegetation and structures can help reduce the noise of traffic outside. At the same time, previous studies have shown that structures can also affect the transmission of traffic sound, and the impact of traffic can be effectively reduced by placing structures on the side closest to the traffic road [51]. Increasing the density of façade elements near the road area helps to reduce the impact of traffic sound on the interior of the park.

The sounds of recreation during the day were mainly distributed at monitoring sites with high canopy closure. Since summer is different from the other seasons, light intensity and temperature were higher, and the spaces with higher canopy closure were most suitable for public activities [52]. Our study found that fountains and other water features did not bring significant attraction to recreational sounds during the opening hours of park's water features. However, past studies have shown that the public has higher expectations for water sound [2], and it is speculated that the waterscape at the node does not produce pleasant soundscapes, such as water flow sound and water droplet sound [15]. At the same time, the waterscape in the park did not have hydrophilic functions, and the area of the venue provided for public activities was reduced. There were also fewer recreational facilities in the surrounding areas, which resulted in the lower frequency and duration of recreational sounds and the mechanical sound of the fountain in the square muffled the sound of water, therefore making the area less attractive to the public.

The SDI in the open space and the park entrance is higher, the recreational sounds in the two types of spaces is diverse, and the vegetation structure is rich around the lake to attract more natural sound. If the facilities in the site are strongly attractive to the public, there will be fewer recreational activities in the site, and the types of recreational sounds will be reduced. For example, the public will be attracted by the artificial bird nest in the field, and conduct birdwatching behavior, and only generate conversation sounds

4.3. Influence of Landscape Space Elements on the Soundscape

Among all landscape spatial elements, the larger the proportion of hard ground, the larger the soundscape power (the higher the PSD) and the higher the soundscape diversity (the higher the SDI). In parks, there are more types of recreational sounds than natural sounds, management sounds and traffic sounds, and the size of hard ground is an important factor related to public activities. In the two types of spaces with low proportion of trees and shrubs and low proportion of sky, the soundscape power is larger and the PSD is higher. It is found that spaces with low sky coverage (mostly covered by vegetation) will attract more recreational activities in summer, resulting in more recreational sounds and greater sound power. At the same time, in open spaces with large hard ground and low enclosed degree of trees and shrubs, it is more conducive to the public to carry out various activities and sound transmission, and the public will have a stronger perception of sound in this type of space. However, areas with large amounts of hard ground faced a reduction in natural sounds, and the increase in number of buildings and tourists will cause a decrease in natural sounds, especially birdsong. The lager the ground cover, the longer the natural, this kind of space was conducive to insect vocalizations. It was areas that were rich in ground-covering plants and shrubs that were conducive to insect habitat [39], and, as the park restricted visitors from entering the grassland, recreational behavior had less impact on insects. We found that recreation sounds increase when the proportion of trees and ground cover were larger, and the proportion of sky was smaller. We also found that the proportion of hard ground had no significant effect on the duration and frequency of recreational sounds, which further indicated that shaded environments and ventilation of the site were the main factors for the public to consider during summer recreation. We noticed that buildings, vegetation, and sky were commonly ignored factors in previous studies, but they, especially the proportion of sky, should be considered as key factors for future soundscape construction [23]. More space enclosed by buildings will reduce the frequency of recreational sound, and such space will be distributed in the boundary area of the park, which will restrict the recreation behavior of the public.

5. Conclusions

Our study drew the following main conclusions: (1) a total of 52 types of sound sources were identified in the forest environment of the urban core area in summer, and the park soundscape was dominated by natural sounds and recreational sounds. (2) Sound sources presented a "U"-shaped diurnal variation characteristic, and the variation characteristic was opposite to the temperature change. The best time for the public to perceive bird song is between 6:00 and 9:00; and after 18:00, the natural sound of the park is dominated by insect chirps. (3) The soundscape power and the soundscape diversity of the park are greatly affected by public recreation behaviors, and some recreation behaviors occupy the high-frequency band of biological vocalizations, which may affect the vocal behavior of organisms such as birds (4) Building spaces with high canopy density can attract more birdsong and recreational sounds in summer, and the combination of "tree + lake" can attract more birdsong. As a negative sound, traffic sound can affect the space within 140 m, while horn sound can reach a further area. Enriching vegetation structure and increasing vegetation width can effectively shorten the influence range of traffic sound in urban forest. (5) In all kinds of landscape space, spaces with large hard ground and low degree of containment will attract public recreation, in such space the public perception of the soundscape will be more intense. In summer, the shade and ventilation condition of site is the main factor of public leisure to consider, this kind of space will attract more birds, at the same time, increase the cover area as well as forbid public access to the lawn is also conducive to insect habitat, increasing the proportion of insects chirping in the park. Compared to previous studies, the grid method was used to set monitoring points to refine the spatial variation characteristics of the soundscape in urban parks. The landscape spatial factors that affected the soundscape were determined based on factor analysis,

48. Stagoll, K.; Lindenmayer, D.B.; Knight, E.; Fischer, J.; Manning, A.D. Large trees are keystone structures in urban parks. *Conserv. Lett.* **2012**, *5*, 115–122. [CrossRef]
49. Yang, G.; Wang, Y.; Xu, J.; Ding, Y.Z.; Wu, S.Y.; Tang, H.M.; Li, H.Q.; Wang, X.M.; Ma, B.; Wang, Z.H. The influence of habitat types on bird community in urban parks. *Acta Ecol. Sin.* **2015**, *35*, 4186–4195.
50. Le Roux, D.S.; Ikin, K.; Lindenmayer, D.B.; Manning, A.D.; Gibbons, P. Single large or several small? Applying biogeographic principles to tree-level conservation and biodiversity offsets. *Biol. Conserv.* **2015**, *191*, 558–566. [CrossRef]
51. Meng, Q.; Kang, J. Study on Soundscapes in Urban Fringe Areas: Taking Tangchang Community Planning as an Example. *City Plan. Rev.* **2018**, *42*, 94–99.
52. Zhang, R.N.; Zhang, Y.; Liu, Y. Relationship between the landscape characteristics and the distribution of the soundscape in community parks: A case study of Lu Xun Park, Shenyang. *J. Appl. Acoust.* **2022**, *41*, 207–215.

Article

Psychophysiological Impacts of Traffic Sounds in Urban Green Spaces

Boya Yu *, Jie Bai, Linjie Wen and Yuying Chai

School of Architecture and Design, Beijing Jiaotong University, Beijing 100044, China; 18311002@bjtu.edu.cn (J.B.); 18311054@bjtu.edu.cn (L.W.); 21126381@bjtu.edu.cn (Y.C.)
* Correspondence: boyayu@bjtu.edu.cn

Abstract: The goal of this study is to investigate the psychophysiological effects of traffic sounds in urban green spaces. In a laboratory experiment, psychological and physiological responses to four traffic sounds were measured, including road, conventional train, high-speed train, and tram. The findings demonstrated that traffic sounds had significant detrimental psychological and physiological effects. In terms of psychological responses, the peak sound level outperformed the equivalent sound level in determining the psychological impact of traffic sounds. The physiological effects of traffic sounds were shown to be significantly influenced by sound type and sound level. The physiological response to the high-speed train sound differed significantly from the other three traffic sounds. The physiological effects of road traffic sounds were found to be unrelated to the sound level. On the contrary, as for the railway sounds, the change in sound level was observed to have a significant impact on the participants' physiological indicators.

Keywords: road traffic noise; railway noise; electrodermal activity; heart rate; soundscape; acoustic comfort; Beijing

Citation: Yu, B.; Bai, J.; Wen, L.; Chai, Y. Psychophysiological Impacts of Traffic Sounds in Urban Green Spaces. *Forests* **2022**, *13*, 960. https://doi.org/10.3390/f13060960

Academic Editors: Xin-Chen Hong, Jiang Liu and Guang-Yu Wang

Received: 15 May 2022
Accepted: 16 June 2022
Published: 19 June 2022

Publisher's Note: MDPI stays neutral with regard to jurisdictional claims in published maps and institutional affiliations.

1. Introduction

Urban green spaces provide flora and fauna for cities [1,2], offering important spaces for relaxation, recreation, social interaction, and sports [3,4]. There is accumulating research indicating that the acoustic environment plays a key role as a component of a positive visitor experience in urban green areas [5–7]. The concept of soundscape, defined by the International Organization for Standardization as the "acoustic environment as perceived or experienced and/or understood by a person or people, in context", has been widely used to investigate the interaction between the acoustic environment and people in urban green spaces [8,9].

The sound source has been proven to be the dominant factor that decides how people understand and respond to the soundscape [10–12]. Numerous studies have been carried out to investigate how individual sound sources affect the perception of green space soundscapes. In field measurement and a questionnaire survey, Bani and Paulo found that traffic sound was the most influential sound source in urban parks [13]. Similar results were also reported in Europe, where traffic sound was identified as the dominant component of the green space soundscape [14–16]. Compared with other sound sources, traffic sounds were found to be the least preferred but to have a dominant position in terms of perceived occurrences or loudness [16]. The annoyance caused by traffic sounds further led to a strong impact on the perception of the overall environment, which increased with the sound level [17]. To achieve a good soundscape quality in green spaces, a limit of 50 dBA was suggested by Nilsson et al. [5]. In addition to the perceived environmental quality, increasing evidence has proven that traffic sounds could also have significant negative effects on recreational activities, restorativeness, and stress recovery [18]. By combining the sound level and the visual contextual features, the TRPT (Tranquility Rating Prediction Tool) was suggested to be efficient in predicting the relaxation effect of urban green spaces [19].

Besides perceptual attributes, exposure to traffic noise has been linked to an increased risk of negative health outcomes [20–25]. Various physiological indicators have been proposed to measure the physiological effects of traffic sounds. Through laboratory experiments, Reinhard et al. revealed that road traffic noise led to higher heart rate (HR) compared to neutral phases [26]. Meanwhile, Basner et al. found this phenomenon was similar for air, road, and rail traffic sounds and that the increase in traffic sound level could cause significant increases in electroencephalographic (EEG) and heart rate (HR) indicators in indoor soundscapes [27]. Through a field survey, Lee et al. further revealed the positive correlation between blood pressure (BP) and traffic noise exposure [28]. Besides, Li et al. found that traffic sound exposure was related to many physiological indicators, including HR, R-wave amplitude, respiration rate (RR), and skin conductance level (SCL) [29].

In the field of urban soundscape research, physiological responses have drawn increasing attention recently in investigating the potential effect of traffic sounds on the overall soundscape [30]. By comparing the physiological effects of various sound elements, Hume and Ahtamad found significant correlations between physiological indicators and perceptual attributes [31]. A more pleasant sound source led to a lower HR and a higher RR, and the traffic sound clip was voted to be the most unpleasant sound. Using functional magnetic resonance imaging, Irwin et al. identified that an unpleasant sound engaged an additional neural circuit including the right amygdala [32]. Besides, the road traffic noise was found to have a significant negative effect on the SCL recovery when compared with the natural sound [33]. Meanwhile, a significant interaction effect of traffic sounds and bird sounds on the SCL in a mixed soundscape was found by Suko et al. [34]. The presence of traffic sounds in a natural soundscape was found to be associated with an increase in SCL. In addition, significant audio–visual interactions of traffic sounds in urban soundscapes were found by Li et al. [35], which indicated that the physiological effects of traffic sound varied in different urban contexts. Focusing on the urban green space soundscape, Matilda et al. found that sound stimuli had a strong impact on physiological responses, including cortisol, sympathetic T-wave amplitude, HR, and HRV [36]. Focusing on children, Shu et al. found that noises produced less physiological restorativeness (EDA) compared with natural sounds [37]. In an in-site experiment, Li et al. also found that sound type had a significant effect on physiological responses in mountainous urban parks. A less remarkable restorative EEG rhythm was found at the traffic-sound-dominant site than at the birdsong-dominant site [38].

Collectively, the negative psychophysiological effects of traffic sound on urban green space soundscapes were outlined by the above studies. However, a simultaneous investigation of the sound type and the sound level was not detailed, to the authors' knowledge. Furthermore, most existing literature focused on road traffic sounds, with little attention paid to rail traffic sounds. However, rail traffic sound is the second most dominant traffic noise, which was found to be significantly different from road traffic sound in not only the physical characteristics but also in the impact on people [39,40].

Therefore, the aim of this study was to reveal the psychophysiological impacts of road traffic and rail traffic sounds on the urban green space soundscape via laboratory experiments, addressing the following questions:

(a) Are there differences between different traffic sounds in the psychophysiological impact on the green space soundscape?
(b) What is the relationship between the sound level and the psychophysiological responses? Does it vary in different traffic groups?
(c) Are there correlations between psychological responses and physiological responses when exposed to traffic sounds?

To answer such questions, four traffic sound sources (road, conventional train, high-speed train, and tram) at three different sound levels (45, 55, and 65 dBA) were used as experiment stimuli in this study. Specifically, there were two sessions for each acoustic stimulus. First, EDA and HR were measured to investigate the physiological impact of traffic sounds. Second, four perceptual attributes were measured to assess the psychological response to each sound stimulus.

2. Methodology

2.1. Participants

Thirty subjects participated in the experiment, including 15 males and 15 females. All subjects were university students (ages from 18 to 24, mean age = 20.3). The participants were recruited via social media and snowball sampling in Beijing Jiaotong University, and they voluntarily participated in the experiment. All of the participants reported normal hearing and corrected vision. None of them were taking prescription medication. All of the participants were informed about the aim and protocol of the experiment.

2.2. Experiment Stimuli

The experiment was carried out in an experiment chamber with a controlled physical environment (temperature = 21–23 °C; background sound level < 25 dBA). Virtual reality (VR) equipment was used to present a complete and realistic visual environment of an urban green space. As shown in Figure 1, an omnidirectional image was captured in a real-world urban park and displayed on a head-mounted display system (HTC VIVE Pro EYE; Resolution: 2880 × 1600; Refresh rate: 90 Hz).

Figure 1. The panoramic view of an urban green space.

Four traffic sounds were considered in this study, including road traffic, conventional train, high-speed train, and tram. The four traffic sounds were selected for the following reasons: (1) they are very common urban traffic sounds that are frequently considered in relevant studies [41–44], and (2) they have been reported to have different influences on subjective evaluation and may have potential differences in psychophysiological effects [28].

Field recordings were collected to extract the experiment stimuli in Beijing, China. The traffic sounds were recorded in a single channel by a sound meter (6228+, Aihua, Hangzhou, China) together with the measurement. To avoid the influences of surrounding environmental sounds, all recording sites were selected in quiet areas (background sound level < 45 dBA). Figure 2 shows the site and implementation of sound recording collection. For each site, recordings were conducted in positions close to the traffic lines (5 to 15 m). The sound meter was placed 1.25 m above the ground and at least 5 m away from the surrounding building facade. The sound recording for each site was conducted on two days (from 8 a.m. to 8 p.m.), including a weekday and a weekend in November 2021. All sound recordings were then manually rechecked to find a clear traffic sound clip to make the experiment stimuli.

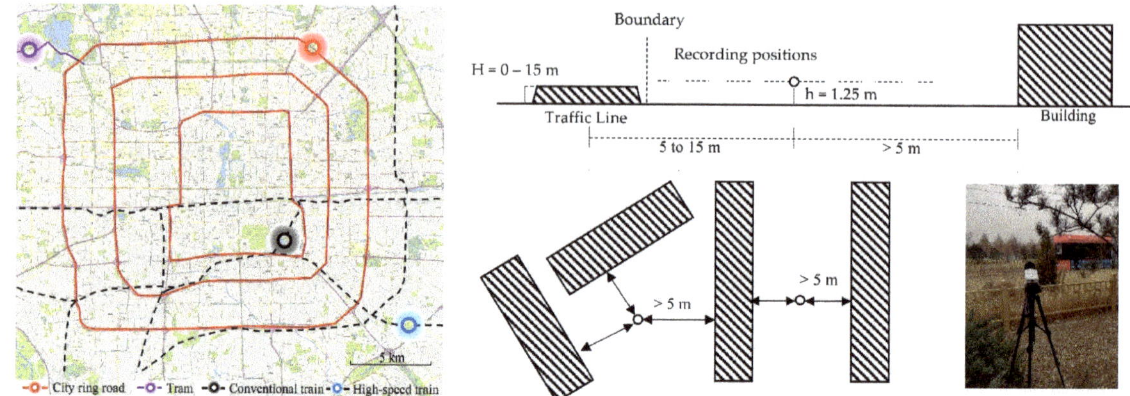

Figure 2. Site and implementation of field recording.

As shown in Figure 3, there were noticeable differences in temporal characteristics between traffic sounds, though the spectral characteristics were rather similar. The road traffic sounds were mainly continuous and steady, while the railway sounds were intermittent and fluctuating. Because of the limitations of physiological measurement, a two-minute stimulus was needed for each traffic sound group. For road traffic sound, a continuous field recording collected near a city highway was used. For each rail traffic sound, a one-minute clip that contained one train passing by was extracted and then repeated to produce the two-minute experimental stimuli.

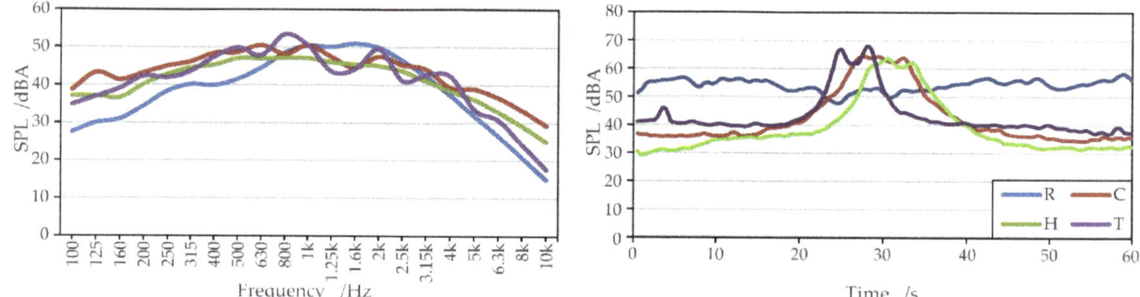

Figure 3. Spectral and temporal characteristics of four traffic noises, including road traffic (R), conventional train (C), high-speed train (H), and tram (T).

The sound stimuli were reproduced in the laboratory through the combination of a computer, a power amplifier, and a reference class headphone (Sennheiser 650HD, Sennheiser, Germany). The single-channel signals were first copied to produce the dual-channel stimuli. A sound level normalization was then conducted to produce the formal experiment stimuli. Four traffic sound clips were normalized to three sound levels, including 45, 55, and 65 dBA (L_{Aeq}, 2 min). To calibrate the sound levels, a sound level meter (6228+, Aihua, Hangzhou, China) was placed 1 cm away from the headphone, which was in position on a participant's ear during the experiment. Then, the sound level calibration was conducted for each channel separately by the Cooledit software (Version 2.0, Syntrillium Software Corporation, Phoenix, AZ, USA).

In addition to traffic sounds, a two-minute silence clip was also used in the experiment as the control stimulus to conduct the baseline measurement. Altogether, there were 13 sound stimuli in the experiment.

2.3. Measures

In this study, two of the most widely used physiological parameters were used: electrodermal activity (EDA) and heart rate (HR), which have been proven to be effective in evaluating the physiological impact of acoustic stimuli [30,37,45,46]. To achieve the simultaneous measurement, the finger electrode method and the photoplethysmography method were used for EDA and HR measurement, respectively. During the experiment, two electrodes (HKR-11, range: 100 to 2500 kΩ; accuracy: 2.5 kΩ; sample frequency: 50 Hz) were attached to the subject's index and middle fingers of the non-dominant hand to record EDA signals. A photoplethysmography (PPG) sensor (HKG-07, measuring range: 30 to 250 bpm; accuracy: 1 bpm; sample frequency: 16 Hz) was attached to the ring finger to measure the HR index [45]. It is suggested that data processing is necessary because the physiological parameters strongly depend on personal characteristics and exposure time [37,46]. The original physiological data were first divided into 20 s segments in this study to reveal the temporal variations. Then, the mean value of each 20 s segment under traffic sounds was compared with the baseline level (2 min silence) to remove the effect of individual differences. Therefore, the change percentages of the physiological parameters were used in further analyses [37,46].

A questionnaire was presented in the VR scene after each sound stimulus for the perceptual evaluation of the green space soundscape, as shown in Figure A1 (Appendix A). Four perceptual attributes, including annoyance, comfort, arousal, and pleasantness, were used to measure the psychological responses to traffic sounds in a green space soundscape. Noise annoyance and acoustic comfort are the most widely used perceptual attributes in relevant research for evaluating the impact of traffic sounds and the soundscape quality, respectively. A five-point verbal scale was used with the verbal marks: (1) "not annoyed at all/very uncomfortable"; (2) "slightly annoyed/uncomfortable"; (3) "moderately annoyed/neither comfortable nor comfortable"; (4) "very annoyed/comfortable"; and (5) "extremely annoyed/very comfortable" [10,47]. In addition, the Self-Assessment Manikin (SAM) was also used to assess the evoked state associated with experiencing sound stimuli [48,49]. As suggested by Hume [31], a 9-point numerical scale was used in this study for the SAM measurement with descriptions from (1) no arousal at all /completely unpleasant to (9) complete arousal/completely pleasant.

2.4. Experimental Procedure

As shown in Figure 4, the experiment was generally divided into two sessions: (1) an adaptation session and (2) a measurement session. In the first 10 min, the participants came into the laboratory to adapt to the experiment environment while the informed consent was read and signed. Then, the VR equipment was put on to adapt to the VR scene for 4 min. During this adaptation session, the physiological sensors were attached, and a practice session for psychological and physiological measurement was conducted. Then, the participants were asked to rest for 1 min with their eyes closed to remove the impact caused by the practice session.

In the measurement session, a silent clip was first used to conduct the baseline measurement. Then, 12 traffic sound stimuli were presented in a random order for each participant. For each experimental stimulus, there were three periods: (1) a 2 min continuous measurement of EDA and HR during the exposure to traffic sound; (2) psychological measurement with the questionnaire, which lasted approximately 30 s; and (3) a 1 min rest to remove the effect of the former stimulus. The total time for the formal experiment was approximately 45 min.

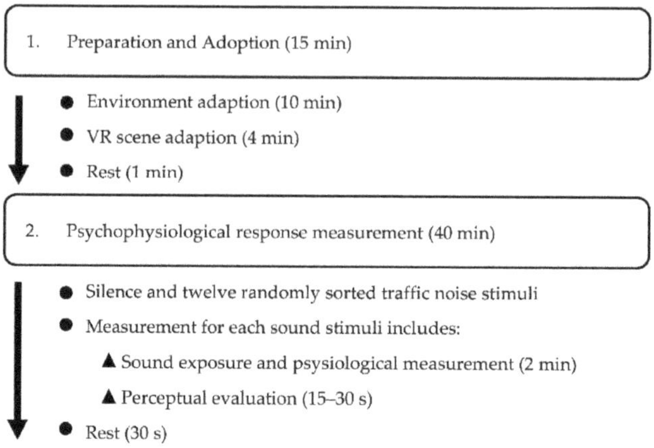

1.   Preparation and Adoption (15 min)

- Environment adaption (10 min)
- VR scene adaption (4 min)
- Rest (1 min)

2.   Psychophysiological response measurement (40 min)

- Silence and twelve randomly sorted traffic noise stimuli
- Measurement for each sound stimuli includes:
  - ▲ Sound exposure and psysiological measurement (2 min)
  - ▲ Perceptual evaluation (15–30 s)
- Rest (30 s)

Figure 4. Schematic diagram of the experiment process.

2.5. Data Analysis

In this study, all statistical analyses were conducted using the SPSS 20.0 software. The multivariate analysis of variance (MANOVA) was first used to identify the influential factors for psychophysiological responses, with the independent variables being sound level, sound type, gender, and exposure time. In addition, the interaction effect of sound type and sound level was also included in the MANOVA analysis. Before the MANOVA analysis, the normality assumptions of the measured responses for each level of the independent variables were examined with the Kolmogorov–Smirnov test. Some of the dataset violated the normality assumption (acoustic comfort and noise annoyance). However, it was suggested that an ANOVA analysis could still yield robust and valid results with non-normally distributed data [50]. The homogeneity of variance was verified with Levene's test (acoustic comfort: $p = 0.0.564$; annoyance: $p = 0.958$; arousal: $p = 0.925$; pleasantness: $p = 0.0.337$; EDA: $p = 0.952$; HR: $p = 0.287$). A pairwise comparison was also applied to show where the differences lay. The least significant difference test (LSD) and the Mann–Whitney U test were applied for variance homogeneity and heterogeneity cases, respectively. Finally, a Spearman correlation analysis was applied to investigate the relationship between the physiological responses and psychological responses. In all analyses, a p-value less than 0.05 was used as the criterion to determine significant differences.

3. Result

3.1. Effect of Traffic Sounds on Psychological Responses

The MANOVA analysis was applied to investigate the effects of the experimental factors on the subjects' psychological responses, including acoustic comfort, annoyance, arousal, and pleasantness. Four factors were used in the ANOVA analysis, including gender, sound type, sound level, and the interaction of sound type and sound level. As discussed in Section 2, there were significant temporal differences between the traffic sound stimuli. Therefore, two different sound level indicators, the equivalent sound level (L_{Aeq}) and the peak level (L_{Afmax}), were used in two independent ANOVA analyses, as shown in Table 1. The results show that gender only showed significant effects on arousal and pleasantness evaluations. Acoustic factors, including sound type and sound level, showed significant effects on all four evaluation dimensions. Using L_{Aeq} as the sound level index (configuration 1), both the sound type and the sound level showed significant influences on participants' psychological responses. However, when replacing L_{Aeq} with L_{Afmax} (configuration 2), only the sound level showed a significant influence on psychological responses. In both ANOVA analyses, no significant interaction effects of sound type and sound level were found.

Table 1. Results of multivariate test of psychological evaluations of traffic sounds. * and ** represent significant differences at 0.05 and 0.01 levels, respectively. Sig. and η_p^2 are the significance coefficient and effect size factor, respectively.

Model	Factor	Comfort		Annoyance		Arousal		Pleasantness	
		Sig.	η_p^2	Sig.	η_p^2	Sig.	η_p^2	Sig.	η_p^2
Model 1: using equivalent sound level	Gender	0.321	0.00	0.09	0.01	0.006 **	0.02	0.006 **	0.02
	Noise type	0.002 **	0.04	0.038 *	0.02	0.050 *	0.02	0.047 *	0.02
	L_{Aeq}	0.000 **	0.37	0.000 **	0.34	0.000 **	0.32	0.000 **	0.31
	Noise type * L_{Aeq}	0.938	0.01	0.585	0.01	0.983	0.00	0.74	0.01
Model 2: using peak sound level	Gender	0.319	0.00	0.089	0.01	0.006 **	0.02	0.006 **	0.02
	Noise type	0.579	0.01	0.118	0.02	0.562	0.01	0.296	0.01
	L_{AFmax}	0.000 **	0.37	0.000 **	0.34	0.000 **	0.32	0.000 **	0.31
	Noise type * L_{AFmax}	0.584	0.01	0.111	0.02	0.767	0.00	0.379	0.01

As shown in Figure 5, the sound type also had a significant influence on the subjects' psychological evaluation. Compared with the baseline condition (silence), all four traffic sounds showed strong negative psychological impacts (lower comfort, higher annoyance, higher arousal, and lower pleasantness). By a further pairwise comparison, the psychological impact of tram sound was found to be significantly stronger than that of the other three traffic sounds. A rank could be assigned based on the negative effects, such as: tram > high-speed train > conventional train > road, which is contrary to the duration order shown in Figure 3.

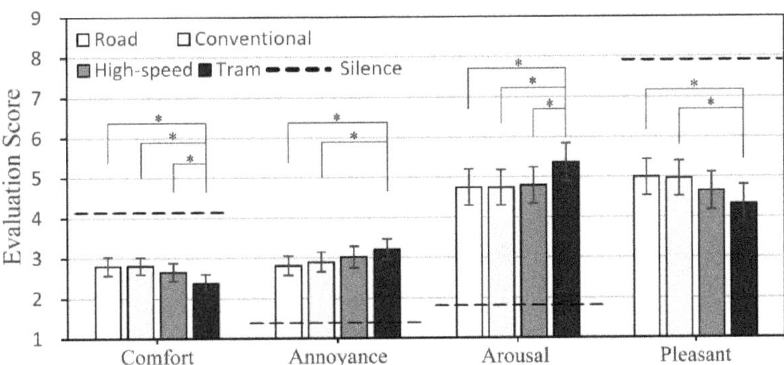

Figure 5. Effect of sound type on psychological attributes. The vertical bars represent 95% confidence intervals. * represents significant differences in pair-wise comparison (LSD) at 0.05 level.

This result explains the difference between the two ANOVA analyses with different sound level indicators, in which the peak sound level showed superior performance compared to the equivalent level in explaining the psychological impacts of different traffic sounds on people. In this experiment, the major difference between the four traffic sounds was temporal duration (road > conventional > high-speed > tram). As discussed in Section 2, this led to the difference in the peak sound level when the equivalent level was equalized. Therefore, the peak sound level described not only the overall sound level but also the temporal characteristics, which led to superior performance in explaining the psychological responses of the participants.

As shown in Figure 6, strong linear correlations were found between the psychological responses and L_{Afmax} ($R^2 = 0.88$–0.9). The negative effects of traffic sounds continued to increase with the increase in L_{Afmax}. According to the regression equation, a 5 dB increase in L_{Afmax} led to changes of 0.35, 0.35, 0.65, and 0.65 for comfort, annoyance, arousal, and pleasantness evaluation, respectively. Meanwhile, the sound levels for achieving neutral evaluations in each psychological dimension were different. By the regression equation, the upper limit of L_{Afmax} for avoiding negative evaluations could be recognized as 60, 69, 65, and 63 dB for comfort, annoyance, arousal, and pleasantness evaluation, respectively. These results show that people have a greater tolerance for traffic noise annoyance than the other three evaluation dimensions. Therefore, relatively weak traffic sounds could have significant negative effects on the psychological state (arousal and pleasantness) and the overall soundscape quality (comfort), although they were perceived as "not annoying". These results indicate the insufficiency of the questionnaire survey with annoyance as the only evaluation dimension to evaluate the impact of traffic sounds.

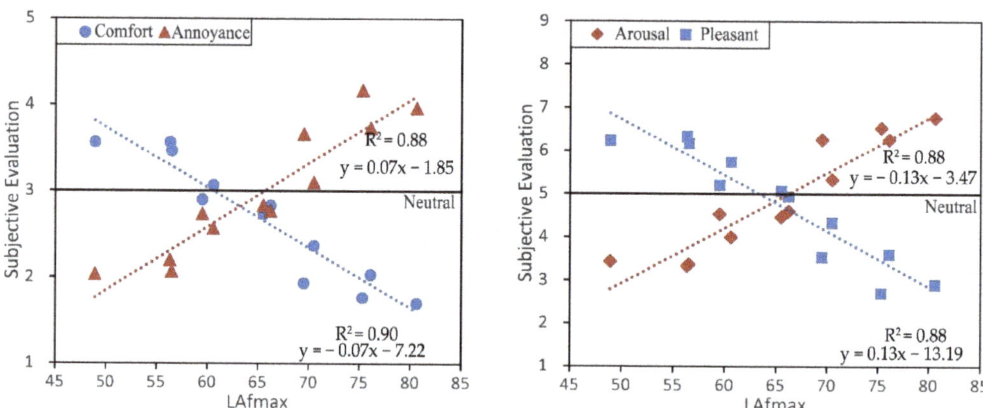

Figure 6. Effect of maximum sound level on psychological responses.

Besides acoustical factors, significant main effects of gender on psychological state were also found, including arousal and pleasantness. As shown in Figure 7, under the impact of traffic sounds, the evaluations of female participants were significantly more negative than those of male participants, with higher arousal evaluations and lower pleasantness evaluations. This result shows that being exposed to traffic sounds has a stronger psychological impact on females than on males in urban green spaces.

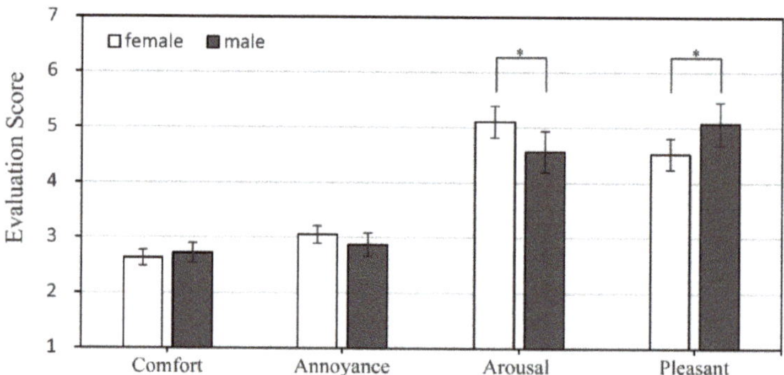

Figure 7. Effect of gender on psychological attributes. The vertical bars represent 95% confidence intervals. * represents significant differences in pair-wise comparison (LSD) at 0.05 level.

3.2. Effect of Traffic Sounds on Physiological Responses

Table 2 shows the results of the MANOVA analysis for physiological responses (EDA and HR). It revealed that both the acoustic factors and the non-acoustic factors had significant effects on the physiological impact of traffic sounds. The main effect of gender was found to be significant on both EDA and HR. Besides, a significant main effect of the exposure time on HR was also found. As for the acoustic factors, the main effects of sound type and sound level were found to be significant on EDA and HR, respectively. In addition, the interaction between the sound type and sound level was observed to be influential on EDA.

Table 2. MANOVA analysis of physiological responses to traffic sounds. * and ** represent significant differences at 0.05 and 0.01 level, respectively.

Factor	EDA			HR		
	F	Sig.	η_p^2	F	Sig.	η_p^2
Gender	92.27	0.000 **	0.041	18.77	0.000 **	0.009
Time	0.07	0.996	0.000	4.92	0.000 **	0.011
Noise Type	3.11	0.025 *	0.004	2.02	0.109	0.003
SPL	1.56	0.209	0.001	3.98	0.019 *	0.004
Noise Type * SPL	3.52	0.002 **	0.010	1.17	0.321	0.003

Figure 8 shows the effects of the sound type and sound level on the physiological responses. By further pairwise comparisons, five significant differences were identified. For the sound type, all four significant differences were between the high-speed train sound and the other traffic sounds, including: (1) the EDA under the high-speed train sound was significantly higher than those under the road traffic sound ($p = 0.010$) and conventional train sound ($p = 0.008$) and (2) the HR under the high-speed train sound was significantly lower than those under the conventional train sound ($p = 0.035$) and tram sound ($p = 0.038$). These results indicate that the physiological impact of the high-speed train sound was significantly different from other traffic sounds. As for the sound level, only one pairwise comparison was found to be significant. A significant increase in HR was found when the sound level increased from 55 dB to 65 dB. This result indicates that the main effect of SPL on the physiological responses was relatively limited, especially when the sound level was low (45 to 55 dB).

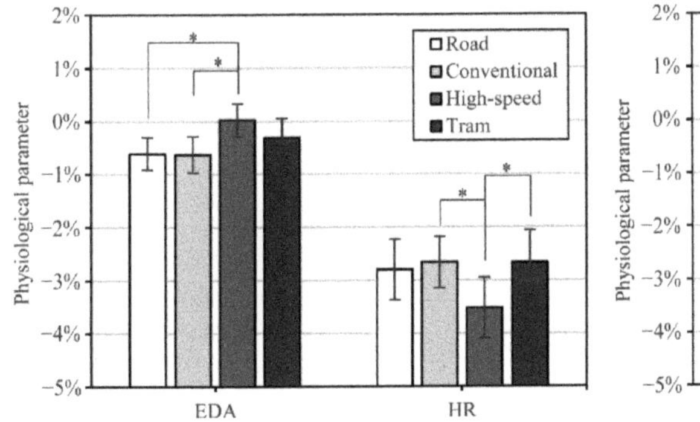
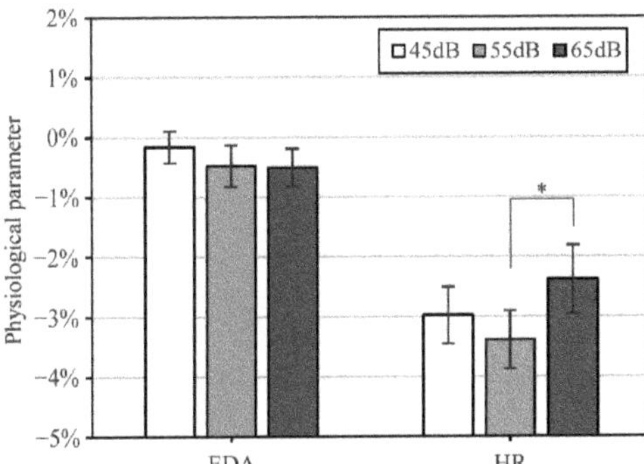

Figure 8. Main effect of noise type and noise level on physiological responses. * represents significant differences in LSD pairwise comparison at the 0.05 level.

An explanation for this phenomenon is that the relationship between the SPL and physiological responses varied in different sound groups, as indicated by the significant interaction effect between SPL and sound type in Table 2. Further pairwise comparisons were carried out to investigate how physiological responses varied with the increase in sound level in each sound group. As shown in Figure 9, the results varied wildly in different traffic groups. As for the road traffic sound, the change in SPL from 45 dB to 65 dB could hardly affect the participant's physiological responses, including EDA and HR, while significant changes in physiological responses were found with the increase in SPL in the railway sound groups. As for the high-speed train sound, when the sound level increased from 45 dB to 65 dB, a significant decrease in EDA was found ($\chi^2 = -3.397$, $p = 0.001$). For the conventional train sound and tram sound, the increase in sound level led to significant increases in HR: (1) when the conventional train sound increased from 45 to 65 dB ($\chi^2 = 2.155$, $p = 0.031$) and 55 to 65 dB ($\chi^2 = 2.035$, $p = 0.042$) and (2) when the tram sound increased from 55 to 65 dB ($\chi^2 = 2.313$, $p = 0.021$). This result indicated that the effect of sound level on the physiological responses depended on the sound type. As for the steady road traffic sound, the decrease in sound level was insufficient to reduce the physiological impact on participants. On the contrary, the control of sound levels is still an effective treatment for controlling the impact of intermittent railway sounds.

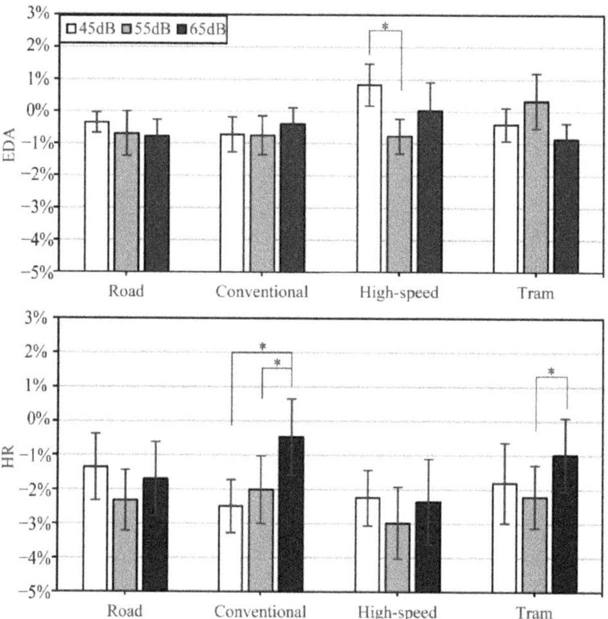

Figure 9. Interaction effect of sound level and sound type on EDA and HR. The vertical bars represent 95% confidence intervals. * represents significant differences in the Mann–Whitney U test at the 0.05 level.

Besides acoustic factors, non-acoustic factors (gender and exposure time) were also found to be influential on physiological responses, as shown in Figure 10. According to the effect factor, gender was found to be more influential on EDA and HR than the acoustic factors. When exposed to traffic sounds, the EDA and HR of males were significantly higher (F = 92.27, $p = 0.000$) and lower (F = 18.77, $p = 0.000$), respectively. Meanwhile, HR was also significantly affected by the exposure time (F = 4.92, $p = 0.000$). With the increase in exposure time, the HR of participants decreased rapidly in the first 60 s and gradually stabilized after 60 s. This result reveals that there was a strong adoption effect on HR when exposed to traffic sounds.

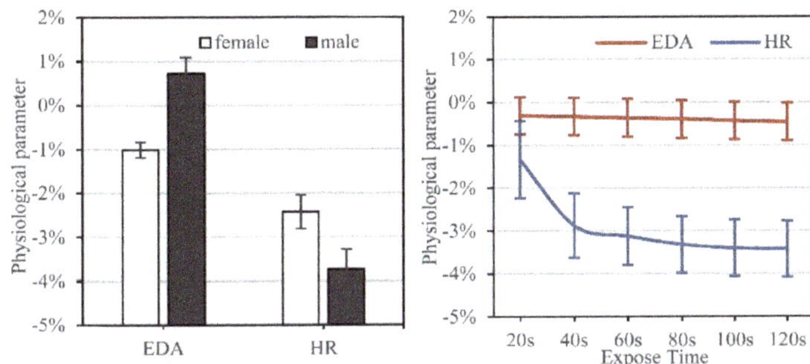

Figure 10. Effects of gender and exposure time on physiological responses to traffic sounds. The vertical bars represent 95% confidence intervals.

3.3. Relationship between Psychological and Physiological Responses

A Spearman correlation analysis was applied to reveal the relationship between the psychological responses and physiological responses, as shown in Table 3. Strong correlations were found between the psychological responses ($|r| = 0.84 - 0.94$, $p < 0.01$). The EDA parameters were discovered to be independent of the HR parameters ($r = -0.057$). Only one weak correlation was found between the physiological responses and psychological responses: that between HR and annoyance ($r = 0.107$, $p < 0.05$). It indicates that more annoying traffic sounds lead to a higher HR level when exposed to traffic sounds. These results revealed that the physiological responses and psychological responses were relatively independent from each other when exposed to traffic sounds. Therefore, it might be insufficient to assess the physiological impact of traffic sounds with the psychological evaluations of the sound environment through self-reported surveys. The measurement of physiological responses might be necessary to have a comprehensive investigation of the impact of traffic sounds on the urban environment.

Table 3. Correlation analysis between psychological and physiological responses. ** represent significant difference at 0.01 level.

Spearman		Perceptual				Physiological	
		Comfort	Annoyance	Arousal	Pleasantness	EDA$_{2min}$	HR$_{2min}$
Perceptual	Comfort	1					
	Annoyance	−0.837 **	1				
	Arousal	−0.858 **	0.863 **	1			
	Pleasantness	0.866 **	−0.853 **	−0.939 **	1		
Physiological	EDA$_{2min}$	0.034	0.049	0.037	−0.005	1	
	HR$_{2min}$	0.101	0.107*	0.082	0.088	−0.057	1

Figure 11 shows the relationship between noise annoyance and physiological responses in each traffic group. In all four groups, EDA was independent of the noise annoyance, while in the conventional train group, a significant positive correlation was identified. In the road traffic and tram groups, the correlation coefficient between HR and annoyance was also positive. However, in the high-speed train group, a weak negative correlation between HR and annoyance was found, which was different from the other three traffic sounds. This result agreed with the result in Figure 8 that the psychophysiological impact of the high-speed train sound was significantly different from the other traffic sounds.

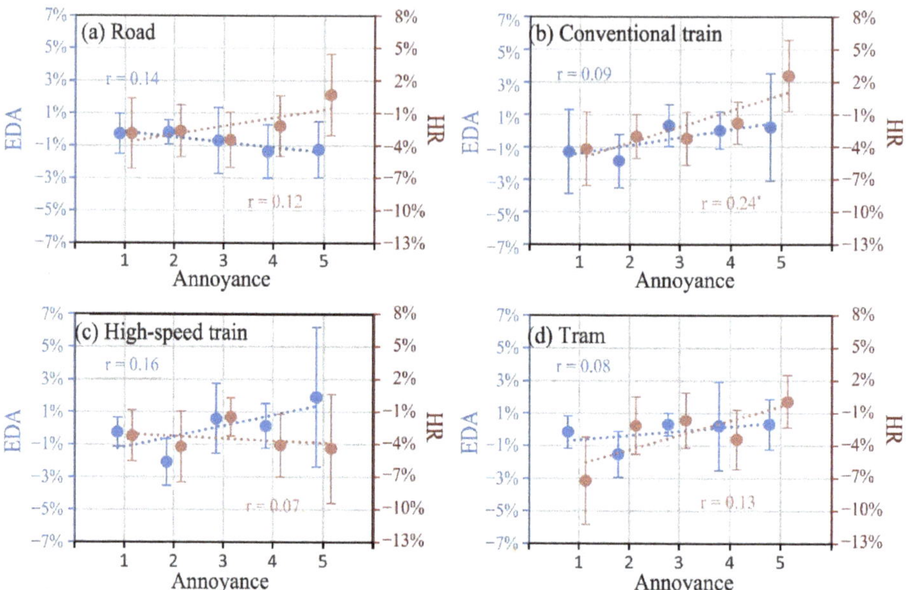

Figure 11. Relationship between psychological response (annoyance) and physiological responses in four traffic sound groups: (**a**) road traffic; (**b**) conventional train; (**c**) High-speed train; (**d**) tram. * represents significant correlation at 0.05 level.

4. Discussion

The findings of this study show that traffic noise has a significant negative impact on the urban green space soundscape, both psychologically and physiologically. Both sound level and traffic type have significant effects on these impacts. However, there were only very weak correlations between the physiological responses and the psychological responses. Therefore, the widely used self-reported questionnaire survey focusing on traffic noise annoyance and soundscape quality might be insufficient for evaluating the physiological impact of traffic sounds. In addition, the common noise level control treatment might be inefficient in controlling the physiological impact of traffic sounds.

As expected, the psychological responses in green space were found to be dominated by the traffic sound magnitude, which agreed with the results of long-term field surveys [51,52]. In addition, the results in this study show that not only the overall energy magnitude but also the temporal variation characteristics are important in determining the psychological impacts. The peak sound level, L_{Afmax}, shows superior performance to the equivalent level. These results reveal that more steady traffic sounds lead to less impact on the green space soundscape when the overall sound energy is controlled. As a result, limiting the carrying capacity and speed of the vehicle may result in a reduction in traffic noise impact, even as the number of vehicles increases.

However, the results of the physiological responses indicated that the decrease in traffic sound level might be inefficient in controlling the physiological impacts because they were strongly affected by the traffic type. First, the impact of the high-speed train sound was found to be significantly different from other traffic sounds. Meanwhile, the relationships between the traffic sound level and physiological parameters varied in four traffic groups. Therefore, the common noise barrier might be effective for railway sounds but ineffective for road traffic sounds. To improve the green space soundscape quality, new strategies should be specially designed according to the traffic type. It was suggested that in the mixed soundscape, the presence of natural sounds, i.e., birds singing and water sounds, had a positive physiological effect in a soundscape exposed to road traffic sound [34]. As a result, the physiological masking effect of pleasant natural sounds could be a potential solution for reducing the impact of traffic noise when noise control treatments are not efficient.

Some limitations of this study should also be noted. First, silence was chosen as the baseline condition, and only traffic sounds were used as the experimental stimuli in this study. However, there are inevitably other sound sources in actual urban green space soundscapes. As suggested by Suko et al., there are significant interactions between different sound sources in mixed soundscapes [34]. Therefore, in a green space soundscape with multiple sound sources, especially when the traffic sounds are relatively weak, the effect of traffic sounds might be different. Future studies need to be conducted to investigate the effects of mixed sounds in such soundscapes. Second, a static picture without traffic vehicles was used in the VR system to establish the visual environment in this study. As suggested by Li et al., there are significant interactions between visual and acoustic information on the psychophysiological response to a soundscape [35]. Therefore, the visual information of the transportation system might also have a potential influence on how people respond to the traffic noises in green space soundscapes. In general, this paper presents the results of an empirical study in which the effect of the two most fundamental factors of traffic sounds were considered: sound type and sound level. However, because of the complexity of the actual urban soundscape, experiments with more factors, including acoustic factors and non-acoustic factors, are needed to have a comprehensive investigation of the psychophysiological effects of traffic sounds in urban green space soundscapes.

5. Conclusions

A laboratory experiment on the psychophysiological impact of traffic sounds on an urban green space soundscape was conducted in this study. The following results were obtained:

In a green space soundscape, significant psychophysiological impacts of traffic sound were observed in the experiment. However, the psychological responses to traffic sounds were found to be independent from the physiological responses. Only one weak correlation between HR and annoyance was found to be significant in the experiment. Higher HR levels were observed when exposed to more annoying traffic sounds.

Strong negative effects of traffic sounds on the psychological assessment of the green space soundscape were identified, which were determined by both the sound type and the sound level. The peak sound level was found to be superior to the equivalent level in describing the psychological impacts of the various traffic sounds.

The physiological effects of the traffic sounds were also discovered to be determined by sound type and sound level. As for the sound type, the physiological response to the high-speed train sound was significantly different from the other three traffic sounds. Exposure to the high-speed train sound led to lower EDA (compared with conventional train and tram sound) and higher HR (compared with road and tram sound). The relationship between the sound level and physiological parameters also depended on the sound type. The physiological impacts of road traffic sound were found to be irrelevant to the sound level. On the contrary, the change in sound level was found to be influential on the participants' physiological parameters in the railway sound group.

The findings in this study highlight that not only the sound level but also the sound type determine the psychophysiological impact of traffic sounds in urban green spaces. Especially for the physiological effects, controlling the sound level does not necessarily lead to an improvement in the environmental quality. Therefore, common noise treatments, such as noise barriers, might be less efficient at masking traffic sounds. Soundscape treatments, such as introducing natural sounds, for instance, might be more effective in masking the negative effect of traffic sounds. To offer more efficient and practical methodologies for architects, planners, and city managers, we will focus on the performance of applying soundscape treatments in reducing the impact of traffic sounds in future studies.

Author Contributions: Conceptualization, B.Y.; methodology, B.Y., J.B. and L.W.; formal analysis, B.Y. and Y.C.; investigation, J.B. and L.W.; data curation, B.Y.; writing—original draft preparation, B.Y.; writing—review and editing, B.Y.; supervision, B.Y.; project administration: B.Y.; funding acquisition, B.Y. All authors have read and agreed to the published version of the manuscript.

Funding: This research was funded by the National Natural Science Foundation of China (Project No. 51808030 and No. 52078026).

Acknowledgments: The authors wish to thank Wei Wu, and Junjie Li for their help with the VR software.

Conflicts of Interest: The authors declare no conflict of interest.

Appendix A

Questionnaire for psychological evaluation:
Introduction:
"Begin by listening through the 12 traffic sounds presented in the headphone, and build your opinion about their character. Thereafter you measure the traffic sounds with the aid of four attribute scales that appears in front of you. The traffic sounds must be measured one at a time on all the attribute scales and in the order presented from up to down.

Your task is to judge to what extent the attributes listed in the protocol are applicable to the traffic sounds in the current environment."

a. Comfort

1-very uncomfortable   **2**-uncomfortable   **3**-Neither comfortable nor uncomfortable   **4**-comfortable   **5**-very comfortable

b. Annoyance

1-not annoyed at all   **2**-slightly annoyed   **3**-moderately annoyed   **4**-very annoyed   **5**-extremely annoyed

c. Arsoual

1-not at all arousal   **9**-complete arousal

1 2 3 4 5 6 7 8 9

d. Pleasantness

1-complete unpleasant   **9**-complete pleasant

Figure A1. Questionnaire used in the experiment.

References

1. De Ridder, K.; Adamec, V.; Bañuelos, A.; Bruse, M.; Bürger, M.; Damsgaard, O.; Dufek, J.; Hirsch, J.; Lefebre, F.; Pérez-Lacorzana, J.M.; et al. An integrated methodology to assess the benefits of urban green space. *Sci. Total Environ.* **2004**, *334–335*, 489–497. [CrossRef] [PubMed]
2. Mörtberg, U.; Wallentinus, H.G. Red-listed forest bird species in an urban environment—Assessment of green space corridors. *Landsc. Urban Plan.* **2000**, *50*, 215–226. [CrossRef]
3. Hillsdon, M.; Panter, J.; Foster, C.; Jones, A. The relationship between access and quality of urban green space with population physical activity. *Public Health* **2006**, *120*, 1127–1132. [CrossRef] [PubMed]
4. Jiang, B.; Larsen, L.; Deal, B.; Sullivan, W.C. A dose-response curve describing the relationship between tree cover density and landscape preference. *Landsc. Urban Plan.* **2015**, *139*, 16–25. [CrossRef]
5. Nilsson, M.E.; Berglund, B. Soundscape quality in suburban green areas and city parks. *Acta Acust. United Acust.* **2006**, *92*, 903–911.
6. Liu, J.; Kang, J.; Behm, H.; Luo, T. Landscape and Urban Planning Effects of landscape on soundscape perception: Soundwalks in city parks. *Landsc. Urban Plan.* **2014**, *123*, 30–40. [CrossRef]
7. Liu, J.; Kang, J. Soundscape design in city parks: Exploring the relationships between soundscape composition parameters and physical and psychoacoustic parameters. *J. Environ. Eng. Landsc. Manag.* **2015**, *23*, 102–112. [CrossRef]
8. Lionello, M.; Aletta, F.; Kang, J. A systematic review of prediction models for the experience of urban soundscapes. *Appl. Acoust.* **2020**, *170*, 107479. [CrossRef]
9. Hong, J.Y.; Jeon, J.Y. Influence of urban contexts on soundscape perceptions: A structural equation modeling approach. *Landsc. Urban Plan.* **2015**, *141*, 78–87. [CrossRef]

10. Yang, W.; Kang, J. Acoustic comfort evaluation in urban open public spaces. *Appl. Acoust.* **2005**, *66*, 211–229. [CrossRef]
11. Botteldooren, D.; De Coensel, B.; De Muer, T. The temporal structure of urban soundscapes. *J. Sound Vib.* **2006**, *292*, 105–123. [CrossRef]
12. Kogan, P.; Gale, T.; Arenas, J.P.; Arias, C. Development and application of practical criteria for the recognition of potential Health Restoration Soundscapes (HeReS) in urban greenspaces. *Sci. Total Environ.* **2021**, *793*, 148541. [CrossRef] [PubMed]
13. Szeremeta, B.; Zannin, P.H.T. Analysis and evaluation of soundscapes in public parks through interviews and measurement of noise. *Sci. Total Environ.* **2009**, *407*, 6143–6149. [CrossRef] [PubMed]
14. Margaritis, E.; Kang, J.; Filipan, K.; Botteldooren, D. The influence of vegetation and surrounding traffic noise parameters on the sound environment of urban parks. *Appl. Geogr.* **2018**, *94*, 199–212. [CrossRef]
15. Brambilla, G.; Gallo, V.; Zambon, G. The Soundscape Quality in Some Urban Parks in Milan, Italy. *Int. J. Environ. Res. Public Health* **2013**, *10*, 2348–2369. [CrossRef] [PubMed]
16. Liu, J.; Wang, Y.; Zimmer, C.; Kang, J.; Yu, T. Factors associated with soundscape experiences in urban green spaces: A case study in Rostock, Germany. *Urban For. Urban Green.* **2019**, *37*, 135–146. [CrossRef]
17. Benfield, J.A.; Bell, P.A.; Troup, L.J.; Soderstrom, N.C. Aesthetic and affective effects of vocal and traffic noise on natural landscape assessment. *J. Environ. Psychol.* **2010**, *30*, 103–111. [CrossRef]
18. Evensen, K.H.; Raanaas, R.K.; Fyhri, A. Soundscape and perceived suitability for recreation in an urban designated quiet zone. *Urban For. Urban Green.* **2016**, *20*, 243–248. [CrossRef]
19. Watts, G.; Miah, A.; Pheasant, R. Tranquillity and soundscapes in urban green spaces- predicted and actual assessments from a questionnaire survey. *Environ. Plan. B Plan. Des.* **2013**, *40*, 170–181. [CrossRef]
20. Jarosińska, D.; Héroux, M.È.; Wilkhu, P.; Creswick, J.; Verbeek, J.; Wothge, J.; Paunović, E. Development of the WHO environmental noise guidelines for the european region: An introduction. *Int. J. Environ. Res. Public Health* **2018**, *15*, 813. [CrossRef]
21. Muzet, A. Environmental noise, sleep and health. *Sleep Med. Rev.* **2007**, *11*, 135–142. [CrossRef] [PubMed]
22. Ndrepepa, A.; Twardella, D. Relationship between noise annoyance from road traffic noise and cardiovascular diseases: A meta-analysis. *Noise Health* **2011**, *13*, 251. [PubMed]
23. Bhatia, P.; Muhar, I. Noise sensitivity and mental efficiency. *Psychol. Int. J. Psychol. Orient* **1988**, *31*, 163–169.
24. Vera, M.N.; Vila, J.; Godoy, J.F. Physiological and subjective effects of traffic noise: The role of negative self-statements. *Int. J. Psychophysiol.* **1992**, *12*, 267–279. [CrossRef]
25. Clark, C.R.; Turpin, G.; Jenkins, L.M.; Tarnopolsky, A. Sensitivity to noise on a community sample: II. Measurement of psychophysiological indices. *Psychol. Med.* **1985**, *15*, 255–263. [CrossRef]
26. Raggam, R.B.; Cik, M.; Höldrich, R.R.; Fallast, K.; Gallasch, E.; Fend, M.; Lackner, A.; Marth, E. Personal noise ranking of road traffic: Subjective estimation versus physiological parameters under laboratory conditions. *Int. J. Hyg. Environ. Health* **2007**, *210*, 97–105. [CrossRef]
27. Basner, M.; Samel, A.; Elmenhorst, E.M.; Maab, H.; Müller, U.; Quehl, J.; Vejvoda, M. Single and combined effects of air, road and rail traffic noise on sleep. In Proceedings of the 35th International Congress and Exposition on Noise Control Engineering (INTER-NOISE 2006), Honolulu, HI, USA, 3–6 December 2006; Volume 7, pp. 4629–4637.
28. Lee, P.J.; Park, S.H.; Jeong, J.H.; Choung, T.; Kim, K.Y. Association between transportation noise and blood pressure in adults living in multi-storey residential buildings. *Environ. Int.* **2019**, *132*, 105101. [CrossRef] [PubMed]
29. Li, Z.; Kang, J.; Ba, M. Influence of distance from traffic sounds on physiological indicators and subjective evaluation. *Transp. Res. Part D Transp. Environ.* **2020**, *87*, 102538. [CrossRef]
30. Erfanian, M.; Mitchell, A.J.; Kang, J.; Aletta, F. The psychophysiological implications of soundscape: A systematic review of empirical literature and a research agenda. *Int. J. Environ. Res. Public Health* **2019**, *16*, 3533. [CrossRef]
31. Hume, K.; Ahtamad, M. Physiological responses to and subjective estimates of soundscape elements. *Appl. Acoust.* **2013**, *74*, 275–281. [CrossRef]
32. Irwin, A.; Hall, D.A.; Peters, A.; Plack, C.J. Listening to urban soundscapes: Physiological validity of perceptual dimensions. *Psychophysiology* **2011**, *48*, 258–268. [CrossRef]
33. Alvarsson, J.J.; Wiens, S.; Nilsson, M.E. Stress recovery during exposure to nature sound and environmental noise. *Int. J. Environ. Res. Public Health* **2010**, *7*, 1036–1046. [CrossRef]
34. Suko, Y.; Saito, K.; Takayama, N.; Sakuma, T.; Warisawa, S. Effect of faint road traffic noise mixed in birdsong on the perceived restorativeness and listeners' physiological response: An exploratory study. *Int. J. Environ. Res. Public Health* **2019**, *16*, 4985. [CrossRef] [PubMed]
35. Li, Z.; Ba, M.; Kang, J. Physiological indicators and subjective restorativeness with audio-visual interactions in urban soundscapes. *Sustain. Cities Soc.* **2021**, *75*, 103360. [CrossRef]
36. Annerstedt, M.; Jönsson, P.; Wallergård, M.; Johansson, G.; Karlson, B.; Grahn, P.; Hansen, Å.M.; Währborg, P. Inducing physiological stress recovery with sounds of nature in a virtual reality forest—Results from a pilot study. *Physiol. Behav.* **2013**, *118*, 240–250. [CrossRef]
37. Shu, S.; Ma, H. Restorative effects of urban park soundscapes on children's psychophysiological stress. *Appl. Acoust.* **2020**, *164*, 107293. [CrossRef]
38. Li, H.; Xie, H.; Woodward, G. Soundscape components, perceptions, and EEG reactions in typical mountainous urban parks. *Urban For. Urban Green.* **2021**, *64*, 127269. [CrossRef]

39. Hygge, S. Classroom experiments on the effects of different noise sources and sound levels on long-term recall and recognition in children. *Appl. Cogn. Psychol.* **2003**, *17*, 895–914. [CrossRef]
40. Brown, A.L.; van Kamp, I. WHO environmental noise guidelines for the European region: A systematic review of transport noise interventions and their impacts on health. *Int. J. Environ. Res. Public Health* **2017**, *14*, 873. [CrossRef] [PubMed]
41. Yano, T.; Sato, T.; Björkman, M.; Rylander, R. Comparison of community response to road traffic noise in Japan and Sweden—Part II: Path analysis. *J. Sound Vib.* **2002**, *250*, 169–174. [CrossRef]
42. Tetsuya, H.; Yano, T.; Murakami, Y. Annoyance due to railway noise before and after the opening of the Kyushu Shinkansen Line. *Appl. Acoust.* **2017**, *115*, 173–180. [CrossRef]
43. Kuwano, S.; Namba, S.; Okamoto, T. Psychological evaluation of sound environment in a compartment of a high-speed train. *J. Sound Vib.* **2004**, *277*, 491–500. [CrossRef]
44. Trollé, A.; Marquis-Favre, C.; Klein, A. Short-term annoyance due to tramway noise: Determination of an acoustical indicator of annoyance via multilevel regression analysis. *Acta Acust. United Acust.* **2014**, *100*, 34–45. [CrossRef]
45. Lee, Y.; Nelson, E.C.; Flynn, M.J.; Jackman, J.S. Exploring soundscaping options for the cognitive environment in an open-plan office. *Build. Acoust.* **2020**, *27*, 185–202. [CrossRef]
46. Park, S.H.; Lee, P.J.; Jeong, J.H. Effects of noise sensitivity on psychophysiological responses to building noise. *Build. Environ.* **2018**, *136*, 302–311. [CrossRef]
47. Brink, M.; Schreckenberg, D.; Vienneau, D.; Cajochen, C.; Wunderli, J.M.; Probst-Hensch, N.; Röösli, M. Effects of scale, question location, order of response alternatives, and season on self-reported noise annoyance using icben scales: A field experiment. *Int. J. Environ. Res. Public Health* **2016**, *13*, 1163. [CrossRef] [PubMed]
48. McManis, M.H.; Bradley, M.M.; Berg, W.K.; Cuthbert, B.N.; Lang, P.J. Emotional reactions in children: Verbal, physiological, and behavioral responses to affective pictures. *Psychophysiology* **2001**, *38*, 222–231. [CrossRef] [PubMed]
49. Bradley, M.M.; Lang, P.J. Measuring emotion: The self-assessment manikin and the semantic differential. *J. Behav. Ther. Exp. Psychiatry* **1994**, *25*, 49–59. [CrossRef]
50. Schmider, E.; Ziegler, M.; Danay, E.; Beyer, L.; Bühner, M. Is It Really Robust?: Reinvestigating the robustness of ANOVA against violations of the normal distribution assumption. *Methodology* **2010**, *6*, 147–151. [CrossRef]
51. Morihara, T.; Sato, T.; Yano, T. Comparison of dose-response relationships between railway and road traffic noises: The moderating effect of distance. *J. Sound Vib.* **2004**, *277*, 559–565. [CrossRef]
52. Fidell, S.; Mestre, V.; Schomer, P.; Berry, B.; Gjestland, T.; Vallet, M.; Reid, T. A first-principles model for estimating the prevalence of annoyance with aircraft noise exposure. *J. Acoust. Soc. Am.* **2011**, *130*, 791–806. [CrossRef] [PubMed]

forests

Article

Effects of Forest on Birdsong and Human Acoustic Perception in Urban Parks: A Case Study in Nigeria

Mary Nwankwo [1], Qi Meng [1,2,*], Da Yang [1] and Fangfang Liu [1]

[1] Key Laboratory of Cold Region Urban and Rural Human Settlement Environment Science and Technology, Ministry of Industry and Information Technology, School of Architecture, Harbin Institute of Technology, 66 West Dazhi Street, Nan Gang District, Harbin 150001, China; blessedmaryc@gmail.com (M.N.); da.yang@hit.edu.cn (D.Y.); liufangfang@hit.edu.cn (F.L.)

[2] Key Laboratory of Architectural Acoustic Environment of Anhui Higher Education Institutes, Anhui Jianzhu University, Hefei 230601, China

* Correspondence: mengq@hit.edu.cn

Abstract: The quality of the natural sound environment is important for the well-being of humans and for urban sustainability. Therefore, it is important to study how the soundscape of the natural environment affects humans with respect to the different densities of vegetation, and how this affects the frequency of singing events and the sound pressure levels of common birds that generate natural sounds in a commonly visited urban park in Abuja, Nigeria. This study involves the recording of birdsongs, the measurement of sound pressure levels, and a questionnaire evaluation of sound perception and the degree of acoustic comfort in the park. Acoustic comfort, which affects humans, describes the fundamental feelings of users towards the acoustic environment. The results show that first, there is a significant difference between the frequency of singing events of birds for each category of vegetation density (low, medium, and high density) under cloudy and sunny weather conditions, but there is no significant difference during rainy weather. Secondly, the measured sound pressure levels of the birdsongs are affected by vegetation density. This study shows a significant difference between the sound pressure levels of birdsongs and the vegetation density under cloudy, sunny, and rainy weather conditions. In addition, the frequency of singing events of birds is affected by the sound pressure levels of birdsongs with respect to different vegetation densities under different weather conditions. Thirdly, the results from the respondents (N = 160) in this study indicated that the acoustic perception of the park was described as being pleasant, vibrant, eventful, calming, and not considered to be chaotic or annoying in any sense. It also shows that the human perception of birdsong in the park was moderately to strongly correlated with different densities of vegetation, and that demographics play an important role in how natural sounds are perceived in the environment under different weather conditions.

Keywords: soundscape; urban park; birdsong; natural vegetation

Citation: Nwankwo, M.; Meng, Q.; Yang, D.; Liu, F. Effects of Forest on Birdsong and Human Acoustic Perception in Urban Parks: A Case Study in Nigeria. *Forests* **2022**, *13*, 994. https://doi.org/10.3390/f13070994

Academic Editors: Guang-Yu Wang, Xin-Chen Hong, Jiang Liu and Chi Yung Jim

Received: 2 April 2022
Accepted: 21 June 2022
Published: 24 June 2022

Publisher's Note: MDPI stays neutral with regard to jurisdictional claims in published maps and institutional affiliations.

1. Introduction

Various studies in Nigeria have explored the benefits of green areas and how urban areas can be improved. Likewise, different approaches have been adopted to further tackle the challenges that are faced in improving urban areas. Results have shown that the improvement of urban areas in Nigeria is negatively affected by the rapid rate of urbanization, less attention being paid to continuous development, and a limited budget for the physical planning and maintenance of green areas [1]. The importance of green areas in cities has been identified in various ways since the 19th century. Green areas reduce urban air pollution and provide environmental, social, and economic value to society [2]. Because of its importance, the green city concept was initiated in Nigeria to focus on achieving sustainable, green, and environmentally friendly cities, and to reduce the negative effects caused by deforestation, especially the emission of CO_2 into the atmosphere [3]. Another important aspect of a functioning city is its urban forest [4,5] and green spaces [6]. This is because green spaces provide an extensive range of benefits, such as climate adaptation [7], climate mitigation [8,9], erosion control, and physical and psychological comfort [10], and because they affect human senses such as sight, smell, sound, taste, and touch [11]. Research by [12] using a quantitative method studied the variance between urban greenery, urban development, and the quest for environmental sustainability in northern Nigerian cities, and the scientific findings indicated that the allocation of open green spaces has not been harmonized with the urban population. This is influenced by a low percentage of urban greenery. This research further suggested the need for a strict adherence to sustainable urban planning that would integrate physical development and environmental considerations in order to enhance urban greenery. Further strategies that support architects and urban planners in developing guidelines required for city planning and design were suggested by [13,14]. These strategies are of high value to the inhabitants, as they support social meetings for all ages and promote the continuous growth and development of urban areas. Although the importance of green spaces and the need for the proper articulation and implementation of planning policies has been constantly emphasized in Nigeria, more studies need to be conducted on individual green spaces, urban parks, and forest and recreational centers that contribute to the physical, social, and health development of urban areas.

Urban parks are recognized as being major contributors to the physical and aesthetic qualities of urban centers, and they play a role in improving the quality of living in urban areas [15–17]. Their uses are good, and generally, humans are free to engage in healthy exercise or other recreational activities. This in turn promotes health, restorative life [18], and better social interactions [19,20]. Urban parks, relaxation gardens, and healing gardens are perfect scenes where natural soundscapes can be perceived [18], and they can offer a reduced degree of exposure to the adverse effects of anthropogenic noise in urban areas [21]. The benefits of urban parks possess high societal values that serve as pathways for economic growth and locations for the complex network of recreational activities that are essential to human function and standards of living. In addition, they are dynamic places with changeable environments that can transform man's idea of nature, and they fulfill a variety of human needs, such as better air quality and noise reduction [22]. They also serve as a useful environmental source for urban dwellers to improve their physical, mental, cognitive, and social wellbeing [23,24]. Parks can consist of natural vegetation, which is a source of natural sounds, and they can function as an important part of the natural ecosystem. When natural scenes are dominated by green vegetation, it has been studied as a restorative environment [25] where exposure provides restoration from stress and mental fatigue [26]. As the human population increases, especially in urban areas, urban dwellers are constantly faced with stress and the need to immerse themselves daily in nature, and to perceive natural sounds. It is the right moment to address factors that affect common avian species in Nigeria and to consider the utmost need for urban bird conservation. Perceived natural sounds are important and useful signals that improve our daily activities. They are part of the natural landscape and play a huge role in the ecosystem [27]. Natural sounds are vital cues that are used by humans to communicate with one another and to perceive environmental conditions [28]. The study of soundscapes has evolved through various disciplines, such as anthropology, acoustics, architecture, ecology, psychology, and landscaping [29,30]. However, it was originally rooted in music, and was first defined by Murray Schafer as any acoustic field of study [31]. Recently, the International Standard Organization (ISO) defined a soundscape as the acoustic environment as perceived, experienced, or understood by a person or people in context [32]. As the study of sound expanded to other disciplines, research into its relationship with the landscape [33] emerged, where it is used to denote the overall sonic environment. The soundscape and the acoustic comfort of an open area or a closed space, such as a classroom, can be designed, measured, and evaluated [34,35]. This can be achieved by using various methods, including laboratory experiments and questionnaires [5], sound walks [36], simulations and virtual modeling [37,38], and psychoacoustic parameters [36,39]. Soundscape research has shown that natural sounds are commonly perceived as being pleasant by humans [40,41] and they have a positive effect on quick recovery from psychological and physiological stress [19]. Researchers such as [5] have contributed to the growing knowledge of soundscapes by evaluating soundscape perception and preferences amongst different users in an urban recreational forest park in Xi'an. The studies revealed that natural sounds were perceived more positively than other artificial sounds that appeared to be dominating in the park, with age and gender also playing important roles as to how certain sounds were tolerated or perceived with different levels of sensitivity.

According to some previous studies, birds are referred to as being one of the most important types of animals that generate pleasant sounds [42,43]. These sounds allow for individual experiences in nature, and they are exceptionally rich in semantic values and associations [44]. Bird sounds possess symbolic values, which affect how they are cognitively appraised and how restorative they are perceived to be [45,46]. Most of the research on the restoration of natural soundscapes in parks and green areas has been focused on birdsong [47]. For instance, research by [48] explained that the soundscape of a park with rich bird sounds can minimize the adverse effects of traffic noise, provide nature-based solutions to human health, and improve general wellbeing in urban areas. In addition, a field experiment by [49] with 70 participants in Shenyang, China found that natural environments with natural sounds have positive effects on the restoration of individual attention. In situations where a space's soundscape is created by birds, it will have peaks of intensity, quantity, and diverse frequencies. Additionally, the time of the year, the hour of the day, meteorological conditions, climate dynamics [27,29], landscape structure, the structure of vegetation [50,51], and human disturbances are factors that could affect bird activities [52]. It has been found that birdsong increases positive perceptions in humans, as well as reducing psychological stress, and it can also be affected by natural factors [19,45,53]. Thus, this paper also provides its own perspective on the factors that affect birds and the generation of birdsongs in urban parks.

Despite research on the promotion of sustainable urban environments and the evaluation of users' perceptions of green spaces in Nigeria, few studies have been conducted specifically regarding natural soundscapes in green spaces, how the natural vegetation can affect natural sounds, human acoustic comfort, and the effects of change in weather conditions on humans and birds in commonly visited urban parks in Nigeria. Thus, this paper aims at exploring natural soundscapes through the study of birdsongs in different vegetation densities and how it affects humans. It focuses on the following research questions: (1) whether there is a significant difference between the density of natural vegetation, the frequency of singing events, and the singing duration of the birds in an urban park under different weather conditions; (2) whether the density of natural vegetation has a significant difference in the sound pressure level of the birdsongs under different weather conditions in an urban park; and (3) whether human acoustic perception in the park is appropriate, especially when listening to the birdsongs. In addition, what are humans' preferences for weather conditions when visiting the park, and the preferences for natural vegetation density? Therefore, this study characterized different vegetation densities in an urban park into three groups: low density, medium density, and high density, before selection. In addition to habitat, preferences such as the presence of evergreen trees for nesting and the presence of food for the birds was taken into consideration over a period of time. The study also analyzed the frequency of singing events of five common bird species in the park, and furthermore, involved in situ sound recordings, measurements, and a questionnaire survey.

2. Methodology

2.1. Study Area

This study area is located within Abuja city in Nigeria (Figure 1). The park was selected due to current urban development activities around it, with Abuja being one of the fastest-growing cities on the African continent. Abuja city lies within a latitude of 8°25″–9°25″ N and longitude of 6°45″–7°45″ E, and is 180 feet above sea level and 8000 km^2 in land mass [54]. The territory is bounded by Niger to the west and northwest, Kaduna to the northeast, Nasarawa to the east, and Kogi to the southwest, with an estimated population of 6 million as of 2016 [55].

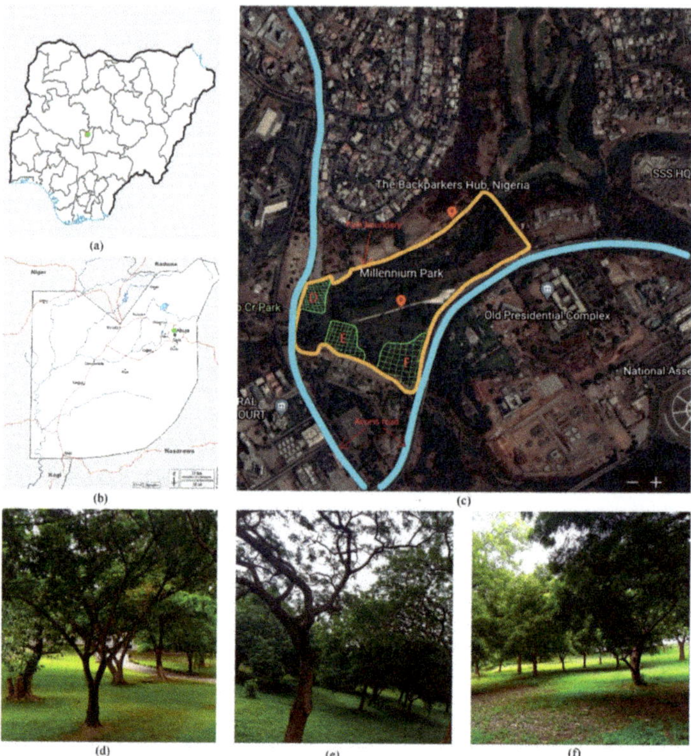

Figure 1. Map and aerial photographs of the study area. (**a**) Map of Nigeria showing Abuja; (**b**) map of Abuja showing Maitama; (**c**) Google Earth map of the study area; (**d**) low-density vegetation; (**e**) medium-density vegetation; (**f**) high-density vegetation.

The study park is Millennium Park in the Federal Capital Territory, Abuja. It is known as the largest public park located in the Maitama district in the capital city of Nigeria. The park was established in 2003, has a total area of 80 acres (32 ha), and is bounded by two major traffic roads in the city. It is the largest multifunctional place of entertainment, allowing for leisure, physical fitness, and other forms of social activities. It attracts a large number of visitors due to its size and natural landscapes; the park is separated by a river along its main rectilinear axis. Its untouched natural vegetation, comprising large evergreen trees that provide shelter and that create a serene environment, distinguishes it from other parks, and it houses different kinds of tropical birds and other natural creatures. Considering that the potential effects of urban noise may influence the birds' sounds [56], the case study site was therefore selected to be distant from traffic roads and residential areas to avoid urban noise. This site is characterized by mountainous vegetation, deciduous forest, savanna, and brushwood vegetation types, with species of flower-bearing plants such as Japanese jasmine (*Jasminum mesnyi*), Arabian jasmine (*Jasminum sambac*), trumpet jasmine (*Jasminum bignoniaceum*), dwarf poinciana (*Caesalpinia pulcherrima*), maidenhair fern (*Adiantum* spp.), and coleus (*Solenostemon*). It also consists of capa de obispo (*Acalypha wilkesiana*), zedoary (*Curcuma zedoaria*), canna (*Canna indica*), broadleaf palm-lily (*Cordyline fruticosa* L. A. Chev), and pigeon berry (*Duranta repens*). Trees include the southern catalpa (*Catalpa bignonioides*), pacara earpod tree (*Enterolobium contortisiliquum*), Malayan banyan (*Ficus microcarpa*), masquerade tree (*Polyalthia longifolia*), sea randa (*Guettarda speciosa*), neem tree (*Azadirachta indica*), flamboyant tree (*Delonix regia*), Broome raintree (*Albizia lebbeck*), Lombardy poplar (*Populus nigra*), weeping willow (*Salix babylonica*), common spruce (*Picea*), Siberian elm (*Ulmus pumila*), and grasses include Bermuda grass (*Cynodon dactylon*).

2.2. Measurement of Sound Pressure Levels

Five different species of birds were identified in each density level of vegetation (low density, medium density, and high density), namely: 1. African reed warblers (*Acrocephalus baeticatus*); 2. African melodious warblers (*Hippolais polyglotta*); 3. wild doves (*Zenaida macroura*); 4. sparrows (*Spizella passerina*); and 5. black kites (*Milvus migrans*). In situ measurements were conducted in the three vegetation densities with respect to the time of the day and the three different weather conditions (cloudy, sunny, and rainy). The different temperatures of the day were recorded for each day, with cloudy days at 26 °C, sunny days at 32 °C, and rainy days at 24 °C. The sound pressure level was measured at specific times of the day between the morning hours (8:00 a.m.) and the evening hours (18:00 p.m.) at 15 min intervals. The sound pressure level was measured with a digital sound level meter (AS824; China), an instrument comprising a microphone that captures and accesses sound by measuring sound pressure levels in decibels (dBA). Slow, fast, and impulse are the three types of sampling settings, depending on the intended results. Since bird sounds usually change quickly and can occur within a short period of time, the digital sound level meter was set to fast sampling [57]. The instruments were placed at different locations in each habitat below the singing birds, at a vertical distance of 3 m, and measurements were recorded with respect to the different times of the day.

2.3. Sound Recording

The birdsongs were recorded to determine the frequencies of singing events for the five different species of birds identified. Each recording was conducted using a H2n Handy Recorder (Zoom Corporation) in the low-density, medium-density, and high-density vegetation areas of the park. The recordings were made under cloudy, sunny, and rainy weather conditions in the park, between the hours of 8:00 and 18:00, at one-hour intervals. In order to record better quality bird sounds from different directions, an omnidirectional microphone was attached to the recording instrument (H2n Handy Recorder) [58], which was placed vertically below the singing birds at 3 m above ground level with the aid of a tripod. Raptor birds such as black kites could be seen hovering in the sky and perching at intervals in the trees. However, the birds could also perch in the trees for calls or for rest, and this was when their songs were comfortably recorded at a height of 3 m. All of the recordings were conducted simultaneously with H2n recorders in the low-density, medium-density, and high-density vegetation areas. Previous studies have shown that crowd density may have an influence on the vocalizations of birds [59]. Measurements of crowd density in the low-, medium-, and high-density vegetation areas were conducted, and the change in crowd density had no significance difference in the vocalizations of the birds among the three categories of vegetation. Therefore, a change of crowd density may not affect the comfortable vocalizations of the birds.

2.4. Questionnaire Design

A questionnaire survey is a tool that can be used to describe the perceived or experienced acoustic environment [60]. A web-based and physical questionnaire containing 15 questions in total was designed for data collection. A total of 171 questionnaires were sent out, adopting an evaluation scale from 1 to 5 (a 5-point Likert scale). A total of 160 effective questionnaires were retrieved, with an effective rate of 93.57%, and a total of 83 males and 77 females participated. The questionnaire questions were divided into four sections:

(1) demographics and social information; (2) the perception of acoustic comfort when listening to birdsongs in the park; (3) the preference for different weather conditions to visit; and (4) the preference for vegetation density areas to remain in. The first section was designed to capture demographic data, including the age, gender, and educational qualifications of the respondents. The second section was designed to obtain data on the perception of acoustic comfort using six indicators [61]: pleasant, calming, annoying, eventful, chaotic, and vibrant. The third section was designed to obtain a database of the preferred weather conditions for visiting in the park, and the fourth question was designed to obtain data based on the preferred density of vegetation in the park. The ages of the respondents were divided into six groups [62,63], and Table 1 shows the questionnaire questions. It took approximately 5–10 min for each respondent to complete the questionnaire.

Table 1. Demographic and social data, and acoustic perceptions determined by the questionnaire.

Demographic and Social Indicators	Categorization and Scale
Gender	1: male; 2: female
Age	1: <18; 2: 19–30; 3: 31–40; 4: 41–50; 5: 51–60; 6: >60
Educational level	1: primary; 2: secondary; 3: indergraduate; 4: graduate; 5: postgraduate
Questions on the acoustic perception of the sound environment and birdsongs in the park	Strongly disagree, disagree, neither disagree nor agree, agree, strongly agree
To what extent do you perceive the current sound environment and birdsongs as being pleasant?	
To what extent do you perceive the current sound environment and birdsongs as being calming?	
To what extent do you perceive the current sound environment and birdsongs as being annoying?	
To what extent do you perceive the current sound environment and birdsongs as being eventful?	
To what extent do you perceive the current sound environment and birdsongs as being chaotic?	
To what extent do you perceive the current sound environment and birdsongs as being vibrant?	
Questions on weather conditions when visiting the park	<1 h, 1 h, 2 h, 3 h, >3 h
How long can you visit the park on a cloudy day?	
How long can you visit the park on a sunny day?	
How long can you visit the park on a rainy day?	
Questions on the density of vegetation when relaxing in the park	1 h, 1 h, 2 h, 3 h, >3 h
How long can you be in the park in a low-density vegetation area?	
How long can you be in the park in a medium-density vegetation area?	
How long can you be in the park in a high-density vegetation area?	

2.5. Data Analysis

The software used for the statistical analysis of the collected data was SPSS 26.0 (IBM, Armonk, NY, USA) and Microsoft Excel. This study adopted common statistical methods. A one-way analysis of variance (ANOVA) was used to ascertain whether there were any significantly significant differences between the frequency of singing events and the sound pressure levels of the birdsongs in low, medium, and high densities of vegetation. The null hypothesis H_0 states that there is no significant difference between each group, and the alternative hypothesis H_A states that there is a significant difference between each group at a 95% confidence level. In addition, Tukey's honest significance test statistically determined whether the low-density, medium-density, and high-density vegetation groups were significantly different to each other, and which were not. Linear regression determined the association between the sound pressure level and the frequency of singing events, the chi-squared test determined the relationships and significance between the age of the respondents and the preference for weather conditions, and Spearman's rank correlation showed the strength and correlation between each of the perceived acoustic indicators and each density level of the vegetation.

3. Results

3.1. Effects of Different Densities of Vegetation on the Singing Events of Birds

A one-way analysis of variance (ANOVA) based on the stated null and alternative hypothesis showed whether there were any statistical differences between the frequency of singing events of the birds in low-density, medium-density, and high-density vegetation under the three different weather conditions at a 95% confidence level. The analysis of variance results were as follows: the frequency of birdsong in low-density, medium-density, and high-density vegetation on a cloudy day was $F\,(2, 27) = 3.47$, $p = 0.045$; on a sunny day, $F\,(2, 27) = 3.40$, $p = 0.048$; and on rainy day, $F\,(2, 27) = 2.43$, $p = 0.146$. The results indicated that there is a significant difference between low-, medium-, and high-density vegetation and the frequency of singing events (frequency). Although this difference was only obtainable on a cloudy day and on a sunny day, with p-values of $p = 0.045$ and $p = 0.048$, respectively, it showed a non-significant effect on a rainy day, with a p-value of $p = 0.146$.

Since the one-way ANOVA is an omnibus test, Tukey's honest significance test (Table 2) was performed on variables that showed significant differences to identify the actual vegetation density that was statistically significant to each other and that which was not. The results indicate that there was a significant difference between the frequency of the birdsongs in low-density vegetation and high-density vegetation on a cloudy day, with $p = 0.036$, but no significant difference between the medium- and high-density vegetation areas, with $p = 0.036$ on the same day. In addition, there was a significant difference between low-density and high-density vegetation on a sunny day, with $p = 0.042$, but there was no significant difference between the medium-density and high-density categories.

Table 2. Measures of frequency of birdsong events in the three densities of vegetation in the park, where LDV = low–density vegetation, MDV = medium–density vegetation, and HDV = high–density vegetation.

Density of Vegetation		Mean Difference	*p*-Value	Upper Bound	Lower Bound
			Cloudy Day		
MDV	LDV	19.20	0.339	−14.04	52.44
	HDV	−16.10	0.463	−49.34	17.14
HDV	LDV	35.30 *	0.036 *	2.06	68.54
			Sunny Day		
MDV	LDV	10.50	0.668	−19.67	40.67
	HDV	−20.70	0.223	−50.87	9.47
HDV	LDV	31.20 *	0.042 *	1.03	61.37

* Correlation is significant at $p < 0.05$.

The five species of birds identified in the low-, medium-, and high-density vegetation areas were: African reed warbler (*Acrocephalus baeticatus*), African melodious warbler (*Hippolais polyglottal*), black kite (*Milvus migrans*), wild dove (*Zenaida macroura*), and sparrow (*Spizella passerina*). The songs of the above listed birds were heard in the low-, medium-, and high-density vegetation areas on cloudy, sunny, and rainy days at different times of the day, from 8:00 to 17:00 at 1 h intervals, as shown in Figure 2a–e. The observed non-polygynous African reed warbler birds sang for a longer period (8 h) on a cloudy day in low-density vegetation, and they had no singing time in medium-density vegetation on the same day. However, it was observed that the African reed warblers had frequent singing times within all densities of vegetation on sunny and rainy days, as shown in Figure 2a. Likewise, the African melodious warbler birds experienced their longest singing time of a duration of 5 h in low-density vegetation on a rainy day, but they experienced no singing time in medium-density vegetation on a sunny day or in high-density vegetation on a rainy day, as shown in Figure 2b. Black kites are normally medium-sized prey birds that are generally dark in color and they are observed around areas with water bodies, where they have access to their prey. This also explains their presence in the park, as the park is characterized by the presence of a river. Since they are raptor birds, they usually hover in the sky, perching at intervals and monitoring the ground prey. The longest singing period of a black kite was recorded on a rainy day, for a duration of 9 h, in medium-density vegetation. Additionally, it was observed that most of the calls of the black kites were recorded toward afternoon and evening, with few calls being recorded in the morning, as shown in Figure 2c, and few calls being recorded on a sunny day in medium-density vegetation. No calls were recorded on a rainy day in low-density vegetation. The black kite call duration was between 8 and 10 times within 30 s.

The wild dove calls (Figure 2d) were recorded in all three densities of vegetation, on cloudy, sunny, and rainy days, with the highest singing duration of 9 h recorded in medium-density on a sunny day, and the lowest singing duration of 2 h recorded in medium-density on a rainy day. One of the characteristics of wild doves in Nigeria is that they are commonly observed birds, especially in evergreen natural vegetation and in areas that are characterized with flamboyant and neem trees, which were present in the study area. Wild doves are easily spotted in urban or rural areas of Nigeria. The wild dove call duration was between seven and eight times within 20 s. The chipping sparrows are found in shrubs or in trees in houses where they are free to nest. It was observed that the same singing period of 5 h was recorded for sparrows on sunny and rainy days in high-density vegetation, and the lowest singing duration of 1 h was recorded in low-density vegetation on a sunny day, as shown in Figure 2e. There were no calls in low-density vegetation on cloudy and rainy days, respectively. From the observations, it can be seen that each species of bird experienced a different duration of calls in different densities of vegetation. From this study, it can be seen that the wild doves had the highest singing period, followed by black kites, African reed warblers, sparrows, and African melodious warblers. In addition, more birdcalls were recorded in higher density vegetation, followed by medium-density vegetation and low-density vegetation.

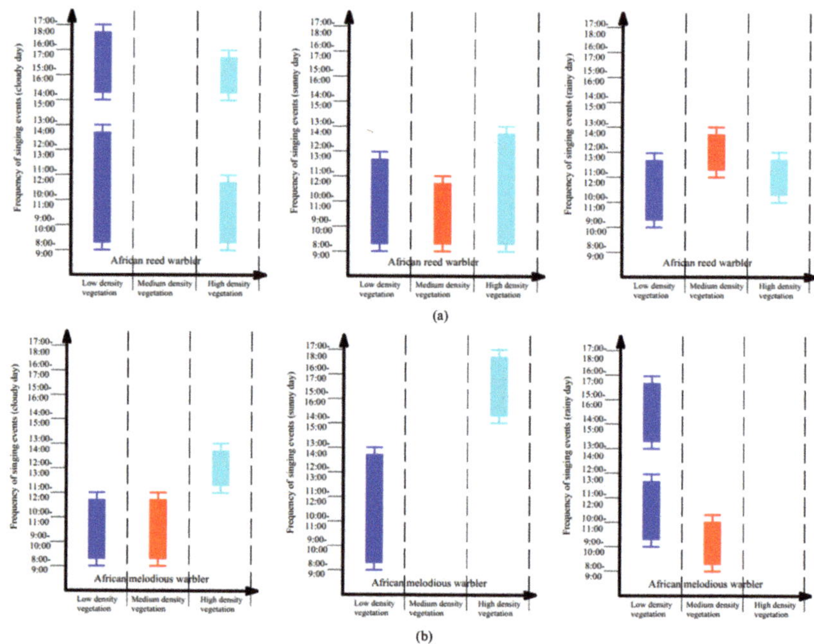

Figure 2. *Cont.*

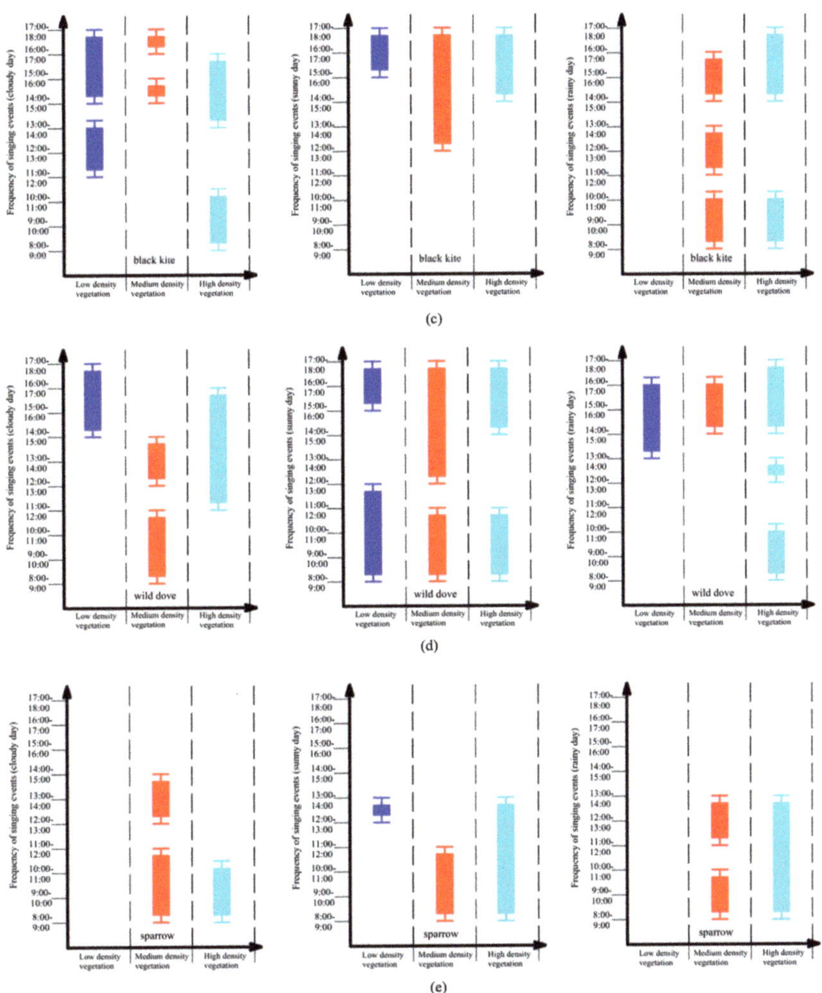

Figure 2. Specific singing duration of the birds in low–, medium–, and high–density vegetation. (**a**) African reed warbler; (**b**) African melodious warbler; (**c**) black kite; (**d**) wild dove; and (**e**) sparrows.

3.2. Effect of Frequency of Singing Events on Sound Pressure Levels of Birdsongs

The measured sound pressure levels of the birdsongs were measured in each density of vegetation on a cloudy day, sunny day, and rainy day, and the data were analyzed using a one-way analysis of variance. The findings showed a significant difference between the sound pressure levels of the birdsongs in each density of vegetation. The statistical results were as follows: for low-density vegetation, $F_{(2, 27)} = 11.15$, $p = 0.000$ ($p < 0.01$), for medium-density vegetation, $F_{(2, 27)} = 4.49$, $p = 0.021$, and for high-density vegetation, $F_{(2, 27)} = 3.44$, $p = 0.047$ ($p < 0.05$). It can be explained that the density of vegetation in which the birds are located can affect their sound pressure levels because they are responsive to their environment. It can also be explained that they also react based on the weather conditions of the day. For a further analysis and for an in-depth study, the relationship between the frequency of the singing events of the birds and the sound pressure level of the birdsongs was analyzed. Figure 3a–i shows the graphical representation of the results obtained. Figure 3a shows that on a cloudy day in low-density vegetation, there was an excellent relationship between the frequency of singing events and the sound pressure levels of the birds, with a correlation coefficient of $R = 0.75$ and a coefficient of determination of $R^2 = 0.57$. In total, 57% was affected by the sound pressure level of the birdsong, whereas 43% could be attributed to other factors that could affect the birds. However, on the same cloudy day in high-density vegetation, the correlation coefficient ($R = 0.64$) implies a moderate relationship, with 41% attributed to sound pressure levels and 59% attributed to factors other than the frequency of singing events, as shown in Figure 3c. The highest correlation coefficient ($R^2 = 0.82$) was obtained in medium-density vegetation on a sunny day, indicating an excellent relationship between the frequency of singing events and the sound pressure levels of the birdsongs shown in Figure 3e, with only 18% being attributed to factors other than the frequency of singing events. Figure 3g shows that in low-density vegetation on a rainy day, the correlation coefficient ($R = 0.89$) obtained also implies that there is an excellent relationship between the frequency of singing events of birds and the sound pressure level of the birdsongs. A correlation coefficient of $R^2 = 0.79$ showed that 79% was attributed to the sound pressure level and 21% was attributed to other factors. In addition, Figure 3h also shows that a coefficient of determination of $R^2 = 0.47$ was obtained, indicating that 47% of the sound pressure level affects the singing times of birds, whereas 53% is attributed to other factors. However, Figure 3i showed a weak relationship between the sound pressure level and the frequency of singing events in the birds. This indicated ($R^2 = 0.10$) that 10% was affected by the sound pressure level, whereas 90% was attributed to factors other than the frequency of singing events in high-density vegetation on rainy days.

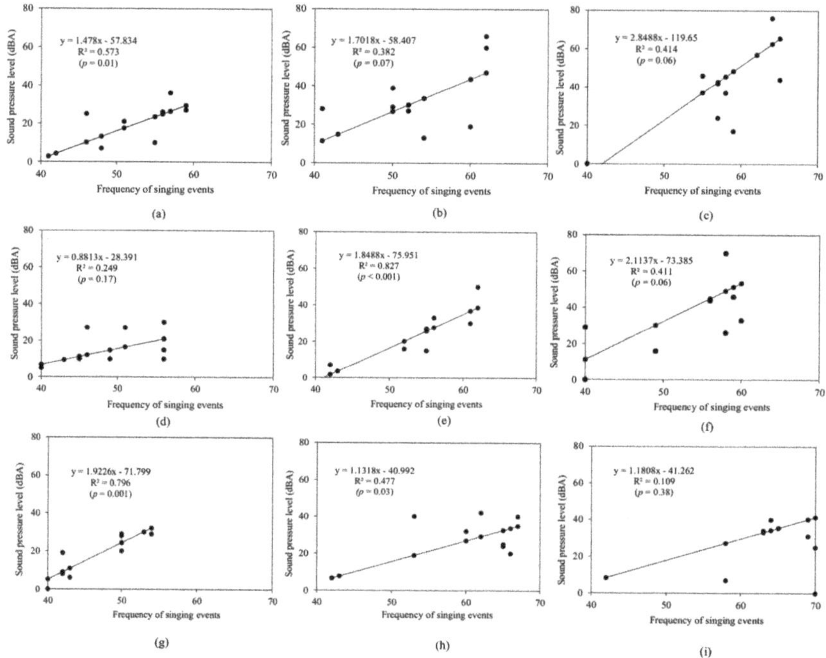

Figure 3. Relationship between the frequency of singing events and the sound pressure levels of birdsongs. (**a**) Low–density vegetation on cloudy days; (**b**) medium–density vegetation on cloudy days; (**c**) high–density vegetation on cloudy days; (**d**) low–density vegetation on sunny days; (**e**) medium–density vegetation on sunny days; (**f**) high–density vegetation on sunny days; (**g**) low–density vegetation on rainy days; (**h**) medium–density vegetation on rainy days; and (**i**) high–density vegetation on rainy days.

3.3. Effects of Densities of Vegetation and Weather Conditions on Acoustic Perception

The arithmetic mean values of the perceived acoustic comfort when listening to birdsongs in a park, the preference for visiting under different weather conditions, and the preference of density of vegetation to stay in were analyzed and are shown in Table 3. In addition, the results from the demographic information show that 51.9% of the respondents were male and 48.1% were female, and the ages of the respondents were <18 (6.9%), 19–30 (53.8%), 31–40 (20.6%), 41–50 (10%), 51–60 (7.5%), and >60 (1.3%), as shown in Table 4. The results of the mean and standard deviation of the ages of the respondents were(mean = 2.61; SD = 1.087, where N = 160.

Table 3. Arithmetic mean values of human acoustic perception and preferences for weather conditions and vegetation densities.

Perceived Acoustic Comfort Indicators	Arithmetic Mean Values		
	Mean	SD	SE Mean
Pleasant	3.55	1.17	0.09
Calming	3.55	1.09	0.86
Annoying	2.07	0.79	0.63
Eventful	4.03	0.74	0.05
Chaotic	2.09	0.66	0.83
Vibrant	3.82	0.74	0.58
Preference of Different Weather Conditions	**Arithmetic Mean Values**		
	Mean	SD	SE Mean
Cloudy day	2.37	1.23	0.09
Sunny day	2.80	1.28	0.10
Rainy	1.45	1.20	0.09
Preference of Different Densities of Vegetation	**Arithmetic Mean Values**		
	Mean	SD	SE Mean
Low density	3.03	1.31	0.10
Medium density	2.90	1.20	0.09
High density	2.68	1.73	0.13

Table 4. Demographic information and social factors of the respondents.

Demographics		Percentage (%)
Gender	Male	51.9
	Female	48.1
Age	<18	6.9
	19–30	53.8
	31–40	20.6
	41–50	10
	51–60	7.5
	>60	1.3
Educational background	Secondary school	2.5
	Undergraduate	28.8
	Graduate	50
	Postgraduate	18.7

The results of the mean and standard deviation for perceived acoustic comfort were pleasant, M = 3.55; SD = 1.17, calming M = 3.55; SD = 1.09), annoying M = 2.07; SD = 0.79, eventful M = 4.03; SD = 0.74, chaotic M = 2.09; SD = 0.66, and vibrant M = 3.82; SD = 0.74. Based on the scale of the analysis of the result, the mean values for pleasant, calming, eventful, and vibrant indicated that the respondents positively agreed with the degree of acoustic comfort in the park when listening to the birdsongs. In addition, the mean values of annoying and chaotic indicated that the respondents disagreed that the birdsongs in the park were annoying sources of noise, or were perceived as unpleasant. However, the mean values for the degree of human comfort in different weather conditions, which were cloudy, M = 2.37, sunny, M = 2.80, and rainy, M = 1.45, indicate that the respondents preferred to spend more time in the park on sunny days, followed by cloudy days, and the least time on rainy days. Likewise, the mean values of the degree of comfort in different densities of vegetation were: low density, M = 3.03, medium density, M = 2.90, and high density, M = 2.68. This indicated that the respondents preferred to stay or spend more time in low-density vegetation, followed by medium-density vegetation, and the least time in high-density vegetation within the park. Figure 4 shows the duration of stay for the respondents under different weather conditions (cloudy, sunny, and rainy) in the park. For periods of less than one hour (<1 h), 83.3% of the respondents indicated that they spent the least time in the park on a rainy day. Additionally, 45% of the respondents spent less than one hour on a cloudy day, and 12.5% on a sunny day. For longer durations of stay (>3 h), the least percentage of respondents recorded was on rainy days with 2.5%, followed by cloudy days (8.8%) and sunny days (10%). Generally, more respondents spent longer hours on a sunny day, followed by cloudy days, and the shortest duration of stay was on a rainy day.

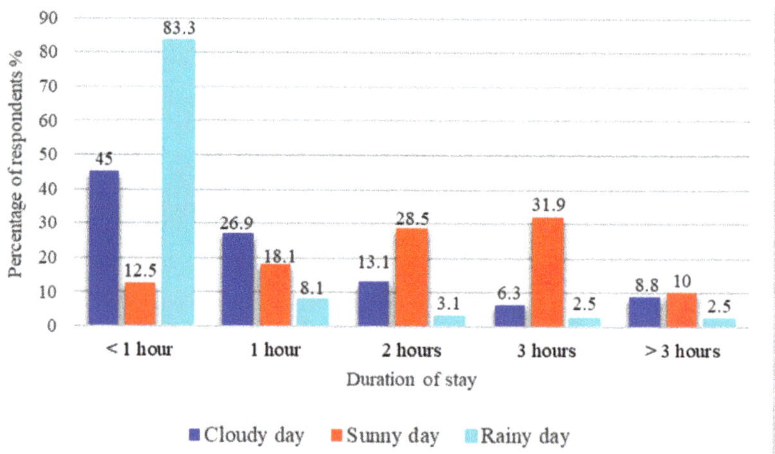

Figure 4. Respondent percentages of durations of stay in the park under cloudy, sunny, and rainy weather conditions.

Figure 5a–c shows the Spearman's correlation coefficient (r_s) of the densities of vegetation and each perceived acoustic comfort indicator. From the results, there was a strong positive correlation between low-density vegetation and pleasant acoustic comfort indicators, which were statistically significant ($r_s = 0.954$, $p < 0.01$). Calming acoustic indicators showed a strong positive correlation with low-density vegetation, which was statistically significant ($r_s = 0.925$, $p < 0.01$). There was a moderate positive correlation between high-density vegetation and the perceived acoustic indicators of annoying and chaotic, with both correlations having the same coefficients ($r_s = 0.712$, $p < 0.01$). There was a similar occurrence in medium density, where the correlation coefficients of the perceived acoustic indicators of annoying and chaotic were the same ($r_s = 0.815$, $p < 0.01$). The smallest correlation coefficient obtained was between the high-density vegetation and vibrant, which was moderately correlated and significant ($r_s = 0.617$, $p < 0.01$). In addition, there were moderate to strong positive correlations between the correlated variables.

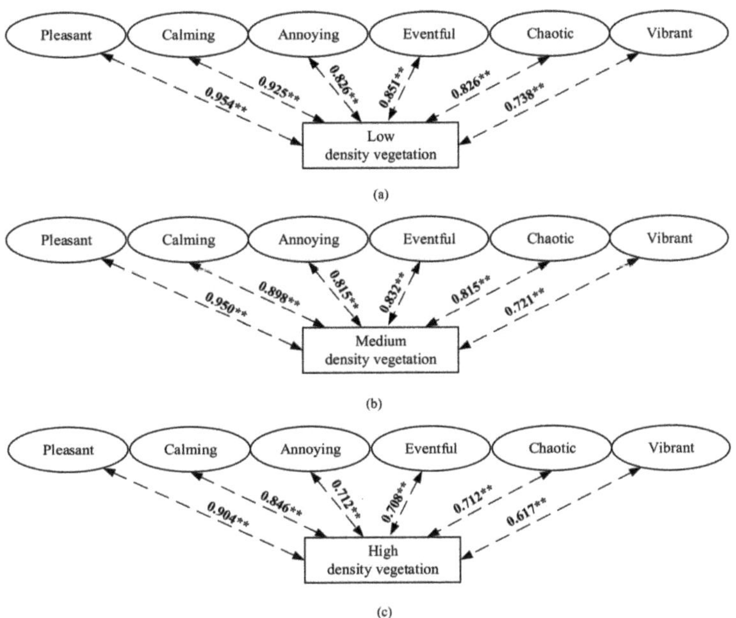

Figure 5. Correlations between densities of vegetation and perceived acoustic comfort indicators. (**a**) Low-density vegetation; (**b**) medium-density vegetation; and (**c**) high-density vegetation (** $p < 0.01$).

A chi-squared test was conducted between the ages of the respondents, the preference for weather conditions to visit the park, and the preference for vegetation density. The results indicated there was a strong relationship between age, preference for weather conditions (X^2 (15, N = 160) = 152.59, $p < 0.001$), and preference for the density of vegetation (X^2 (10, N = 160) = 199.69, $p < 0.001$). In addition, the results showed that there was a relationship between gender and preference for weather conditions (X^2 (12, N = 160) = 147.77, $p < 0.001$). The results also showed that 20.6% of the male respondents preferred cloudy days, 30% preferred sunny days, and 1.3% preferred rainy days. A total of 45.6% of female respondents preferred sunny days, and 2.5% preferred all weather conditions. Table 5 shows the influence of demographics on the respondents' preferences for the density of vegetation to be in and the weather conditions when visiting the park.

Table 5. Influence of the demographics of the respondents regarding preferences to weather condition and vegetation density.

Age of Respondents	Percentage of Response (%)	Preferred Weather Condition
(<18, 19–30)	20.6	Cloudy day
(19–30)	1.3	Rainy day
(19–30, 31–40, 41–50, 51–60)	75.6	Sunny day
(>60)	2.5	All weather
Age of Respondents	**Percentage of Response (%)**	**Preferred Density of Vegetation**
(<18, 19–30)	20.6	Low density
(19–30)	40	Medium density
(31–40, 41–50, 51–60, >60)	39.4	High density

4. Discussion

Nature and green spaces contribute to public health by reducing stress and psychological disorders [64–66]. They attract and function as a living habitat for living organisms, especially birds that produce sounds [67]. For these reasons, there has been a recent establishment of programs and agencies that are dedicated to tree planning, afforestation, and the significant control and ban of deforestation in Nigeria. As research into soundscapes continues to gain a wider spectrum in other diciplines, especially in urban planning and noise control engineering [68], it is paramount that established agencies that are responsible for urban growth continually avoid undermining the positive and health-related roles of urban forests and green spaces on the environment.

For this same reason, the National Environmental Standards and Regulations Enforcement Agency (NESREA) of Nigeria is a targeted body to which this research should be applied. The agency established in 2007 that it was committed to ensuring that Nigerians gain a feeling of being in a cleaner and healthier environment. The evaluation of avian species, as highlighted in this study, gives a better insight into the richness and density of vegetation and its effects on the sound pressure level and singing times of the birds in an environment that is meant to be conserved and protected. This study has also shown that bird calls are affected by their environment and by the weather conditions of the day. The NESREA should ensure that urban green spaces are continually conserved, as avian species could be threatened if natural vegetation in the environment becomes low. The smart and sustainable planning of a city's urban forest and green parks should be promoted using policies and guidelines which favor the economy and wellbeing of the citizens [69]. For these reasons, the Ministry of Housing and Urban Development is responsible for the creation of urban parks and recreational centers in Nigeria, and the maintenance and management of these parks is the responsibility of the Ministry of Environment. This study has shown the need for the creation of sub-agencies in various states in the country, led by the Ministry of Environment in Nigeria, to ensure that urban parks, recreational centers, and other green spaces that are responsible for providing an environment for healthy activities are constantly being evaluated. This will ensure that the acoustic environment meets international standards. In addition, the agencies should ensure that park users' perceptions of the natural environment are considered as being important and are not neglected. This study has evaluated users' perceptions of natural sounds, especially birdsongs in the park, and human preferences for weather conditions for visiting the park and the density of natural vegetation to be in. The results from this study should inform the Ministry of Environment and other agencies in Nigeria, which are responsible for providing a greener and healthier environment, to consider the provision of facilities that protect humans from harsh weather conditions in the park, especially during rain. The park's vegetation, irrespective of density, should be conserved and nurtured so that its positive environmental benefits can be reaped. Policies that allow for the provision and maintenance of a restorative environment through natural vegetation should be implemented and revised if need be, in order to meet the current needs of citizens.

This study also has its limitations, as previous studies have shown that breeding seasons affect the vocalizations of birds [70,71]. However, this study focused on vegetation densities, weather conditions, and acoustic perceptions, and thus, it did not take into consideration the breeding seasons of the birds. This will be captured in future research, where birds' behaviors during different climatic seasons of the year will be studied. In addition, this study classified five common birds in the three different vegetation densities, and the sample locations were grouped into three densities of vegetation. Future studies would entail the classification and identification of more birds in the park and would include more sample locations. The spectrogram of the birdsongs is important for determining the perceived acoustic perceptions of birdsongs [72]. Since the main purpose of the current study was to determine the effects of forest density on birdsong and human acoustic perception in an urban park, and the results were correlated with perceived acoustic comfort, a spectrogram was therefore not conducted, but this approach will be considered in future studies.

5. Conclusions

This article focused on different densities of vegetation and how these affect the sound pressure level, the frequency of singing events, and the duration of singing by individual bird species, and how this affects the human perception of sound in the natural environment.

In the context of whether different densities of natural vegetation have effects on the frequency of singing events for birds in cloudy, sunny, and rainy weather conditions, there was a significant difference between the frequency of singing events of birds on cloudy and sunny days only, with the results showing that there were no effects of vegetation densities on the frequency of singing events of birds on rainy days. With reference to the duration of singing for each species of bird, the vegetation density and the weather conditions affect the duration of singing for each bird. Some species of birds did show a different length of singing when heard in different vegetation densities under different weather conditions.

The vegetation density also affects the sound pressure levels of the birdsongs. As shown in this study, there is a significant difference between the sound pressure level of birdsongs in different densities of vegetation on a cloudy day, a sunny day, and a rainy day. In addition, the frequency of bird singing events is affected by the sound pressure levels of birdsongs, indicating that there is a significant positive correlation, as shown in Figure 3, and that the human perception of birdsongs is not affected by the sound pressure levels of the singing birds.

The respondents showed that they perceived the sound environment in the park as being pleasant, calming, eventful, and vibrant, and they did not perceive the sound environment as being chaotic or annoying. The respondents also experienced better comfort in low-density vegetation, and the least comfort in high-density vegetation. In addition, the respondents also most preferred to visit the park on sunny days, and least preferred to visit the park on rainy days. It can be seen that demographics were key factors in determining how the acoustic environment was perceived and the preferred vegetation density to be in. Regarding the length of time that people spent in the park, the respondents spent a longer time in the park on sunny days and the least time on rainy days.

The findings from this research are beneficial for the design and furnishing of public parks. It is vital to factor in different weather conditions as a design consideration when designing new parks or upgrading existing parks. In addition, the park landscape should be of high-density vegetation if birdsong is to be more frequent. In addition, parks should be furnished with facilities that would encourage people to visit and to enjoy nature, irrespective of the weather conditions, especially on rainy days where the shortest visiting time and lowest preference was recorded.

Author Contributions: Conceptualization, M.N. and Q.M.; data curation, M.N.; methodology, M.N., Q.M. and D.Y.; formal analysis M.N., Q.M. and D.Y.; funding acquisition, Q.M., D.Y. and F.L.; software, M.N. and Q.M.; writing—original draft preparation, M.N., writing—review and editing, M.N., Q.M., D.Y. and F.L. All authors have read and agreed to the published version of the manuscript.

Funding: This study was supported by the National Natural Science Foundation of China (NSFC) (grant numbers 52178070, 51878210, 51678180, and 51608147), the Open Projects Fund of the Key Laboratory of Architectural Acoustic Environment of Anhui Higher Education Institutes, Anhui Jianzhu University, Hefei, China (AAE2021ZD03), and the Natural Science Foundation of Heilongjiang Province (YQ2019E022).

Institutional Review Board Statement: Not applicable.

Informed Consent Statement: Not applicable.

Data Availability Statement: Not applicable.

Conflicts of Interest: The authors declare no conflict of interest.

References

1. Adegun, O.B.; Ikudayisi, A.E.; Morakinyo, T.E.; Olusoga, O.O. Urban Green Infrastructure in Nigeria: A Review. *Sci. Afr.* **2021**, *14*, e01044. [CrossRef]
2. Swanwick, C.; Dunnett, N.; Woolley, H. Nature, Role and Value of Green Space in Towns and Cities: An Overview. *Built Environ.* **2003**, *29*, 94–106. [CrossRef]
3. Di Lallo, G.; Mundhenk, P.; López, S.E.Z.; Marchetti, M.; Köhl, M. REDD+: Quick Assessment of Deforestation Risk Based on Available Data. *Forests* **2017**, *8*, 29. [CrossRef]
4. Barron, S.; Sheppard, S.R.J.; Condon, P.M.; Escobedo, F.; Livesley, S.J.; Morgenroth, J. Urban Forest Indicators for Planning and Designing Future Forests. *Forests* **2016**, *7*, 208. [CrossRef]
5. Fang, X.; Gao, T.; Hedblom, M.; Xu, N.; Xiang, Y.; Hu, M.; Chen, Y.; Qiu, L. Soundscape Perceptions and Preferences for Different Groups of Users in Urban Recreational Forest Parks. *Forests* **2021**, *12*, 468. [CrossRef]
6. Nguyen, P.Y.; Astell-Burt, T.; Rahimi-Ardabili, H.; Feng, X. Green Space Quality and Health: A Systematic Review. *Int. J. Environ. Res. Public Health* **2021**, *18*, 11028. [CrossRef]
7. Hall, J.M.; Handley, J.F.; Ennos, A.R. The Potential of Tree Planting to Climate-Proof High Density Residential Areas in Manchester, UK. *Landsc. Urban Plan.* **2012**, *104*, 410–417. [CrossRef]
8. Akbari, H. Shade Trees Reduce Building Energy Use and CO_2 Emissions from Power Plants. *Environ. Pollut.* **2002**, *116*, S119–S126. [CrossRef]
9. Nowak, D.J.; Crane, D.E. Carbon Storage and Sequestration by Urban Trees in the USA. *Environ. Pollut.* **2002**, *116*, 381–389. [CrossRef]
10. Ezeabasili, A.; Iloghalu, I.; Okoro, B.; Manafa, I. Sustainable Urban Forestry in Nigerian Built Environments. *J. Sci. Res. Rep.* **2015**, *5*, 524–531. [CrossRef]
11. Yamada, Y. Soundscape-Based Forest Planning for Recreational and Therapeutic Activities. *Urban For. Urban Green.* **2006**, *5*, 131–139. [CrossRef]
12. Zakka, S.D.; Permana, A.S.; Majid, M.R.; Danladi, A.; Bako, P.E. Urban Greenery a Pathway to Environmental Sustainability in Sub Saharan Africa: A Case of Northern Nigeria Cities. *Int. J. Built Environ. Sustain.* **2017**, *4*, 180–189. [CrossRef]
13. Madu, I.E.; Innocent, E.O. Urban Planning Problems in Nigeria: A Case of Onitsha Metropolis of Anambra State. *Singap. J. Bus. Econ. Manag. Stud.* **2013**, *1*, 41–59. [CrossRef]
14. Lamond, J.; Awuah, J.; Lewis, E.; Bloch, R.; Falade, J. *Urban Land, Planning and Governance Systems in Nigeria*; ICF International: Lexington, KY, USA, 2015.
15. Tempesta, T. Benefits and Costs of Urban Parks: A Review. *Aestimum* **2015**, *67*, 127–143. [CrossRef]
16. Ahmed, N.; El Aziz, A. Pocket Park Design in Informal Settlements in Cairo City, Egypt. *Landsc. Archit. Reg. Plan.* **2017**, *2*, 51–60. Available online: https://www.sciencepublishinggroup.com/journal/paperinfo?journalid=241&paperId=10020954 (accessed on 9 June 2022).
17. Buxton, R.T.; Pearson, A.L.; Allou, C.; Fristrup, K.; Wittemyer, G. A Synthesis of Health Benefits of Natural Sounds and Their Distribution in National Parks. *Proc. Natl. Acad. Sci. USA* **2021**, *118*, 14. [CrossRef]
18. Payne, S.R. Are Perceived Soundscapes within Urban Parks Restorative. *J. Acoust. Soc. Am.* **2008**, *123*, 3809. [CrossRef]
19. Ulrich, R.S.; Simons, R.F.; Losito, B.D.; Fiorito, E.; Miles, M.A.; Zelson, M. Stress Recovery during Exposure to Natural and Urban Environments. *J. Environ. Psychol.* **1991**, *11*, 201–230. [CrossRef]
20. Chiesura, A. The Role of Urban Parks for the Sustainable City. *Landsc. Urban Plan.* **2004**, *68*, 129–138. [CrossRef]
21. Dzhambov, A.; Tilov, B.; Markevych, I.; Dimitrova, D. Residential Road Traffic Noise and General Mental Health in Youth: The Role of Noise Annoyance, Neighborhood Restorative Quality, Physical Activity, and Social Cohesion as Potential Mediators. *Environ. Int.* **2017**, *109*, 1–9. [CrossRef]
22. Shashua-Bar, L.; Hoffman, M.E. Geometry and Orientation Aspects in Passive Cooling of Canyon Streets with Trees. *Energy Build.* **2003**, *35*, 61–68. [CrossRef]
23. Hartig, T.; Mitchell, R.; De Vries, S.; Frumkin, H. Nature and Health. *Annu. Rev. Public Health* **2014**, *35*, 207–228. [CrossRef] [PubMed]
24. Berman, M.G.; Jonides, J.; Kaplan, S. The Cognitive Benefits of Interacting with Nature. *Psychol. Sci.* **2008**, *19*, 1207–1212. [CrossRef] [PubMed]
25. Kaplan, S.; Berman, M.G. Directed Attention as a Common Resource for Executive Functioning and Self-Regulation. *Perspect. Psychol. Sci.* **2010**, *5*, 43–57. [CrossRef] [PubMed]
26. Hartig, T.; Evans, G.W.; Jamner, L.D.; Davis, D.S.; Gärling, T. Tracking Restoration in Natural and Urban Field Settings. *J. Environ. Psychol.* **2003**, *23*, 109–123. [CrossRef]
27. Pijanowski, B.C.; Villanueva-Rivera, L.J.; Dumyahn, S.L.; Farina, A.; Krause, B.L.; Napoletano, B.M.; Gage, S.H.; Pieretti, N. Soundscape Ecology: The Science of Sound in the Landscape. *Bioscience* **2011**, *61*, 203–216. [CrossRef]
28. Yang, W.; Kang, J. Acoustic Comfort Evaluation in Urban Open Public Spaces. *Appl. Acoust.* **2005**, *66*, 211–229. [CrossRef]
29. Truax, B.; Barrett, G.W. Soundscape in a Context of Acoustic and Landscape Ecology. *Landsc. Ecol.* **2011**, *26*, 1201–1207. [CrossRef]
30. Van den Bosch, K.A.M.; Welch, D.; Andringa, T.C. The Evolution of Soundscape Appraisal through Enactive Cognition. *Front. Psychol.* **2018**, *9*, 1129. [CrossRef]
31. Schafer, R.M. *The Soundscape: Our Sonic Environment and the Tuning of the World*; Destiny Books: Rochester, VT, USA, 1993; p. 301.

32. ISO—ISO 12913-1:2014—Acoustics—Soundscape—Part 1: Definition and Conceptual Framework. Available online: https://www.iso.org/standard/52161.html (accessed on 9 June 2022).
33. Viollon, S.; Lavandier, C.; Drake, C. Influence of Visual Setting on Sound Ratings in an Urban Environment. *Appl. Acoust.* **2002**, *63*, 493–511. [CrossRef]
34. Kang, J.; Aletta, F.; Gjestland, T.T.; Brown, L.A.; Botteldooren, D.; Schulte-Fortkamp, B.; Lercher, P.; van Kamp, I.; Genuit, K.; Fiebig, A.; et al. Ten Questions on the Soundscapes of the Built Environment. *Build. Environ.* **2016**, *108*, 284–294. [CrossRef]
35. Zhu, P.; Tao, W.; Lu, X.; Mo, F.; Guo, F.; Zhang, H. Optimisation Design and Verification of the Acoustic Environment for Multimedia Classrooms in Universities Based on Simulation. *Build. Simul.* **2021**, *15*, 1419–1436. [CrossRef]
36. Liu, J.; Kang, J. Soundscape Design in City Parks: Exploring the Relationships between Soundscape Composition Parameters and Physical and Psychoacoustic Parameters. *J. Environ. Eng. Landsc. Manag.* **2015**, *23*, 102–112. [CrossRef]
37. Wang, C.; Kang, J. Development of Acoustic Computer Simulation for Performance Spaces: A Systematic Review and Meta-Analysis. *Build. Simul.* **2022**, 1–17. [CrossRef]
38. Alonso, A.; Suárez, R.; Sendra, J.J. Virtual Reconstruction of Indoor Acoustics in Cathedrals: The Case of the Cathedral of Granada. *Build. Simul.* **2017**, *10*, 431–446. [CrossRef]
39. Çakır, O.; İlal, M.E. Utilization of Psychoacoustic Parameters for Occupancy-Based Acoustic Evaluation in Eating Establishments. *Build. Simul.* **2021**, *15*, 729–739. [CrossRef]
40. Brown, A.L.; Muhar, A. An Approach to the Acoustic Design of Outdoor Space. *J. Environ. Plan. Manag.* **2004**, *47*, 827–842. [CrossRef]
41. Nilsson, M.E. Soundscape Quality in Suburban Green Areas and City Parks. *Acta Acust. United Acust.* **2006**, *92*, 903–911.
42. Cox, D.T.C.; Gaston, K.J. Likeability of Garden Birds: Importance of Species Knowledge & Richness in Connecting People to Nature. *PLoS ONE* **2015**, *10*, e0141505. [CrossRef]
43. Curtin, S. Wildlife Tourism: The Intangible, Psychological Benefits of Human–Wildlife Encounters. *Curr. Issues Tour.* **2009**, *12*, 451–474. [CrossRef]
44. Mynott, J. *Birdscapes: Birds in Our Imagination and Experience*; Princeton University Press: Prinston, NJ, USA, 2009.
45. Ratcliffe, E.; Gatersleben, B.; Sowden, P.T. Associations with Bird Sounds: How Do They Relate to Perceived Restorative Potential? *J. Environ. Psychol.* **2016**, *47*, 136–144. [CrossRef]
46. Ratcliffe, E.; Gatersleben, B.; Sowden, P.T. Predicting the Perceived Restorative Potential of Bird Sounds Through Acoustics and Aesthetics. *Environ. Behav.* **2020**, *52*, 371–400. [CrossRef]
47. Zhu, X.; Gao, M.; Zhao, W.; Ge, T. Does the Presence of Birdsongs Improve Perceived Levels of Mental Restoration from Park Use? Experiments on Parkways of Harbin Sun Island in China. *Int. J. Environ. Res. Public Health* **2020**, *17*, 2271. [CrossRef] [PubMed]
48. Uebel, K.; Marselle, M.; Dean, A.J.; Rhodes, J.R.; Bonn, A. Urban Green Space Soundscapes and Their Perceived Restorativeness. *People Nat.* **2021**, *3*, 756–769. [CrossRef]
49. Zhang, Y.; Kang, J.; Kang, J. Effects of Soundscape on the Environmental Restoration in Urban Natural Environments. *Noise Health* **2017**, *19*, 65. [CrossRef] [PubMed]
50. Briefer, E.; Osiejuk, T.S.; Rybak, F.; Aubin, T. Are Bird Song Complexity and Song Sharing Shaped by Habitat Structure? An Information Theory and Statistical Approach. *J. Theor. Biol.* **2010**, *262*, 151–164. [CrossRef]
51. Farina, A.; Lattanzi, E.; Malavasi, R.; Pieretti, N.; Piccioli, L. Avian Soundscapes and Cognitive Landscapes: Theory, Application and Ecological Perspectives. *Landsc. Ecol.* **2011**, *26*, 1257–1267. [CrossRef]
52. Bucur, V. *Urban Forest Acoustics*; Springer: Berlin/Heidelberg, Germany, 2006; pp. 1–181. [CrossRef]
53. Hedblom, M.; Heyman, E.; Antonsson, H.; Gunnarsson, B. Bird Song Diversity Influences Young People's Appreciation of Urban Landscapes. *Urban For. Urban Green.* **2014**, *13*, 469–474. [CrossRef]
54. Chibuike, E.M.; Ibukun, A.O.; Abbas, A.; Kunda, J.J. Assessment of Green Parks Cooling Effect on Abuja Urban Microclimate Using Geospatial Techniques. *Remote Sens. Appl. Soc. Environ.* **2018**, *11*, 11–21. [CrossRef]
55. State Population, 2006—Nigeria Data Portal. Available online: https://nigeria.opendataforafrica.org/ifpbxbd/state-population-2006 (accessed on 10 June 2022).
56. Luther, D.; Baptista, L. Urban Noise and the Cultural Evolution of Bird Songs. *Proc. R. Soc. B Biol. Sci.* **2010**, *277*, 469–473. [CrossRef]
57. Buckland, S.T.; Anderson, D.R.; Burnham, K.P.; Laake, J.L.; Borchers, D.L.; Thomas, L. *Introduction to Distance Sampling: Estimating Abundance of Biological Populations*; Oxford University Press: New York, NY, USA, 2001. Available online: https://global.oup.com/academic/product/introduction-to-distance-sampling-9780198509271?cc=us&lang=en& (accessed on 9 June 2022).
58. Efford, M.G.; Dawson, D.K.; Borchers, D.L. Population Density Estimated from Locations of Individuals on a Passive Detector Array. *Ecology* **2009**, *90*, 2676–2682. [CrossRef]
59. Slabbekoorn, H.; Dooling, R.J.; Popper, A.N.; Fay, R.R. *Effects of Anthropogenic Noise on Animals*; Springer: New York, NY, USA, 2018. [CrossRef]
60. Maffei, L.; Masullo, M.; Aletta, F.; Di Gabriele, M. The Influence of Visual Characteristics of Barriers on Railway Noise Perception. *Sci. Total Environ.* **2013**, *445*, 41–47. [CrossRef] [PubMed]
61. ISO—ISO/TS 12913-2:2018—Acoustics—Soundscape—Part 2: Data Collection and Reporting Requirements. Available online: https://www.iso.org/standard/75267.html (accessed on 9 June 2022).

62. Gao, T.; Song, R.; Zhu, L.; Qiu, L. What Characteristics of Urban Green Spaces and Recreational Activities Do Self-Reported Stressed Individuals Like? A Case Study of Baoji, China. *Int. J. Environ. Res. Public Health* **2019**, *16*, 1348. [CrossRef] [PubMed]
63. Zhao, W.; Li, H.; Zhu, X.; Ge, T. Effect of Birdsong Soundscape on Perceived Restorativeness in an Urban Park. *Int. J. Environ. Res. Public Health* **2020**, *17*, 16–5659. [CrossRef] [PubMed]
64. Annerstedt, M.; Östergren, P.O.; Björk, J.; Grahn, P.; Skärbäck, E.; Währborg, P. Green Qualities in the Neighbourhood and Mental Health—Results from a Longitudinal Cohort Study in Southern Sweden. *BMC Public Health* **2012**, *12*, 337. [CrossRef]
65. Ward Thompson, C.; Roe, J.; Aspinall, P.; Mitchell, R.; Clow, A.; Miller, D. More Green Space Is Linked to Less Stress in Deprived Communities: Evidence from Salivary Cortisol Patterns. *Landsc. Urban Plan.* **2012**, *105*, 221–229. [CrossRef]
66. Kellert, S.R.; Wilson, E.O. *The Biophilia Hypothesis*; Island Press: Washington, DC, USA, 1993; Volume 15, pp. 52–53, ISBN 1-55963-148-1. [CrossRef]
67. Van Renterghem, T. Towards Explaining the Positive Effect of Vegetation on the Perception of Environmental Noise. *Urban For. Urban Green.* **2019**, *40*, 133–144. [CrossRef]
68. Kang, J.; Aletta, F. The Impact and Outreach of Soundscape Research. *Environments* **2018**, *5*, 58. [CrossRef]
69. Sacchelli, S.; Favaro, M. A Virtual-Reality and Soundscape-Based Approach for Assessment and Management of Cultural Ecosystem Services in Urban Forest. *Forests* **2019**, *10*, 731. [CrossRef]
70. Wilson, D.M.; Bart, J. Reliability of Singing Bird Surveys: Effects of Song Phenology during the Breeding Season. *Condor* **1985**, *87*, 69–73. [CrossRef]
71. Dawson, D.K.; Efford, M.G. Bird Population Density Estimated from Acoustic Signals. *J. Appl. Ecol.* **2009**, *46*, 1201–1209. [CrossRef]
72. Dowling, J.L.; Luther, D.A.; Marra, P.P. Comparative Effects of Urban Development and Anthropogenic Noise on Bird Songs. *Behav. Ecol.* **2012**, *23*, 201–209. [CrossRef]

forests

Article

Before Becoming a World Heritage: Spatiotemporal Dynamics and Spatial Dependency of the Soundscapes in Kulangsu Scenic Area, China

Zhu Chen [1,2], Tian-Yuan Zhu [1,3], Jiang Liu [1,*] and Xin-Chen Hong [1]

1 School of Architecture and Urban-Rural Planning, Fuzhou University, Fuzhou 350108, China
2 Institute of Environmental Planning, Leibniz University Hannover, Herrenhäuser Str. 2, 30419 Hanover, Germany
3 Institute of Urban Environment, Chinese Academy of Sciences, Xiamen 361021, China
* Correspondence: jiang.liu@fzu.edu.cn; Tel.: +86-15705958118

Abstract: Kulangsu is a famous scenic area in China and a World Heritage Site. It is important to obtain knowledge with regard to the status of soundscape and landscape resources and their interrelationships in Kulangsu before it became a World Heritage. The objective of this study was to explore the spatial dependency of the soundscapes in Kulangsu, based on the spatiotemporal dynamics of soundscape and landscape perceptions, including perceived sound sources, soundscape quality, and landscape satisfaction degree, and the spatial landscape characteristics, including the distance to green spaces, normalized difference vegetation index, and landscape spatial patterns. The results showed that perception of soundscape and landscape were observed in significant spatiotemporal dynamics, and the dominance of biological sounds in all sampling periods and human sounds in the evening indicated that Kulangsu scenic area had a good natural environment and a developed night-time economy, respectively. The green spaces and commercial lands may contribute to both the soundscape pleasantness and eventfulness. Moreover, the soundscape quality was dependent on the sound dominant degree and landscape satisfaction degree but not on the landscape characteristics. The GWR model had better goodness of fit than the OLS model, and possible non-linear relationships were found between the soundscape pleasantness and the variables of perceived sound sources and landscape satisfaction degree. The GWR models with spatial stationarity were found to be more effective in understanding the spatial dependence of soundscapes. In particular, the data applied should ideally include a complete temporal dimension to obtain a relatively high fitting accuracy of the model. These findings can provide useful data support and references for future planning and design practices, and management strategies for the soundscape resources in scenic areas and World Heritage Sites.

Keywords: soundscape mapping; soundscape quality; spatiotemporal dynamics; landscape satisfaction; landscape pattern; scenic area

Citation: Chen, Z.; Zhu, T.-Y.; Liu, J.; Hong, X.-C. Before Becoming a World Heritage: Spatiotemporal Dynamics and Spatial Dependency of the Soundscapes in Kulangsu Scenic Area, China. *Forests* 2022, 13, 1526. https://doi.org/10.3390/f13091526

Academic Editor: Isabella De Meo

Received: 24 August 2022
Accepted: 11 September 2022
Published: 19 September 2022

Publisher's Note: MDPI stays neutral with regard to jurisdictional claims in published maps and institutional affiliations.

1. Introduction

Rapid urbanization has led to an increasing proportion of the population settling in cities worldwide [1]. Living in high-density built-up areas limits urban residents' access to nature and may expose them to certain environmental hazards, such as noise pollution [2]. Scenic areas possess abundant natural and cultural landscapes as well as infrastructure and facilities for human activities [3], which is a critical vehicle for people to interact with natural environment. The scenic areas located in urban areas are an important and special urban green space (UGS), providing a variety of important ecosystem services to urban residents, such as air purification [4], biodiversity conservation [5], nature-based recreation [6], and noise and quiet reduction [7]. Furthermore, they can improve the quality of urban environments, promote sustainable lifestyles, contribute to human health and well-being [8], and even reduce mortality [9]. However, with the development of tourist industry, the conflicting interests e.g., between noise of tourist activities and experiencing the sounds of nature have become one of the focal issues in the management of scenic areas [10,11], because auditory perception is as important as visual perception in people's visiting experience [12,13].

Currently, scholars no longer equate the good acoustic environment with simply reducing noise levels [14,15], but rather emphasize the importance of people's subjective perception of the acoustic environment following the soundscape concept of *ISO 12913-1* [16]. Against this backdrop, more and more studies have explored and focused on the human perceived acoustic environment [17,18]. However, exploring soundscapes needs a multi-faceted perspective because soundscape perception is associated with not only acoustic components but also non-acoustic factors [19]. Sound sources are the most basic and important acoustic components in creating soundscapes in urban areas [20], because the sound dominance in a soundscape is able to affect the spatiotemporal dynamics and the outcome of people's perception of the soundscape [21,22]. Regarding non-acoustic factors, landscape compositions account for a high proportion of the influence on soundscape perception, with can be summarized as subjective and objective aspects [18]. The former is mainly due to people's perception varied from landscape elements, such as naturalness [23], visual quality [24], urban contexts [25], features of landscape and architecture [17], audio-visual coherence [26], and infrastructure services [27]. The latter is related to the landscape characteristics, for example, accessibility [28], vegetation coverage [29], landscape spatial pattern [7], and biodiversity [30].

These relationships indicate that a soundscape may possess spatial dependencies on such factors. The spatial dependence is seen as a normal extension of the first law in geography [31]—"everything is related to everything else, but near things are more related than distant things". It may occur due to the spatial dimensions with regard to social-cultural contexts and economic factors [32]. Previous studies found that the temporal dimension is also a critical aspect for exploring the spatial dependence, and neglecting the temporal characteristics could lead to a misunderstanding of the "real" measure of spatial dependence over time [33,34]. However, to date only a few studies have investigated spatial dependence of soundscape quality but with some deficiencies. For instance, Hong and Jeon [35] explored the spatial dependence of urban soundscapes, nonetheless, solely on the perceived sound sources. Rice et al. [36] explored the spatial dependence of noise abatement on the features of protected areas, but they did not consider people's perception of the acoustic environment. In general, current studies have neither explored the spatial dependency of soundscape quality with the principal components, i.e., pleasantness and eventfulness [37], nor included variables in terms of landscape perception and green space features. Besides, the temporal characteristics of the spatial dependence of soundscape have not been effectively explored either. Exploring the spatial dependence of soundscape can clarify the interrelationship between soundscape perception and impact factors, which may help planners and managers identify the main disturbances to the acoustic environment in scenic areas and therefore find solutions and protection measures [35].

In 2016, Kulangsu was suggested as a cultural heritage by United Nations Educational, Scientific, and Cultural Organization (UNESCO), and subsequently was successfully listed as a World Heritage at the 41st World Heritage Congress in Krakow, Poland, in July 2017 [38]. Therefore, 2016 was a "landmark year" that represents a turning point for Kulangsu from China to the world. Kulangsu is a unique and historic international settlement, an important UGS, and a famous tourist attraction [39], the natural and cultural resources of which constitutes the unique soundscapes of outstanding universal value [40]. Unfortunately, such soundscapes are suffering from excessive disturbance and destruction by human activities, whether present, past, or even future. Accordingly, exploring the spatiotemporal dynamics of soundscapes and landscapes as well as the spatial dependence of soundscape quality in Kulangsu Island has significant implications for soundscape planning and management, and soundscape resources conservation, not only in UGS but also in the scenic area and even other similar World Heritage Sites. Given this importance, the objective of this study is (1) to visualize and analyze the spatiotemporal dynamics with regard to soundscape and landscape perception, as well as the spatial landscape features in Kulangsu Island; and (2) to examine the spatial dependence of soundscape quality on these compositions. To this end, both global and local spatial regression methods were employed.

2. Materials and Methods

2.1. Study Area

The study area is located on Kulangsu Island (Figure 1a), a World Heritage and one of the National 5A level tourist attractions in China. The area of it is about 1.92 km^2, with a length of 2.3 km from north to south and 1.6 km from east to west. The green spaces occupied more than one-third of the island (Figure 1b), which was the highest percentage (32.36%) of land use type (according to 2017 land use vector data of Kulangsu Island obtained from Xiamen Municipal Natural Resources and Planning Bureau). The types of green space include parks, squares, and woodlands [41].

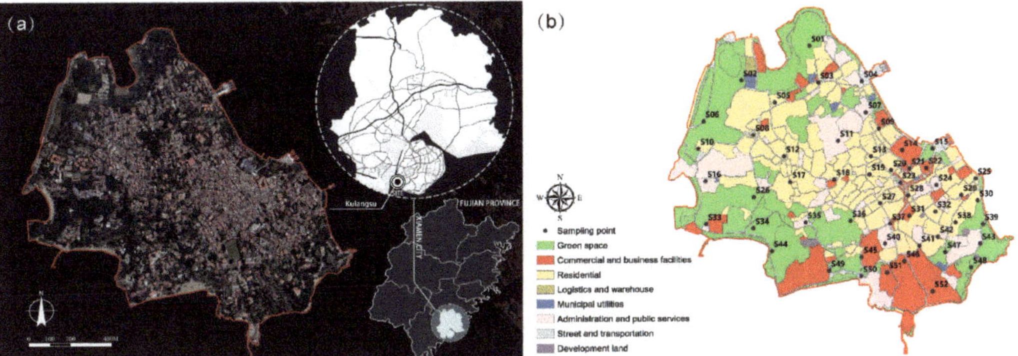

Figure 1. Case study area: (**a**) location of Kulangsu Island in Fujian Province, China; (**b**) land use type with sampling points.

The present study was conducted from 17 July 2016, to 21 July 2016, in the application process of World Heritage. In July 2017, Kulangsu Island was officially inscribed on the World Heritage List. Based on the pilot study and relevant literature [7,12,42], 4 main-categories and 19 sub-categories sound sources were identified in the area (Table 1).

Table 1. Classification of sound sources in the study area.

Main Category (Abbreviation)	Sub-Category
Human activity sound (HS)	Talking, footstep, playing children, hawking, folk activity, live performance
Mechanical sound (MS)	Music radio, broadcast notification, construction, traffic noise, alarm
Biological sound (BS)	Birdsong, insect, cat
Geophysical sound (GS)	Sea wave, wind, tree, water, raining

We selected 52 observation points according to the land use type, function, and accessibility, and interviewed random passersby by questionnaire surveys on site in the morning (8:00–11:00), afternoon (13:00–16:00), and evening (17:00–20:00) of the days. A team of 12 college students from landscape architecture faculty of Fuzhou University got involved to help conduct the field survey and distribute the questionnaire to random people in the study area. They were professionally trained before executing the field survey in order to ensure the quality of study. The data includes personal information and subjective perception ratings for the soundscape and landscape of the interviewees. A total of 703 valid questionnaires were returned, with 10 to 15 questionnaires on each sample site, indicating that enough and accurate results can be achieved [43,44]. The interviewees' information is shown in Figure S1 (Supplementary Material). Analysis of the questionnaire data in SPSS 25.0 showed that the alpha coefficient was 0.86, indicating that the reliability of the questionnaire data was good and suitable for further analysis.

2.2. Data Collection

2.2.1. Soundscape and Landscape Perception

Participants were asked to evaluate the sound source, soundscape quality, and landscape satisfaction degree of the environment, based on perceived indicators using a Likert 5-point scale (Table 2).

Table 2. Detailed information for each perceived indicator.

Category	Indicators (Abbreviation)	Survey Question	Rating Scale or Formula	Reference
Sound source	Perceived occurrences of sound (POS)	To what frequency do you presently hear the following four types of sounds?	1-never, 2-occasionally, 3-normal, 4-frequently, 5-too frequently	[12,17,24]
	Perceived loudness of sound (PLS)	To what intensity do you presently hear the following four types of sounds?	1-too weak, 2-weak, 3-neither weak nor strong, 4-strong, 5-too strong	
	Sound dominant degree (SDD)	/	$SDD_{ij} = POS_{ij} \times PLS_{ij}$	
Soundscape quality	Pleasant Comfort Harmonious Vivid Richness Eventful	To what extent do you agree or disagree that the present surrounding sound environment is … ?	1-strongly disagree, 2-disagree, 3-general, 4-agree, 5-strongly agree	[37]
Landscape satisfaction degree	Satisfaction of natural landscapes (SNL) Satisfaction of landscape design (SLD) Satisfaction of historical building (SHB) Satisfaction of visual-audio experience (SVA) Satisfaction of service facilities (SSF)	To what extent do you satisfy or dissatisfy that the present surrounding landscape with regard to … ?	1-very dissatisfied, 2-not satisfied, 3-general, 4-satisfied, 5-very satisfied	[12,27]

Notes: j is the jth sample, i is the ith source, and n is the sample size.

2.2.2. Analysis of Landscape Characteristics

This study objectively quantifies the landscape features within Kulangsu in two aspects: (1) features of green spaces, including Distance to green spaces (DtGS) and Normalized difference vegetation index (NDVI), and (2) landscape spatial patterns, including patterns of the green space class and overall landscape. DtGS was calculated based on the "Euclidean distance" spatial analysis tool in GIS, which is measured in the projection unit of the raster and calculated from one cell center to the other cell center, and can objectively measure the spatial distance of the global scale [45], with the Equation (1) shows:

$$D = \sqrt{(x_2 - x_1)^2 + (y_2 - y_1)^2} \tag{1}$$

where (x_1, y_1) are the coordinates of one point, (x_2, y_2) are the coordinates of the other point; D is the Euclidean distance between (x_1, y_1) and (x_2, y_2).

We imported the collected Sentinel-2 level 2A images as an ensemble into the Google Earth Engine (GEE) JavaScript-based code editor environment [46], to calculate the value of NDVI for each time series image. Only images with less than 10% cloudiness in the study area were extracted, to ensure data integrity, and the calculated images were between March 2017 (the first Sentinel-2 available image in GEE) and January 2019. All data were top-of-atmosphere (TOA) images and had been atmospherically corrected [47]. The FMask algorithm [48] was used in multispectral instrument (MSI) data processed to identify cloud, cloud shadow, cirrus, and snow/ice observations. The *NDVI* values were calculated for each image element and a time series *NDVI* image collection with a spatial resolution of 10 m is generated. The calculation formula is shown in Equation (2):

$$NDVI = \frac{NIR - RED}{NIR + RED} \tag{2}$$

where *RED* is the TOA values of the red band (630–680 nm); *NIR* is the TOA values of the near infrared band (845–885 nm). The *NDVI* takes values in the range of −1 to 1 [49].

To examine the landscape patterns, we considered landscape features in terms of area, density, shape, diversity, and aggregation on class and landscape levels [7,50]. Details of the calculated landscape spatial indices are shown in Supplementary Material (Table S1).

2.3. Mapping Process

Based on the questionnaire data, the mean values of ratings for soundscape and landscape perception were calculated for each observation point at each sampling time period, and the data were subsequently visualized in ArcGIS 10.7. By comparing the different interpolation methods provided in spatial analyst tools, the Inverse Distance Weighted (IDW) method was chosen to produce the soundscape and landscape perception maps to analyze their spatiotemporal dynamic characteristics. The IDW method is based on the spatial distance of the data points for weighted interpolation, and the closer the point is to the value, the greater the effect is. Conversely, the smaller the effect is [7,20,21].

The landscape index visualization is based on the moving window technique in Fragstats 4.2. The size and shape of the window can be defined by the user. The window is moved over each cell with positive value in the raster data, and the selected landscape index within the window is calculated. These values are then returned to the focal point (mid-point) of the cell, while a new continuous type of grid data is generated for each selected landscape index, where the cell values represent the "local neighborhood structure" [51]. According to the previous study [7], we set 175 m radius as the window size. In addition, we considered that Fragstats software gives negative values for cells near the edges and cells that are not fully included in the input grid window in the calculating process, which may lead to incomplete spatial data. Therefore, we created a buffer of the same size as the moving window (175 m) before inputting the grid data, thus minimizing the effect of boundary effects [52].

2.4. Statistical Analysis

(1) Principal Component Analysis (PCA). Based on the semantic attributes of the soundscape, the PCA method was applied to extract the principal components so as to obtain the determinants of overall soundscape quality. The eigenvalues of the extracted principal components were all greater than 1. The analysis was performed in SPSS 25.0.

(2) Multicollinearity Analysis. Before constructing a spatial regression model, it should be ensured that there is no multicollinearity problem among the independent variables included in the model [53]. The tolerance (TOL) and variance inflation factor (VIF) were used to perform the diagnosis of multicollinearity problems. TOL > 0.1 or VIF < 5 indicates that there is no multicollinearity problem between the analyzed [54]. Equations (3) and (4) were used to calculate TOL and VIF, respectively:

$$TOL = \frac{1}{VIF} \tag{3}$$

$$VIF = \frac{1}{1 - R_j^2} \tag{4}$$

where R_j^2 is the coefficient of determination for the regression analysis on all other variables.

(3) Spatial Regression Model. Spatial regression models capturing spatial dependence were used to examine the effects of perceived sound sources, landscape satisfaction degree, and landscape characteristics on perceived soundscape quality within Kulangsu Island.

Both the global spatial regression model—ordinary least squares (OLS) estimation model, and the local spatial regression model—geographically weighted regression (GWR) model were used (see Supplementary Material for details) [55], and the formulae of them are shown in Equations (5) and (6), respectively [56,57]. The explanatory ability and goodness of fit of these models were examined by the coefficient of determination (R^2) and the Akaike Information Criterion (AIC) [58,59]. All operations were performed in ArcGIS 10.7.

$$Y_i = \beta_0 + \beta_1 x_1 + \beta_2 x_2 + \cdots + \beta_n x_n + \varepsilon_i \frac{1}{2} \tag{5}$$

where Y_i is the dependent variable being explained; $x_1, x_2 \ldots x_n$ are the independent variable; β_0 is constant; $\beta_1, \beta_2 \ldots \beta_n$ are variable coefficients; ε_i is the bias in estimating the coefficients.

$$Y_i = \beta_0(\mu_i, v_i) + \sum_{k=1}^{p} \beta_k(\mu_i, v_i) x_{ik} + \varepsilon_i (i = 1, 2, \ldots, n) \tag{6}$$

where Y_i is the explained dependent variable, x_{ik} is the independent variable at kth sample of I, (μ_i, v_i) denotes the coordinates of the ith sample, $\beta_0 (\mu_i, v_i)$ denotes the intercept of the ith sample, $\beta_k (\mu_i, v_i)$ denotes the regression parameters of the ith sample, and ε_i denotes the residuals of the model at the ith sample in estimating the coefficients. All data were normalized to 0 to 1 prior to analysis.

3. Results

3.1. Spatiotemporal Dynamics of Soundscape and Landscape Perceptions

3.1.1. Soundscape Mapping

Figure 2 shows significant spatial and temporal variability in SDD for all four major sound sources. The SDD-GS is higher in the eastern and western parts, and relatively lower in the central part of the study area. The spatial distribution of SDD for BS, GS, and HS all changed significantly over time. The SDD-HS changed most significantly, and its high values covered almost entire area in the evening. However, the whole area was mainly dominated by BS in the morning. Compared with the SDD of the BS, GS, and HS, SDD-MS only changed slightly over time, and the overall distribution pattern was relatively stable. For the temporal variation of the mean values across land use types, the maximum mean values of SDD-BS in all periods were in the logistics and warehouse land. The mean values of SDD-GS were the always highest in development land, and the lowest in municipal land and logistics and warehouse land. The highest mean values of SDD-HS were found in logistics and warehouse land in P1 and P3, respectively, but in municipal utility land in P2. For SDD-MS, the highest mean values occurred in different land types in each period. The general trend of it was more diverse than the other three sounds.

PCA was used to extract the principal components of the six semantic attributes based on the varimax rotation method, for representing the soundscape quality. The Kaiser-Mayer-Olkin (KMO) test was 0.839 (KMO > 0.60), and the Bartlett's test of sphericity was 0.000 indicating highly significant ($p < 0.001$), which means that the dataset was suitable for PCA. A total of two principal components were extracted: Component 1 (Pleasantness) and Component 2 (Eventfulness), which explained 41.8% and 38.8% of the variance in the semantic attribute dataset, respectively (Table 3). The results of this analysis are in good agreement with the two-dimensional model of perceived affective quality of soundscape (Pleasantness-Eventfulness) proposed in the previous study [37]. This model can provide comprehensive information on soundscape perception.

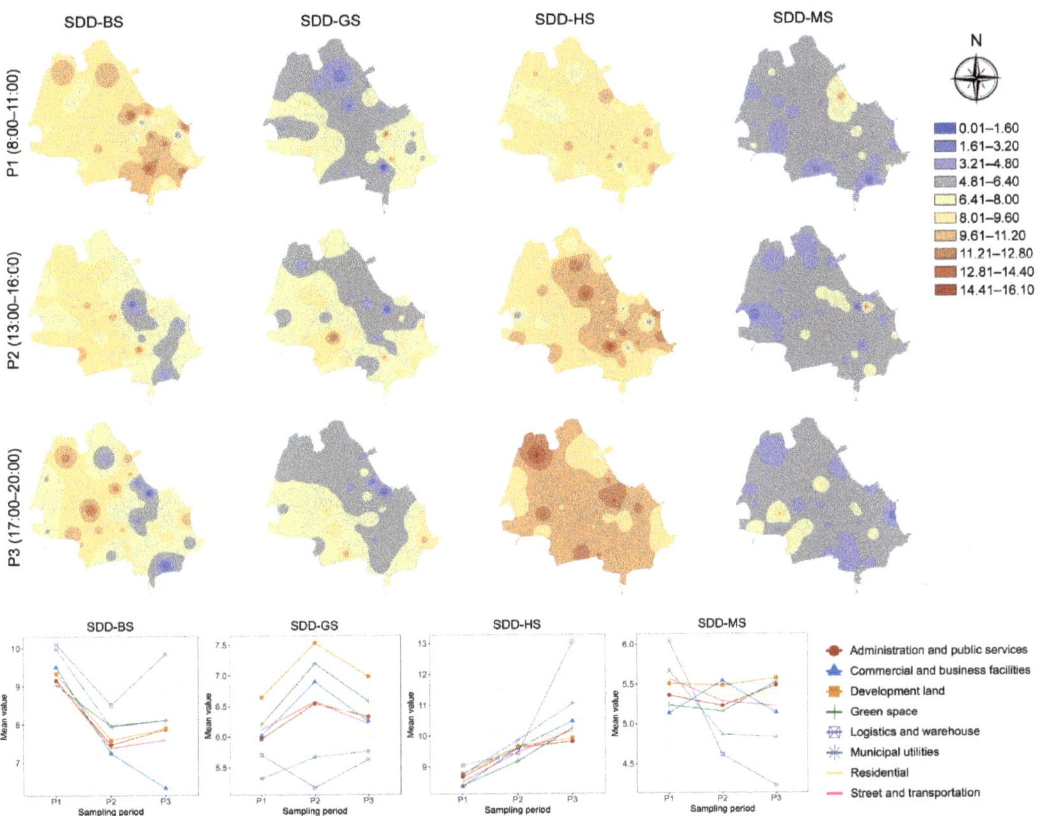

Figure 2. Spatiotemporal distribution and temporal variation of the mean values of sound source dominance of the four main category sound sources.

Table 3. Rotated component matrices of the PCA based on semantic attributes (numbers in parentheses represent explained variance).

Semantic Attribute	Component 1: Pleasantness (41.8%)	Component 2: Eventfulness (38.8%)
Pleasant	0.880	0.250
Comfort	0.906	0.236
Harmony	0.739	0.441
Vivid	0.491	0.703
Richness	0.295	0.867
Eventful	0.195	0.877

The values of pleasantness and eventfulness for each observing point were calculated in SPSS 25.0, and then visualized via ArcGIS 10.7 (Figure 3). Both components had significant spatiotemporal characteristics, and appeared to be similar in their distribution patterns. In general, the values of pleasantness and eventfulness in the study area showed low and high values in the north and south, respectively, in all periods. Regarding mean values of the components in different periods for each land use type, the trends of pleasantness and eventfulness were similar. The maximum and minimum values occurred in the same land use type in all periods. The maximum values were found in development land all the time. The minimum values were found in municipal utility land in P1, and in logistics and warehouse land in P2 and P3. The mean value of pleasantness first reduced and then improved in logistics and warehouse land, nonetheless, it had opposite trends in each of the rest land use types. The trend of eventfulness was different from that of pleasantness only in commercial and business facility land, which continued to decline over time.

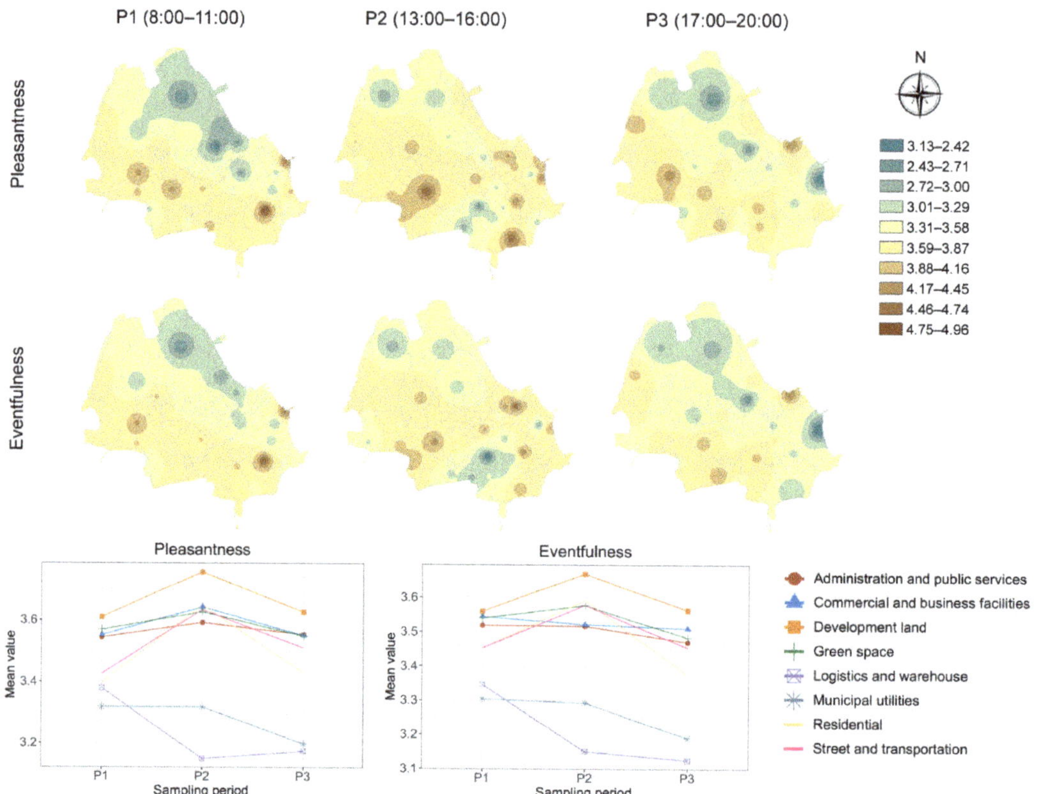

Figure 3. Spatiotemporal distribution and temporal variation of the mean values of pleasantness and eventfulness.

3.1.2. Landscape Satisfaction Degree

Figure 4 indicates the spatiotemporal distribution of each landscape satisfaction degree indicator. Values of SNL, SLD, and SHB were relatively high, while that of SSF was significantly lower. The mapping results show that the spatial differences were pronounced for SLD and SHB, respectively, but that for SSF was relatively small. However, the spatial distribution patterns of them changed significantly with time. In P2, high values of SNL, SLD, and SHB covered almost the whole area. Nonetheless in P3, the high values of the first two weakened in the central and northern areas, but the high values of SHB still covered almost the whole area. The landscape satisfaction degree indicators showed presented the same trend of mean values in some land use types. The mean values of SNL and SLD in all types of land use, except for development land, showed the same trends. The trends of mean values of SHB and SSF were the same in all land use types. Both of their maximum mean values were found in logistics and warehouse land and development land in P2 and P3, respectively. The mean value of SVA increased and then decrease in all land use types, except for commercial and business facility land.

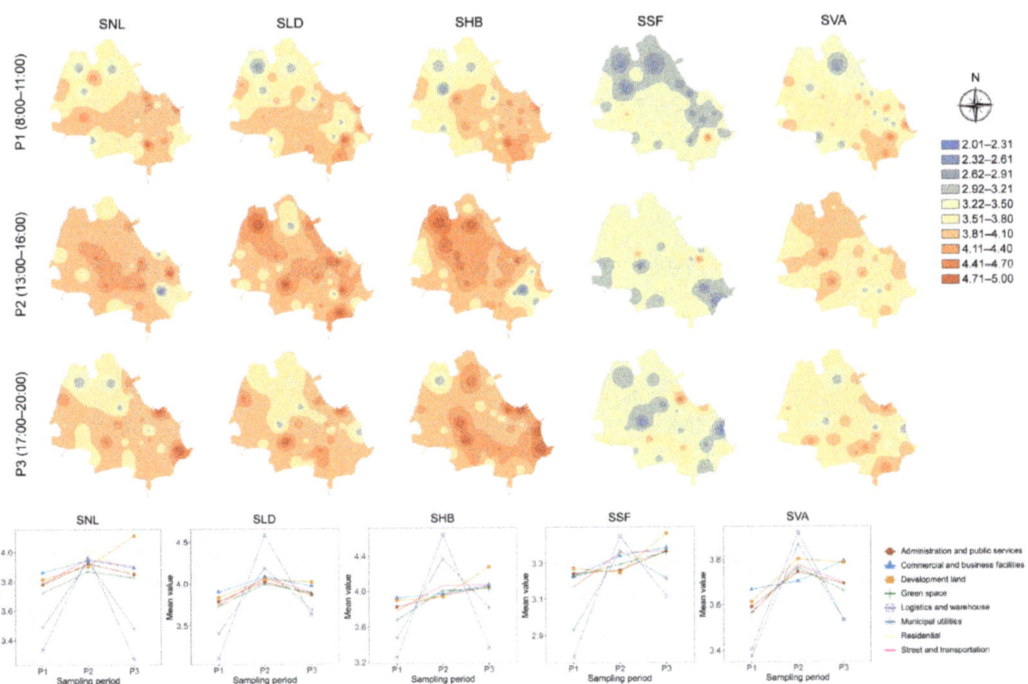

Figure 4. Spatiotemporal variation and temporal variation of the mean values of landscape satisfaction degree.

3.2. Spatial Landscape Characteristics

The DtGS and NDVI of the study area are shown in Figure 5. The result shows that the vast majority of the study area was close to green spaces (Figure 5a). The NDVI values showed low values in the central area (Figure 5b). The low NDVI values were also found in a few marginal areas of the study area. However, the NDVI values were generally at a high level, with more than 50% of the areas having the values over 0.5 [60], indicating a high vegetation coverage in Kulangsu Island.

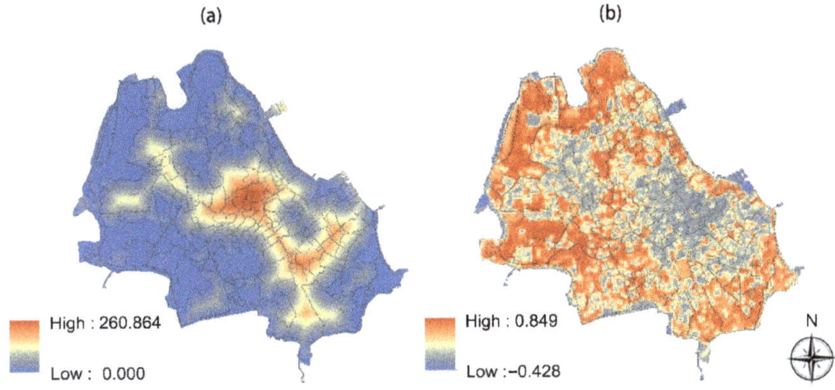

Figure 5. The spatial features of green spaces: (**a**) distance to green spaces (DtGS) and (**b**) normalized difference vegetation index (NDVI).

Figure 6 shows the spatial distribution of the landscape indices at class level and landscape level. Except for SPLIT at the class level, all other landscape indices had significant spatial differences. For the landscape pattern of greenspaces, high values occurred at similar locations for CA and ED, but the latter covered a wider area. The high values of PD occurred only in a few areas near the northern and southern edges. The high values of IJI covered almost the entire area, while MESH and SPLIT showed lower levels within the area. For the landscape pattern of whole area, the high values of PD concentrated in the central and northern areas, and the high values of LSI were mainly found in the central-eastern area. The high values of SHDI were distributed similarly to those of IJI, which mainly gathered in the periphery of the study area, while the middle area showed a band of low values. SPLIT had a large area of low values in the area, and the high values of COHESION were found mainly in the central and southwestern areas.

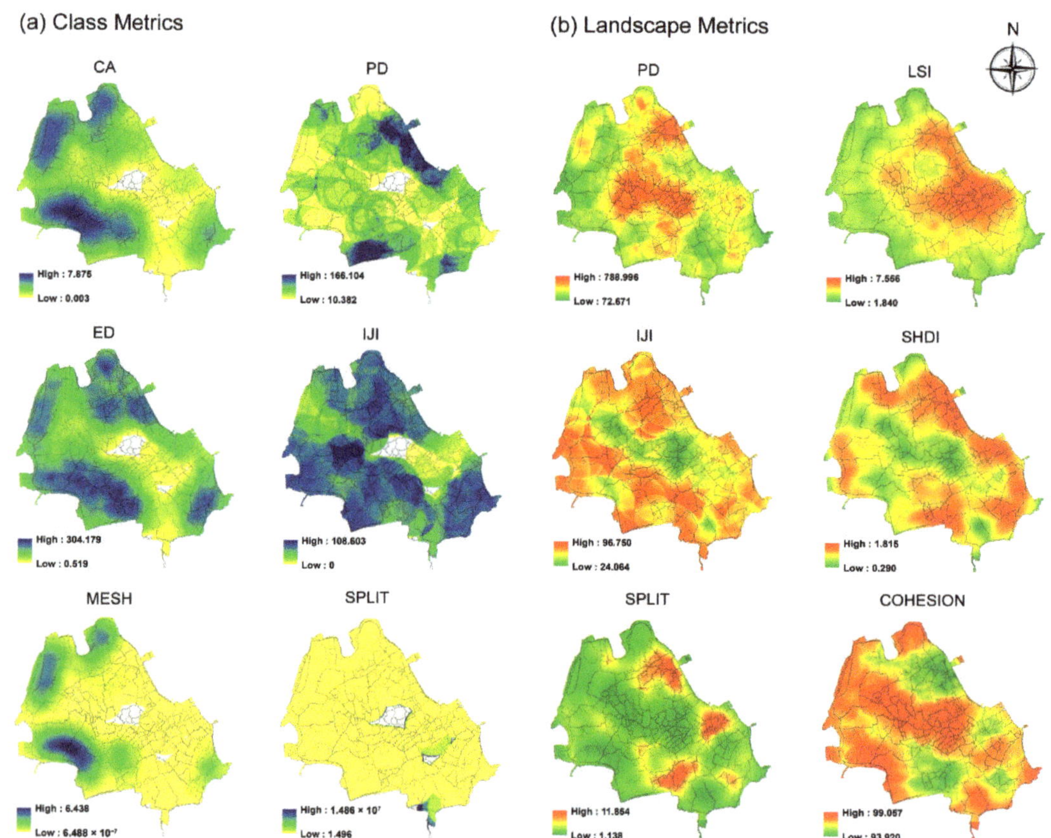

Figure 6. Landscape spatial patterns in (**a**) class metrics and (**b**) landscape metrics.

3.3. Spatial Regression Models of Soundscape Quality

Spatial regression models, including OLS and GWR models, were built to examine the spatial dependence of soundscape quality (dependent variables) in Kulangsu Island. The explanatory (independent) variables include SDD of 4 main sound sources, landscape satisfaction degree, DtGS, NDVI, and landscape pattern indices.

3.3.1. Multicollinearity Diagnostic Results

To avoid possible covariance problems among the independent variables, we performed covariance diagnosis in the independent variables at the three periods and the whole day (Total), respectively, and selected the variables without multicollinearity in each period for the OLS regression. The results of multicollinearity diagnosis are shown in Table S2, Supplementary Material. Most variables passed the multicollinearity test, indicating rationality of the indicator selection. This analysis was performed in SPSS 25.0.

3.3.2. Global Spatial Regression Model

The OLS model used the data on each grid to build relationships between the independent and dependent variables (Table 4). None of the landscape indices significantly influenced the soundscape quality at all periods. Only the variables related to SDD and landscape satisfaction degree were significant for the soundscape quality. Pleasantness was only influenced by SLD in P3, and eventfulness was only significantly influenced by SLD in both P2 and P3. The included independent variables had better performances in explaining the pleasantness (R^2 was 0.653 and 0.649, respectively) and eventfulness (R^2 was 0.591 for both) in P1 and Total, compared with those in P2 and P3. Similarly, the results of the AIC demonstrate that the OLS models in P1 and Total had better fitting accuracy than those in P2 and P3.

Table 4. Results of the OLS model (* $p < 0.05$, ** $p < 0.01$).

Sampling Period	Indicator	Pleasantness		Eventfulness	
P1	Independent variable	SDD-MS	−0.505 **	SDD-BS	0.170 *
		SHB	0.330 *	SDD-MS	−0.345 *
		SVA	0.312 **	SVA	0.407 **
	R^2		0.653		0.591
	AIC		−59.061		−56.234
P2	Independent variable	SDD-GS	0.433 *	SLD	0.295 **
		SLD	0.362 **		
	R^2		0.302		0.328
	AIC		−5.066		−19.834
P3	Independent variable	SLD	0.401 *	SLD	0.497 **
	R^2		0.379		0.418
	AIC		−17.154		−31.585
Total	Independent variable	SDD-BS	0.196 **	SDD-BS	0.238 **
		SDD-GS	0.299 **	SDD-GS	0.234 *
		SLD	0.510 **	SLD	0.504 **
				SSF	0.351 *
	R^2		0.649		0.591
	AIC		−62.988		−50.111

3.3.3. Local Spatial Regression Model

The GWR model was to explore the spatial relationships between soundscape quality and the significant variables derived from the OLS model (Table 4). Before executing the GWR model, a kernel function needs to be selected for performing the geo-weighting algorithm, which is to estimate the local coefficients and their bandwidth sizes. There are two commonly used functions in GIS, namely the Gaussian fixed and adaptive kernel type. Based on the finding of previous study [35], we selected the Gaussian fixed kernel type to perform the GWR model.

The values of AIC of the GWR models in all period were significantly smaller than those of the OLS models (Table 5), indicating that the GWR model had better fitting results than the OLS model. The mean values of R^2 of the GWR models were higher than those of the OLS models for the all periods, nonetheless, except for the pleasantness models in P1 and Total. As shown in Figure 7, the R^2 spatial distributions of pleasantness model in P1, and both pleasantness and eventfulness models in Total were relatively stationary compared with those in P2 and P3. Also, Table 5 indicates that the results of R^2 and AIC in P1 and Total were better than those in P2 and P3. These findings indicate that such spatial stationarity may achieve a better performance of the statistical model in terms of explanation of variance and goodness-of-fit.

Table 5. R^2 and AIC of the GWR model.

Sampling Period	Indicator		Pleasantness	Eventfulness
P1	R^2	Mean	0.596	0.674
		Minimum	0.588	0.577
		Maximum	0.645	0.81
	AIC		−69.975	−84.445
P2	R^2	Mean	0.507	0.380
		Minimum	0.343	0.001
		Maximum	0.58	0.453
	AIC		−46.329	−46.172
P3	R^2	Mean	0.468	0.601
		Minimum	0.212	0.258
		Maximum	0.558	0.773
	AIC		−51.287	−75.251
Total	R^2	Mean	0.598	0.615
		Minimum	0.62156	0.64534
		Maximum	0.62183	0.64564
	AIC		−84.948	−80.86

Note: To account for differences, the maximum and minimum values of the R^2 of the GWR models for the total dataset were retained to five decimal places.

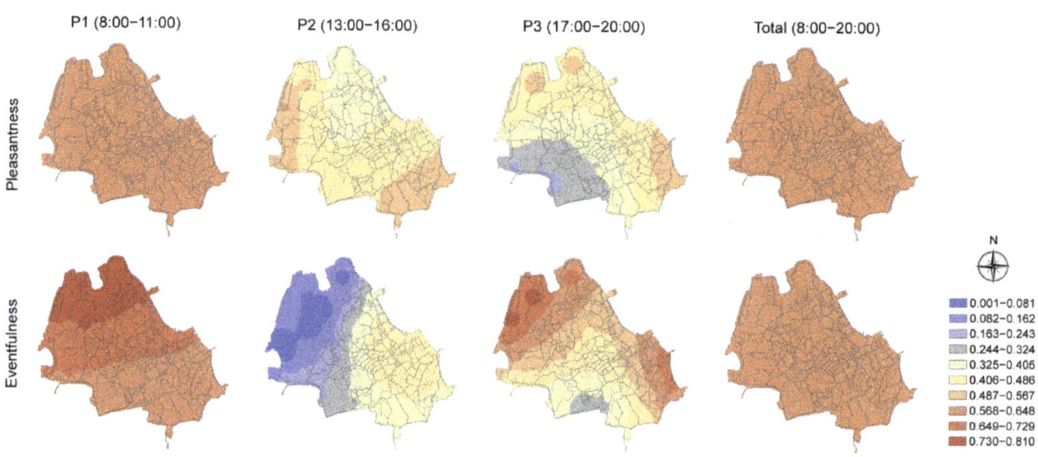

Figure 7. Spatial distribution of local R^2 for pleasantness and eventfulness in the three and total sampling periods.

The spatial distribution of the coefficients of each independent variable are shown in Figure 8, the spatial stationarity also presented in pleasantness model in P1, and both pleasantness and eventfulness models in Total. In P1, SDD-MS and SVA exhibited the strongest negative and positive effects, respectively, on both pleasantness and eventfulness. In P2, SLD was positively related to both pleasantness and eventfulness in most areas, but their negative relationships appeared in few areas in the northwest. In addition, pleasantness was also positively influenced by SDD-GS, and the spatial differences were significant with the coefficients decreasing from northwest to southeast (from 0.808 to

0.139). In P3, both pleasantness and eventfulness were positively related to SLD only, with the coefficients increasing from southwest to northeast. In the model of Total, all included variables had positive relationships with pleasantness and eventfulness, respectively. SLD was the variable with strongest effects on pleasantness and eventfulness, respectively.

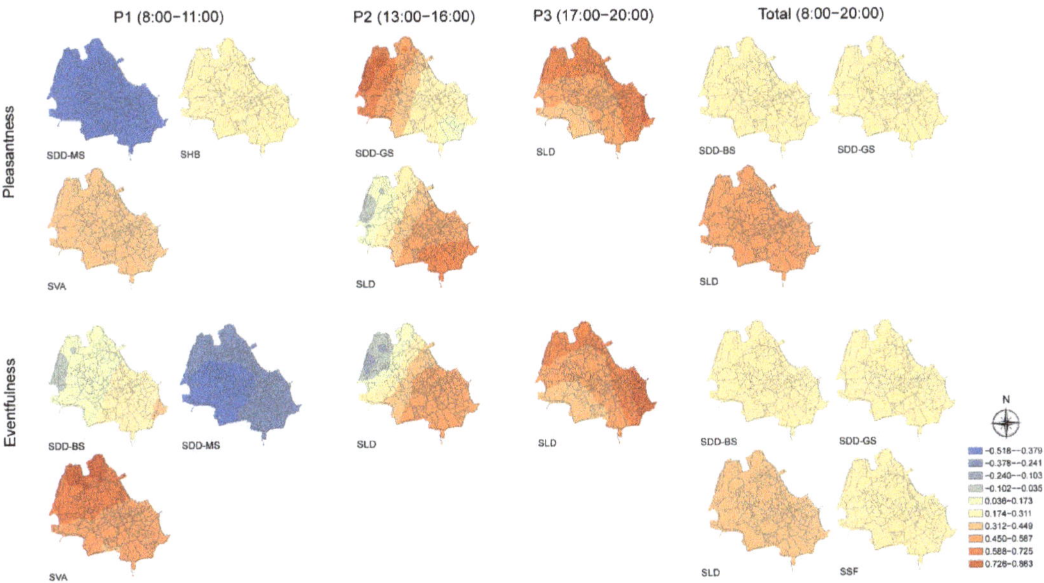

Figure 8. Spatial distribution of local coefficients of significant variables in the three and total sampling period.

4. Discussion

4.1. Spatiotemporal Dynamics of Soundscape and Landscape in Kulangsu

The results of the present study showed that both soundscape and landscape perception in Kulangsu Island were characterized by significant spatiotemporal dynamics. In terms of perceived sound sources, we calculated and visualized the SDD for each main sound source, to present the dominance of them in different temporal and spatial dimensions in Kulangsu (Figure 2). The results demonstrated that BS and HS were the dominant sounds during the three sampling periods. The dominance of BS may imply the good quality of ecological environment within Kulangsu, which can attract various vocal organisms such as birds and insects to congregate and communicate here [61]. This was also reflected the high level of biodiversity in the island. Such abundance of natural resources also provides a good opportunity for tourism development, with many urban residents and tourists choosing to visit the area for e.g., nature-based recreation [62]. We speculate that this maybe one of the reasons why HS was another dominant sound.

We found that the HS-dominated area gradually expanded over time, and covered almost the entire area in the evening (Figure 2), indicating that this period was the most intensive period for people's activities. This survey was conducted during the summer of July, which is generally the hottest period of the year in Xiamen [63]. Accordingly, we speculate that one reason for this phenomenon may be due to the higher thermal comfort in the evening, compared to the built-up area. This is probably due to the good vegetation condition in Kulangsu, which has a significant effect on reducing temperature thus creating a comfortable microclimate. Generally, such cooling effects were most pronounced in the evening [64]. Another reason may be the development of the night-time economy in China in recent years, which has led to a new urban living pattern dominated by night-time entertainment activities [65].

In the analysis of SDD mean values of the sound sources for different land use types (Figure 2), high SDD of BS and HS unexpectedly appeared several times in logistics and warehouse land. We suspect that this may be because this type of land use was generally small in size and surrounded by large areas of green spaces. This also potentially reflects that both animals and visitors were more willing to get close to the natural landscapes. The high GS-SDD mean values were mainly found in development land. This may be due to the harbors and waterfront spaces that were close to the sea in such land use type and therefore the GS especially water and wave sounds were more dominant. Such results are similar to the findings of previous studies located in coastal areas [66].

A similarity was found in the spatiotemporal distribution patterns of pleasantness and eventfulness of the soundscape (Figure 3), which implies they may be potentially interrelated. This finding is the same as previous soundscape studies located in urban built-up areas and urban forests [20,22]. For mean values of pleasantness and eventfulness across land use types (Figure 3), the maximum values of them were always found in development land, and mainly concentrated in the south according to the mapping result. This is possibly because the development land to the south were close to large areas of green spaces and commercial lands, and therefore subject to the "radiation effect" from such land types. The development land, on the other hand, were relatively smaller and thus may receive higher aggregated values of pleasantness and eventfulness. Another possible reason was that the development land encompassed many waterfront spaces where people prefer to stay, to experience nature and response higher values of pleasantness. Moreover, the consequent increase in foot traffic may also lead to high values of eventfulness [67].

As for landscape satisfaction degree, we found that the spatiotemporal distribution patterns of SNL, SLD, SHB, and SVA were similar to those of pleasantness and eventfulness in P1, but they were relatively different in P2 and P3 (Figures 3 and 4). This finding suggests that the harmony or congruency between the soundscapes and landscapes in Kulangsu was still deficient especially in the afternoon and evening, and therefore it has room to improve in future landscape planning and design. We speculate that this audio-visual inconsistency may be due to the increasing spatial dominance of HS in these periods (Figure 2), which negatively affected the audio-visual perception [68]. Regarding the spatiotemporal distribution of SSF, it was significantly lower compared to the other landscape perception indicators. However, the distribution pattern of SSF was also similar to those of pleasantness and eventfulness in the morning and afternoon (Figures 3 and 4). This similarity indicates that people may prefer to use these service facilities especially in these two periods. We suppose that this is because the temperatures in daytime are generally high in Kulangsu. This finding is also in line with the results of the previous study, which found infrastructure and facilities were an important factor contributing to the soundscape quality in scenic areas [27].

The analysis of "objective" landscape characteristics also illustrates the superior natural environmental characteristics of Kulangsu (Figures 5 and 6), which was consistent with the finding derived from the spatiotemporal distribution of SDD-BS (Figure 2). These results indicated that Kulangsu had good green space accessibility and a high proportion of vegetation cover, which allows visitors to easily approach and experience nature and contributes to the comfortable microclimate here [60]. We found that although the high values of SDD-BS appeared in different areas over time, they almost always fell in the areas with low DtGS or high NDVI, indicating the interrelationship between biological sounds and green space features. Regarding the landscape pattern, the results of class metrics indicate that the green space patches were large and morphologically intact, with a high degree of connectivity, and did not have significant fragmentation. In addition to SPLIT, the spatial distributions of CA, PD, ED, IJI, and MESH were more or less similar to those of pleasantness and eventfulness in the three periods, suggesting that diverse and coherent structure of green space patches may contribute to soundscape quality [69]. Based on the results of the landscape metrics, we found that the structure of the patches in the center of Kulangsu was more complex and homogeneous, and the connectivity between patches was good. This may be due to the high density of historical architectures in the central area [70], but such buildings still maintain a good continuity. The edge of the study area, on the other hand, had a richer patch composition and different types of patches may be interspersed between similar patches. We found that this complexity tends to accompany areas where commercial land occurred. Based on such, we suggest that the remediation planning of scenic areas could focus on preventing the fragmentation effects on the landscape especially from dispersed commercial activities [71].

4.2. Spatial Dependencies of Soundscape Quality

Most indicators with regard to perceived sound sources, landscape satisfaction degree, the features of green spaces, and landscape patterns passed the test of multicollinearity on explaining the soundscape quality (Table S2), and then included in the OLS and GWR models. Interestingly, the results of OLS models (Table 4) show that the soundscape quality had no spatial dependence on all objective landscape characteristics, namely DtGS, NDVI, and landscape metrics on class and landscape levels. Both pleasantness and eventfulness of the soundscape were significantly dependent on the perceptual indicators of soundscape and landscape, i.e., SDD-BS, SDD-GS, SDD-MS, SLD, SHB, and SSF, in different periods. This finding was not exactly the same as previous studies on the relationship between soundscape elements and landscape patterns [7,72]. We speculate that this discrepancy may arise from two aspects. On the one hand, the research objectives and methods were different. The present study aimed to explore the spatial dependence of soundscape quality on many types of variables rather than the correlations between two variables. The spatial dependence can indicate the propensity for nearby locations to interact and share similar properties, which is also an essential part of modeling soundscape [35]. Therefore, the regression model accounting for the spatial dependence can explain more about the relationships between the dependent and independent variables, and predict the dependent variable based on such, while the correlation does not necessarily imply causation [73]. On the other hand, the study area was different. Our study was devoted to exploring the soundscape and landscape resources of the Kulangsu scenic area, whereas the previous studies were in a multifunctional urban area [7,20,72]. Kulangsu Scenic Area is rich in natural and cultural resources and locally distinctive sound sources, which is basically different from the urban area.

The result of soundscape spatial dependency on SDD-BS, SDD-GS, and SDD-MS suggests that introducing more birdsongs and vegetation sounds, through enriching the vegetation types and density, may contribute to the biodiversity and reduction of mechanical noise [74], therefore improving the pleasure and structure diversity of the soundscape [37]. Besides, according to the dependence on SLD and SHD, the soundscape quality can also be enhanced by optimizing the spatiotemporal features of landscape design, with regard to the lightscape (especially in the evening) and cultural innovation [24,75]. For instance, adding interesting lightscapes designed according to the natural and cultural environment characteristics of Kulangsu. The type and application of the light sources should be able to integrate into but not affect the natural environment [76]. Furthermore, the historical culture of Kulangsu can be combined with modern elements to revitalize old buildings by injecting new functions, such as cultural and creative business or exhibitions [77]. Also, the "Piano Island", one of the famous titles of Kulangsu, can be used to create e.g., a theme-specific "musical environment" in different public open spaces by combining classical and contemporary featured piano music. The significant dependency of soundscape quality on the landscape perception indicators implies that only considering the effects of sound features on soundscape quality was insufficient in the past study [35]. This result also proves that the landscape compositions are indispensable factors in creating a soundscape [17,43]. The temporal differences of spatial dependence of the soundscape in Kulangsu can contribute specific information to the soundscape planning and design strategies for scenic areas and Heritage Site.

The GWR model was found to have a better goodness-of-fit than the OLS model for all periods according to the results of AIC, which was consistent with the previous study [35]. However, the R^2 results of GWR model performing pleasantness in P1 and Total were lower than those of OLS model (Tables 4 and 5). Interestingly, the spatially stationary relationships were found in the results of R^2 and coefficients of the variables in pleasantness model in the morning and both pleasantness and eventfulness models in total dataset, respectively (Figures 7 and 8). This finding suggests that the pleasantness models had better fitting accuracy if they were spatially stationary, but the explanatory power of the regression line for the dependent variable became weaker, which implies that the nonlinear relationship may be able to better explain the spatial dependence of soundscape pleasantness. This is also a good proof that the spatial stationarity helps to verify and select the suitable statistical methods in predicting variables [78,79]. As shown in Figures 7 and 8, the total dataset models of pleasantness and eventfulness exhibited the best spatial stationarity, indicating the data used should ideally be of a time-varying nature that may contain as complete a temporal dimension as possible. For instance, the collected data should ideally cover morning, afternoon, evening, and even in the night or early morning.

4.3. Limitations and Future Studies

The limitations of the present study may mainly stem from the data collected and the model employed. Although vegetation coverage was considered in this study, the classification of green spaces was not in sufficient detail in terms of the vegetation type. The outcomes may therefore vary due to more detailed classification of vegetation cover. However, a high spatial resolution vegetation type data in China is difficult to obtain in general. Future studies are suggested to utilize data regarding vegetation cover types, such as coniferous forests, deciduous forests, shrubs, and grasslands, to explore the spatial relationship between their features and soundscape quality.

Moreover, the questionnaire data collected in 2016 was not up-to-date, nonetheless such data was still of research interest and significance, because this investigation can provide valuable historical research data for further exploring the environmental changes and soundscape planning and management. Given this importance, we suggest that future research could explore the spatiotemporal evolution of soundscape quality in Kulangsu in conjunction with the results of the present study. Based on e.g., the concept of DPSIR model [80], it is possible to predict the future state of soundscape resources in landscape planning.

The spatial model used in this study was based on a linear regression algorithm. Therefore, although the results indicated that landscape characteristics did not have significant effects on soundscape quality, this does not account for the possible non-linear relationships between these variables [81]. It has been proposed that the non-linear models may be able to explain soundscape perception better than the linear models [82]. However, the non-linear prediction models related to soundscape are still in their infancy, and most of them are only able to predict the values of output variables without spatiotemporal dynamic features [22,83,84]. Based on such, we recommend future studies could use the non-linear models to explore the soundscape spatiotemporal properties and to predict soundscape perception based on such. The models, such as the cellular automata, artificial neural networks, and Fuzzy-logic models [85,86]. Although the linear model is not perfect, the spatial regression model used in this study still presented a relatively better performance in spatially explaining the soundscape quality than the conventional regression model [20].

5. Conclusions

This study provided indispensable data with regard to the spatiotemporal dynamics of soundscape and landscape perception as well as the spatial landscape characteristics in Kulangsu. Furthermore, the spatial dependence of soundscape quality was examined. We found that sound dominant degree, soundscape quality, and landscape satisfaction degree all showed high spatiotemporal dynamics, and the distance to greenspaces, NDVI, and landscape patterns presented obviously spatial variations. Moreover, soundscape quality had spatial dependence on the sound dominant degree and landscape satisfaction degree, but not on the objective landscape characteristics. Such spatial dependence of soundscape quality provides useful suggestions for the design of soundscape and landscape. The GWR model had better goodness-of-fit than the OLS model, and the results with spatial stationarity of the GWR models suggest that applied data should consider as complete a time dimension as possible in exploring the spatial dependence of soundscape quality, which also guidance for soundscape modeling. The findings of this study will allow landscape planners to determine what factors the quality of a soundscape depends on, which gives a relatively comprehensive consideration of selected indicators for modeling soundscapes. They are also meaningful for the bringing soundscape evaluation into planning and design practices, especially for the development of specific planning goals and design strategies, which aim to protect, restore, and optimize the soundscape resources in the scenic areas and World Heritage Sites.

Supplementary Materials: The following supporting information can be downloaded at: https://www.mdpi.com/article/10.3390/f13091526/s1, Figure S1: Social/demographical/behavioral information of the interviewees; Table S1: Calculated landscape spatial indices at class and landscape levels; Table S2: Results of multicollinearity diagnosis for pleasantness and eventfulness at each sampling period; detailed concepts and definition in the statistical analysis. Reference [87] is cited in the supplementary materials.

Author Contributions: Conceptualization, Z.C., T.-Y.Z. and J.L.; methodology, Z.C. and J.L.; Software, Z.C. and T.-Y.Z.; validation, Z.C., T.-Y.Z., J.L. and X.-C.H.; formal analysis, Z.C. and T.-Y.Z.; investigation, Z.C. and T.-Y.Z.; resources, Z.C., T.-Y.Z. and J.L.; data curation, Z.C. and T.-Y.Z.; writing—original draft preparation, Z.C.; writing—review and editing, Z.C. and J.L.; visualization, Z.C. and T.-Y.Z.; supervision, J.L.; project administration, Z.C., J.L. and X.-C.H.; funding acquisition, J.L. and X.-C.H. All authors have read and agreed to the published version of the manuscript.

Funding: This research was funded by the National Natural Science Foundation of China (51508101 and 52208052), Fujian Provincial Department of Science & Technology (2017J01694), and the Program of Humanities and Social Science Research Program of Ministry of Education of China (21YJCZH038).

Data Availability Statement: Data available on request due to restrictions, e.g., privacy or ethical.

Acknowledgments: The first author, Zhu Chen, would like to thank the China Scholarship Council for the support of a doctoral scholarship (grant number: 202108080105). All the authors would like to thank the anonymous reviewers for their valuable comments, which have greatly improved the quality of this paper. We thank Christina von Haaren and Johannes Hermes for their very helpful suggestions in this paper.

Conflicts of Interest: The authors declare no conflict of interest.

References

1. Zlotnik, H. World urbanization: Trends and prospects. In *New Forms of Urbanization*; Routledge: London, UK, 2017; pp. 43–64. ISBN 1315248077.
2. Margaritis, E.; Kang, J. Relationship between urban green spaces and other features of urban morphology with traffic noise distribution. *Urban For. Urban Green.* **2016**, *15*, 174–185. [CrossRef]
3. Zhang, Y.; Han, M.; Chen, W. The strategy of digital scenic area planning from the perspective of intangible cultural heritage protection. *EURASIP J. Image Video Process.* **2018**, *2018*, 130. [CrossRef]
4. Liu, H.-L.; Shen, Y.-S. The impact of green space changes on air pollution and microclimates: A case study of the Taipei metropolitan area. *Sustainability* **2014**, *6*, 8827–8855. [CrossRef]
5. Gao, T.; Liu, F.; Wang, Y.; Mu, S.; Qiu, L. Reduction of atmospheric suspended particulate matter concentration and influencing factors of green space in urban forest park. *Forests* **2020**, *11*, 950. [CrossRef]
6. Wen, C.; Albert, C.; von Haaren, C. The elderly in green spaces: Exploring requirements and preferences concerning nature-based recreation. *Sustain. Cities Soc.* **2018**, *38*, 582–593. [CrossRef]
7. Liu, J.; Kang, J.; Behm, H.; Luo, T. Landscape spatial pattern indices and soundscape perception in a multi-functional urban area, Germany. *J. Environ. Eng. Landsc. Manag.* **2014**, *22*, 208–218. [CrossRef]

8.   WHO. *Urban Green Spaces: A Brief for Action*; World Health Organization: Geneva, Switzerland, 2017.
9.   Vienneau, D.; de Hoogh, K.; Faeh, D.; Kaufmann, M.; Wunderli, J.M.; Röösli, M.; SNC Study Group. More than clean air and tranquillity: Residential green is independently associated with decreasing mortality. *Environ. Int.* **2017**, *108*, 176–184. [CrossRef]
10.   Herrera-Montes, M.I. Protected area zoning as a strategy to preserve natural soundscapes, reduce anthropogenic noise intrusion, and conserve biodiversity. *Trop. Conserv. Sci.* **2018**, *11*, 1940082918804344. [CrossRef]
11.   Miller, N.P. US National Parks and management of park soundscapes: A review. *Appl. Acoust.* **2008**, *69*, 77–92. [CrossRef]
12.   Liu, J.; Xiong, Y.; Wang, Y.; Luo, T. Soundscape effects on visiting experience in city park: A case study in Fuzhou, China. *Urban For. Urban Green.* **2018**, *31*, 38–47. [CrossRef]
13.   Li, H.; Lau, S.-K. A review of audio-visual interaction on soundscape assessment in urban built environments. *Appl. Acoust.* **2020**, *166*, 107372. [CrossRef]
14.   Aletta, F.; Kang, J.; Axelsson, Ö. Soundscape descriptors and a conceptual framework for developing predictive soundscape models. *Landsc. Urban Plan.* **2016**, *149*, 65–74. [CrossRef]
15.   Hong, J.Y.; Ong, Z.-T.; Lam, B.; Ooi, K.; Gan, W.-S.; Kang, J.; Feng, J.; Tan, S.-T. Effects of adding natural sounds to urban noises on the perceived loudness of noise and soundscape quality. *Sci. Total Environ.* **2020**, *711*, 134571. [CrossRef]
16.   *ISO 12913-1:2014*; Acoustics-Soundscape-Part 1: Definition and Conceptual Framework. ISO Genebra: Vernier, Switzerland, 2014.
17.   Liu, J.; Yang, L.; Xiong, Y.; Yang, Y. Effects of soundscape perception on visiting experience in a renovated historical block. *Build. Environ.* **2019**, *165*, 106375. [CrossRef]
18.   Chen, Z.; Hermes, J.; Liu, J.; von Haaren, C. How to integrate the soundscape resource into landscape planning? A perspective from ecosystem services. *Ecol. Indic.* **2022**, *141*, 109156. [CrossRef]
19.   Hong, J.Y.; Lam, B.; Ong, Z.-T.; Ooi, K.; Gan, W.-S.; Kang, J.; Yeong, S.; Lee, I.; Tan, S.-T. Effects of contexts in urban residential areas on the pleasantness and appropriateness of natural sounds. *Sustain. Cities Soc.* **2020**, *63*, 102475. [CrossRef]
20.   Hong, J.Y.; Jeon, J.Y. Relationship between spatiotemporal variability of soundscape and urban morphology in a multifunctional urban area: A case study in Seoul, Korea. *Build. Environ.* **2017**, *126*, 382–395. [CrossRef]
21.   Liu, J.; Kang, J.; Luo, T.; Behm, H.; Coppack, T. Spatiotemporal variability of soundscapes in a multiple functional urban area. *Landsc. Urban Plan.* **2013**, *115*, 1–9. [CrossRef]
22.   Hong, X.-C.; Wang, G.-Y.; Liu, J.; Song, L.; Wu, E.T.Y. Modeling the impact of soundscape drivers on perceived birdsongs in urban forests. *J. Clean. Prod.* **2021**, *292*, 125315. [CrossRef]
23.   Watts, G.; Marafa, L. Validation of the tranquillity rating prediction tool (TRAPT): Comparative studies in UK and Hong Kong. *Noise Mapp.* **2017**, *4*, 67–74. [CrossRef]
24.   Hong, X.-C.; Wang, G.-Y.; Liu, J.; Dang, E. Perceived Loudness Sensitivity Influenced by Brightness in Urban Forests: A Comparison When Eyes Were Opened and Closed. *Forests* **2020**, *11*, 1242. [CrossRef]
25.   Hong, J.Y.; Jeon, J.Y. Influence of urban contexts on soundscape perceptions: A structural equation modeling approach. *Landsc. Urban Plan.* **2015**, *141*, 78–87. [CrossRef]
26.   Kogan, P.; Gale, T.; Arenas, J.P.; Arias, C. Development and application of practical criteria for the recognition of potential Health Restoration Soundscapes (HeReS) in urban greenspaces. *Sci. Total Environ.* **2021**, *793*, 148541. [CrossRef]
27.   Xiong, Y.; Yang, L.; Wang, X.; Liu, J. Mediating effect on landscape experience in scenic area: A case study in Gulangyu Island, Xiamen City. *Int. J. Sustain. Dev. World Ecol.* **2020**, *27*, 276–283. [CrossRef]
28.   Votsi, N.-E.P.; Drakou, E.G.; Mazaris, A.D.; Kallimanis, A.S.; Pantis, J.D. Distance-based assessment of open country Quiet Areas in Greece. *Landsc. Urban Plan.* **2012**, *104*, 279–288. [CrossRef]
29.   Dzhambov, A.M.; Markevych, I.; Tilov, B.; Arabadzhiev, Z.; Stoyanov, D.; Gatseva, P.; Dimitrova, D.D. Lower noise annoyance associated with GIS-derived greenspace: Pathways through perceived greenspace and residential noise. *Int. J. Environ. Res. Public Health* **2018**, *15*, 1533. [CrossRef]
30.   Gunnarsson, B.; Knez, I.; Hedblom, M.; Sang, Å. Effects of biodiversity and environment-related attitude on perception of urban green space. *Urban Ecosyst.* **2017**, *20*, 37–49. [CrossRef]
31.   Tobler, W.R. A computer movie simulating urban growth in the Detroit region. *Econ. Geogr.* **1970**, *46*, 234–240. [CrossRef]
32.   Anselin, L. *Spatial Econometrics: Methods and Models*; Springer Science & Business Media: New York, NY, USA, 1988; ISBN 9024737354.
33.   Dubé, J.; Legros, D. A spatio-temporal measure of spatial dependence: An example using real estate data. *Pap. Reg. Sci.* **2013**, *92*, 19–30. [CrossRef]
34.   Li, L.; Li, Y.; Li, Z. Efficient missing data imputing for traffic flow by considering temporal and spatial dependence. *Transp. Res. Part C Emerg. Technol.* **2013**, *34*, 108–120. [CrossRef]
35.   Hong, J.Y.; Jeon, J.Y. Exploring spatial relationships among soundscape variables in urban areas: A spatial statistical modelling approach. *Landsc. Urban Plan.* **2017**, *157*, 352–364. [CrossRef]
36.   Rice, W.L.; Newman, P.; Miller, Z.D.; Taff, B.D. Protected areas and noise abatement: A spatial approach. *Landsc. Urban Plan.* **2020**, *194*, 103701. [CrossRef]
37.   Axelsson, Ö.; Nilsson, M.E.; Berglund, B. A principal components model of soundscape perception. *J. Acoust. Soc. Am.* **2010**, *128*, 2836–2846. [CrossRef] [PubMed]
38.   Fourrier, J. Protecting World Heritage. *Beijing Rev.* **2017**, *30*, 32–33.

39. Liang, X.; Coscia, C.; Dellapiana, E.; Martin, J.; Zhang, Y. Complex Social Value-Based Approach for Decision-Making and Valorization Process in Chinese World Cultural Heritage Site: The Case of Kulangsu (China). *Land* **2022**, *11*, 614. [CrossRef]
40. Fengze, L.; Fengming, C.; Mingjian, Z. (Eds.) User Experience Centered Application Design of Multivariate Landscape in Kulangsu, Xiamen. In Proceedings of the 23rd HCI International Conference, HCII 2021, Virtual Event, 24–29 July 2021; Springer: Berlin/Heidelberg, Germany, 2021; pp. 43–59.
41. GB50137; Code for Classification of Urban Land Use and Planning Standards of Development Land. Ministry of Housing and Urban-Rural Development of the People's Republic of China: Beijing, China, 2011.
42. Liu, J.; Kang, J.; Behm, H.; Luo, T. Effects of landscape on soundscape perception: Soundwalks in city parks. *Landsc. Urban Plan.* **2014**, *123*, 30–40. [CrossRef]
43. Hong, X.; Liu, J.; Wang, G.; Jiang, Y.; Wu, S.; Lan, S. Factors influencing the harmonious degree of soundscapes in urban forests: A comparison of broad-leaved and coniferous forests. *Urban For. Urban Green.* **2019**, *39*, 18–25. [CrossRef]
44. Liu, J.; Kang, J. Soundscape design in city parks: Exploring the relationships between soundscape composition parameters and physical and psychoacoustic parameters. *J. Environ. Eng. Landsc. Manag.* **2015**, *23*, 102–112. [CrossRef]
45. Balaji, R.; Bapat, R.B. On Euclidean distance matrices. *Linear Algebra Appl.* **2007**, *424*, 108–117. [CrossRef]
46. Gorelick, N.; Hancher, M.; Dixon, M.; Ilyushchenko, S.; Thau, D.; Moore, R. Google Earth Engine: Planetary-scale geospatial analysis for everyone. *Remote Sens. Environ.* **2017**, *202*, 18–27. [CrossRef]
47. Praticò, S.; Solano, F.; Di Fazio, S.; Modica, G. Machine learning classification of mediterranean forest habitats in google earth engine based on seasonal sentinel-2 time-series and input image composition optimisation. *Remote Sens.* **2021**, *13*, 586. [CrossRef]
48. Foga, S.; Scaramuzza, P.L.; Guo, S.; Zhu, Z.; Dilley Jr, R.D.; Beckmann, T.; Schmidt, G.L.; Dwyer, J.L.; Hughes, M.J.; Laue, B. Cloud detection algorithm comparison and validation for operational Landsat data products. *Remote Sens. Environ.* **2017**, *194*, 379–390. [CrossRef]
49. Coppin, P.; Lambin, E.; Jonckheere, I.; Muys, B. Digital change detection methods in natural ecosystem monitoring: A review. In *Series in Remote Sensing: Analysis of Multi-Temporal Remote Sensing Images, Proceedings of the First International Workshop on Multitemp, Trento, Italy, 13–14 September 2001*; World Scientific: Singapore, 2002; pp. 3–36.
50. Ren, Y.; Wei, X.; Wang, D.; Luo, Y.; Song, X.; Wang, Y.; Yang, Y.; Hua, L. Linking landscape patterns with ecological functions: A case study examining the interaction between landscape heterogeneity and carbon stock of urban forests in Xiamen, China. *For. Ecol. Manag.* **2013**, *293*, 122–131. [CrossRef]
51. McGarigal, K.; Cushman, S.A.; Ene, E. FRAGSTATS v4: Spatial Pattern Analysis Program for Categorical and Continuous Maps. Computer Software Program Produced by the Authors at the University of Massachusetts, Amherst. 2012, Volume 15. Available online: http://www.researchgate.net/publication/259011515_FRAGSTATS_Spatial_pattern_analysis_program_for_categorical_and_maps (accessed on 20 July 2021).
52. Modica, G.; Vizzari, M.; Pollino, M.; Fichera, C.R.; Zoccali, P.; Di Fazio, S. Spatio-temporal analysis of the urban–rural gradient structure: An application in a Mediterranean mountainous landscape (Serra San Bruno, Italy). *Earth Syst. Dyn.* **2012**, *3*, 263–279. [CrossRef]
53. Pourtaghi, Z.S.; Pourghasemi, H.R. GIS-based groundwater spring potential assessment and mapping in the Birjand Township, southern Khorasan Province, Iran. *Hydrogeol. J.* **2014**, *22*, 643–662. [CrossRef]
54. Chen, W.; Li, H.; Hou, E.; Wang, S.; Wang, G.; Panahi, M.; Li, T.; Peng, T.; Guo, C.; Niu, C. GIS-based groundwater potential analysis using novel ensemble weights-of-evidence with logistic regression and functional tree models. *Sci. Total Environ.* **2018**, *634*, 853–867. [CrossRef]
55. McMillen, D.P. *Geographically Weighted Regression: The Analysis of Spatially Varying Relationships*; JSTOR: New York, NY, USA, 2004.
56. Nazeer, M.; Bilal, M. Evaluation of ordinary least square (OLS) and geographically weighted regression (GWR) for water quality monitoring: A case study for the estimation of salinity. *J. Ocean Univ. China* **2018**, *17*, 305–310. [CrossRef]
57. Mennis, J. Mapping the results of geographically weighted regression. *Cartogr. J.* **2006**, *43*, 171–179. [CrossRef]
58. Srivastava, A.K.; Srivastava, V.K.; Ullah, A. The coefficient of determination and its adjusted version in linear regression models. *Econom. Rev.* **1995**, *14*, 229–240. [CrossRef]
59. Hu, S. Akaike information criterion. *Cent. Res. Sci. Comput.* **2007**, *93*. Available online: https://www.researchgate.net/profile/Shuhua-Hu/publication/267201163_Akaike_Information_Criterion/links/599f662aa6fdccf5941f894b/Akaike-Information-Criterion.pdf (accessed on 20 July 2021).
60. Hashim, H.; Abd Latif, Z.; Adnan, N.A. Urban vegetation classification with NDVI threshold value method with very high resolution (VHR) Pleiades imagery. *Int. Arch. Photogramm. Remote Sens. Spat. Inf. Sci.* **2019**, *42*, 237–240. [CrossRef]
61. Shaw, T.; Hedes, R.; Sandstrom, A.; Ruete, A.; Hiron, M.; Hedblom, M.; Eggers, S.; Mikusiński, G. Hybrid bioacoustic and ecoacoustic analyses provide new links between bird assemblages and habitat quality in a winter boreal forest. *Environ. Sustain. Indic.* **2021**, *11*, 100141. [CrossRef]
62. Lackey, N.Q.; Tysor, D.A.; McNay, G.D.; Joyner, L.; Baker, K.H.; Hodge, C. Mental health benefits of nature-based recreation: A systematic review. *Ann. Leis. Res.* **2021**, *24*, 379–393. [CrossRef]
63. Wang, F.; Wang, Y. Potential role of local contributions to record-breaking high-temperature event in Xiamen, China. *Weather Clim. Extrem.* **2021**, *33*, 100338. [CrossRef]
64. Yin, S.; Peng, L.L.H.; Feng, N.; Wen, H.; Ling, Z.; Yang, X.; Dong, L. Spatial-temporal pattern in the cooling effect of a large urban forest and the factors driving it. *Build. Environ.* **2022**, *209*, 108676. [CrossRef]

65. Lin, V.S.; Qin, Y.; Ying, T.; Shen, S.; Lyu, G. Night-time economy vitality index: Framework and evidence. *Tour. Econ.* **2022**, *28*, 665–691. [CrossRef]
66. Nicolosi, V.; Wilson, J.; Yoshino, A.; Viren, P. The restorative potential of coastal walks and implications of sound. *J. Leis. Res.* **2021**, *52*, 41–61. [CrossRef]
67. Axelsson, Ö. (Ed.) How to Measure Soundscape Quality. In Proceedings of the Euronoise 2015 Conference, Maastricht, The Netherlands, 31 May–3 June 2015; pp. 1477–1481.
68. Yimprasert, W.; Teampanpong, J.; Somnam, K. Soundscape Quality in Recreation Areas of Khao Yai National Park in Thailand. *J. Environ. Manag. Tour.* **2021**, *12*, 1324–1334.
69. Zhao, J.; Liu, X.; Dong, R.; Shao, G. *Landsenses Ecology and Ecological Planning toward Sustainable Development*; Taylor & Francis: Oxfordshire, UK, 2016; Volume 23, pp. 293–297.
70. Tan, X.; Han, L.; Li, G.; Zhou, W.; Li, W.; Qian, Y. A quantifiable architecture for urban social-ecological complex landscape pattern. *Landsc. Ecol.* **2022**, *37*, 663–672. [CrossRef]
71. Hakizimana, C.; Goldsmith, P.; Nunow, A.A.; Roba, A.W.; Biashara, J.K. Land and agricultural commercialisation in Meru County, Kenya: Evidence from three models. *J. Peasant Stud.* **2017**, *44*, 555–573. [CrossRef]
72. Liu, J.; Kang, J.; Behm, H. Birdsong as an element of the urban sound environment: A case study concerning the area of Warnemünde in Germany. *Acta Acust. United Acust.* **2014**, *100*, 458–466. [CrossRef]
73. Shi, R.; Conrad, S.A. Correlation and regression analysis. *Ann. Allergy Asthma Immunol.* **2009**, *103*, S35–S41. [CrossRef]
74. Ow, L.F.; Ghosh, S. Urban cities and road traffic noise: Reduction through vegetation. *Appl. Acoust.* **2017**, *120*, 15–20. [CrossRef]
75. Yu, W. (Ed.) Cultural Elements in the Urban Landscape Design Innovation Based on Big Data Era. In *International Conference on Big Data Analytics for Cyber-Physical System in Smart City*; Springer: New York, NY, USA, 2021; pp. 1–8.
76. Valetti, L.; Pellegrino, A.; Aghemo, C. Cultural landscape: Towards the design of a nocturnal lightscape. *J. Cult. Herit.* **2020**, *42*, 181–190. [CrossRef]
77. Li, D.; Li, X. The Improvement Study of Knowledge Management in Mount Laojun Scenic in Lijiang. *Int. J. Cult. Hist.* **2015**, *1*, 29–32. [CrossRef]
78. Rao, S.S. Statistical analysis of a spatio-temporal model with location-dependent parameters and a test for spatial stationarity. *J. Time Ser. Anal.* **2008**, *29*, 673–694. [CrossRef]
79. Kupfer, J.A.; Farris, C.A. Incorporating spatial non-stationarity of regression coefficients into predictive vegetation models. *Landsc. Ecol.* **2007**, *22*, 837–852. [CrossRef]
80. Albert, C.; Galler, C.; Hermes, J.; Neuendorf, F.; Von Haaren, C.; Lovett, A. Applying ecosystem services indicators in landscape planning and management: The ES-in-Planning framework. *Ecol. Indic.* **2016**, *61*, 100–113. [CrossRef]
81. Siddagangaiah, S.; Chen, C.-F.; Hu, W.-C.; Farina, A. The dynamical complexity of seasonal soundscapes is governed by fish chorusing. *Commun. Earth Environ.* **2022**, *3*, 109. [CrossRef]
82. Lionello, M.; Aletta, F.; Kang, J. A systematic review of prediction models for the experience of urban soundscapes. *Appl. Acoust.* **2020**, *170*, 107479. [CrossRef]
83. Giannakopoulos, T.; Orfanidi, M.; Perantonis, S. (Eds.) Athens Urban Soundscape (Athus): A Dataset for Urban Soundscape Quality Recognition. In Proceedings of the International Conference on Multimedia Modeling, Thessaloniki, Greece, 8–11 January 2019; Springer: New York, NY, USA, 2019; pp. 338–348.
84. Romero, V.P.; Maffei, L.; Brambilla, G.; Ciaburro, G. Modelling the soundscape quality of urban waterfronts by artificial neural networks. *Appl. Acoust.* **2016**, *111*, 121–128. [CrossRef]
85. Baig, M.F.; Mustafa, M.R.U.; Baig, I.; Takaijudin, H.B.; Zeshan, M.T. Assessment of land use land cover changes and future predictions using CA-ANN simulation for selangor, Malaysia. *Water* **2022**, *14*, 402. [CrossRef]
86. Rocha, J.; Gutierres, F.; Gomes, P.; Teodoro, A.C. A hybrid CA-ANN-Fuzzy model for simulating coastal changing patterns. In *Beach Management Tools-Concepts, Methodologies and Case Studies*; Springer: New York, NY, USA, 2018; pp. 197–217.
87. Dismuke, C.; Lindrooth, R. Ordinary least squares. *Methods Des. Outcomes Res.* **2006**, *93*, 93–104.

MDPI
St. Alban-Anlage 66
4052 Basel
Switzerland
Tel. +41 61 683 77 34
Fax +41 61 302 89 18
www.mdpi.com

Forests Editorial Office
E-mail: forests@mdpi.com
www.mdpi.com/journal/forests